Communications and Control Engineering

Series Editors
E.D. Sontag • M. Thoma • A. Isidori • J.H. van Schuppen

Published titles include:

Stability and Stabilization of Infinite Dimensional Systems with Applications
Zheng-Hua Luo, Bao-Zhu Guo and Omer Morgul

Nonsmooth Mechanics (Second edition)
Bernard Brogliato

Nonlinear Control Systems II
Alberto Isidori

L_2-Gain and Passivity Techniques in Nonlinear Control
Arjan van der Schaft

Control of Linear Systems with Regulation and Input Constraints
Ali Saberi, Anton A. Stoorvogel and Peddapullaiah Sannuti

Robust and H_∞ Control
Ben M. Chen

Computer Controlled Systems
Efim N. Rosenwasser and Bernhard P. Lampe

Control of Complex and Uncertain Systems
Stanislav V. Emelyanov and Sergey K. Korovin

Robust Control Design Using H_∞ Methods
Ian R. Petersen, Valery A. Ugrinovski and Andrey V. Savkin

Model Reduction for Control System Design
Goro Obinata and Brian D.O. Anderson

Control Theory for Linear Systems
Harry L. Trentelman, Anton Stoorvogel and Malo Hautus

Functional Adaptive Control
Simon G. Fabri and Visakan Kadirkamanathan

Positive 1D and 2D Systems
Tadeusz Kaczorek

Identification and Control Using Volterra Models
Francis J. Doyle III, Ronald K. Pearson and Babatunde A. Ogunnaike

Non-linear Control for Underactuated Mechanical Systems
Isabelle Fantoni and Rogelio Lozano

Robust Control (Second edition)
Jürgen Ackermann

Flow Control by Feedback
Ole Morten Aamo and Miroslav Krstić

Learning and Generalization (Second edition)
Mathukumalli Vidyasagar

Constrained Control and Estimation
Graham C. Goodwin, María M. Seron and José A. De Doná

Randomized Algorithms for Analysis and Control of Uncertain Systems
Roberto Tempo, Giuseppe Calafiore and Fabrizio Dabbene

Switched Linear Systems
Zhendong Sun and Shuzhi S. Ge

Subspace Methods for System Identification
Tohru Katayama

Digital Control Systems
Ioan D. Landau and Gianluca Zito

Multivariable Computer-controlled Systems
Efim N. Rosenwasser and Bernhard P. Lampe

Dissipative Systems Analysis and Control (Second edition)
Bernard Brogliato, Rogelio Lozano, Bernhard Maschke and Olav Egeland

Algebraic Methods for Nonlinear Control Systems
Giuseppe Conte, Claude H. Moog and Anna M. Perdon

Polynomial and Rational Matrices
Tadeusz Kaczorek

Simulation-based Algorithms for Markov Decision Processes
Hyeong Soo Chang, Michael C. Fu, Jiaqiao Hu and Steven I. Marcus

Iterative Learning Control
Hyo-Sung Ahn, Kevin L. Moore and YangQuan Chen

Distributed Consensus in Multi-vehicle Cooperative Control
Wei Ren and Randal W. Beard

Control of Singular Systems with Random Abrupt Changes
El-Kébir Boukas

Nonlinear and Adaptive Control with Applications
Alessandro Astolfi, Dimitrios Karagiannis and Romeo Ortega

Stabilization, Optimal and Robust Control
Aziz Belmiloudi

Control of Nonlinear Dynamical Systems
Felix L. Chernous'ko, Igor M. Ananievski and Sergey A. Reshmin

Periodic Systems
Sergio Bittanti and Patrizio Colaneri

Discontinuous Systems
Yury V. Orlov

Huaguang Zhang · Derong Liu · Zhiliang Wang

Controlling Chaos

Suppression, Synchronization and Chaotification

Springer

Huaguang Zhang, PhD
Zhiliang Wang, PhD

Northeastern University
College of Information Science
and Engineering
Shenyang 110004
PR China
hgzhang@ieee.org

Derong Liu, PhD
Chinese Academy of Sciences
Institute of Automation
Beijing 100190
PR China
and
University of Illinois at Chicago
Department of Electrical
and Computer Engineering
Chicago
Illinois 60607
USA
derongliu@gmail.com

ISBN 978-1-4471-2282-1 e-ISBN 978-1-84882-523-9
DOI 10.1007/978-1-84882-523-9
Springer Dordrecht Heidelberg London New York

British Library Cataloguing in Publication Data
A catalogue record for this book is available from the British Library

Softcover reprint of the hardcover 1st edition 2009

Cover design: le-tex publishing services GmbH, Leipzig, Germany

Printed on acid-free paper

Springer is part of Springer Science+Business Media (www.springer.com)

To

Leon O. Chua

The father of nonlinear circuit theory and cellular neural networks
The inventor of a five-element electronic circuit for generating chaotic signals – the Chua circuit

Foreword

Today, *chaos* is no longer a stranger to many. Instead, chaos theory and technology have gradually become well known as a promising research field with significant impacts on an increasing number of novel, potentially attractive, time- and energy-critical engineering applications.

There are sufficient scientific evidence and practical reasons for studying and utilizing chaos control and synchronization. In a system where irregular responses are undesirable or even harmful, chaos should be reduced as much as possible or even completely suppressed, while if complex dynamics is beneficial and useful, chaos synchronization and generation become desirable. The scope of chaos control for applications, as it has turned out, covers such widespread areas as secure wired and wireless communications and encryption of information data (e.g., texts, images, and videos), biological systems (e.g., understanding of the functioning of human brain and heart, so as to suppress deadly epilepsy and cardiac arrhythmias), liquid mixing (e.g., chemical reactions and medical drug production), crisis management (e.g., critical prevention of cascading fatal voltage collapses of power grids), high-performance circuits and devices (e.g., cellular neural networks, multi-coupled signal modulators, and power converters), and artificial intelligence in decision making and organization of complex multi-agent networks and systems in industries, economics, as well as military alike.

Chaos control, even by linguistic definition, involves both basic chaos theory and automatic control techniques. More than ten years ago, we pointed out that[1] "We subscribe to the idea that any viable theory should be one that is constantly evolving and developing. When merged, the twin virtues of classical control theory and advanced nonlinear science may bring about unexpected benefits. For chaotic systems, if we can incorporate control mechanisms that exploit certain defining characteristics of chaos, the approach becomes more interesting and, hopefully, more efficient and effective. In addition, due to the inherent association of nonlinearity and chaos to various issues, the scope of chaos control is much more diverse. When compared

[1] Preface, G. Chen and X. Dong, *From Chaos to Order: Methodologies, Perspectives and Applications*. World Scientific, Singapore, 1998.

to other conventional approaches, chaos control has its unique features with regard to such aspects as objectives, perspectives, problem formulations, and performance measures." Today, we are especially pleased to welcome the present monograph *Controlling Chaos* that continues to address the important subject of chaos control, including in particular the key topics of chaos suppression, synchronization, and generation (also named chaotification).

This exciting and yet challenging research and development area has continuously been a scientific interdiscipline involving systems and control engineers, theoretical and experimental physicists, computational and applied mathematicians, biologists and physiologists, and electronics specialists, among others. The progress has been very promising and encouraging to date. Nevertheless, achievements notwithstanding,[2] "Chaos control calls for new efforts and endeavors for the coming millennium. In this new era, perhaps today's contemporary concepts of nonlinear dynamics and controls will undergo yet another cycle of rethinking and reorganizing; perhaps the chaos control theories, methodologies, and perspectives that have been drawn together in this book will unleash some other new ideas and valuable applications; perhaps real breakthroughs will begin to take place, bringing enhancement, improvement, and sustainability to the complex living nature – it is only the beginning … ."

Hong Kong
February 2009

Guanrong Chen

---

[2] Epilogue, G. Chen and X. Dong, *From Chaos to Order: Methodologies, Perspectives and Applications*. World Scientific, Singapore, 1998.

Preface

Background of This Book

Chaos theory, once considered to be the third revolution in physics following relativity theory and quantum mechanics, has been studied extensively in the past thirty years. A lot of chaotic phenomena have been found and enormous mathematical strides have been taken. Nowadays, it has been agreed by scientists and engineers that chaos is ubiquitous in natural sciences and social sciences, such as in physics, chemistry, mathematics, biology, ecology, physiology, economics, and so on. Wherever nonlinearity exists, chaos may be found. For a long time, chaos was thought of as a harmful behavior that could decrease the performance of a system and therefore should be avoided when the system is running. One remarkable feature of a chaotic system distinguishing itself from other nonchaotic systems is that the system is extremely sensitive to initial conditions. Any tiny perturbation of the initial conditions will significantly alter the long-term dynamics of the system. This fact means that when one wants to control a chaotic system one must make sure that the measurement of the needed signals is absolutely precise. Otherwise any attempt of controlling chaos would make the dynamics of the system go to an unexpected state. With the development of chaos theory and practice in engineering, more and more people want to know the answers to the following questions:

(1) Can chaos be controlled?
(2) Can chaos be utilized?
(3) Can two chaotic systems be in resonance as in the case of periodic ones?
(4) If the answer to the second question is positive, then how to generate chaos in a nonchaotic system?

These questions have been partly answered by Ott, Pecora, and Chen in the 1990s, which has led a surge in the application study of chaos. From then on, a new research area, chaos control, including suppression, utilization, and generation of chaotic phenomena, came into being. Among these studies, three aspects attract more attention; that is, stabilization of chaos, synchronization of chaos, and anticontrol of chaos.[3]

Why This Book?

Although several monographs on controlling chaos have been published, the present book has unique features which distinguish it from others.

First, the types of chaotic systems studied in this monograph are rather extensive. From the point of view of physics, readers can find not only well-known chaotic systems, such as the Lorenz system, the Rössler system, and the Hénon map, but also some new chaotic systems which appeared in recent years, such as the Liu hyperchaotic system, the Liao chaotic system, the Chen chaotic system, and the Lü chaotic system. From the point of view of models, one can find difference equations, ordinary differential equations, and time-delayed differential equations in this book, which are the main mathematical models describing chaos.

Second, since the monograph is a summary of the authors' previous research, the methods proposed here for stabilizing, synchronizing, and generating chaos in a great degree benefit from the theory of nonlinear control systems, and are more advanced than that appear in other introductory books. One example is that in order to stabilize a chaotic system to one of its equilibria, an inverse optimal control method is developed in this book. The controller designed according to this method not only stabilizes the system but also optimizes a meaningful cost functional. Therefore, the difficulty of solving the Hamilton–Jacobi–Bellman (HJB) equation is avoided. Another example is that in order to synchronize two discrete-time chaotic systems, the exact linearization method is used which provides a unified framework for controller design for both continuous-time and discrete-time chaotic systems. Yet a third example is that in order to chaotify a continuous-time nonchaotic system, a kind of impulsive control method is developed. A mathematical proof shows that the chaos induced by this method satisfies Devaney's definition of chaos.

Last but not least, some rather unique contributions are included in this monograph. One notable feature is the combination of fuzzy logic and chaos. Besides the famous Takagi–Sugeno (T–S) fuzzy model, a novel model, the fuzzy hyperbolic model (FHM), which was initially proposed by one of the authors and whose merits in modeling and control have been illustrated in our book[4] earlier, is also included in this book. In this monograph we combine chaos and fuzzy logic in many aspects: the T–S fuzzy model is used in Chaps. 4 and 8 for suppressing, modeling, and synchronizing chaotic systems, respectively; and the chaotification of the discrete-time FHM and the continuous-time FHM is studied in Chap. 9. Another notable feature is that the methods proposed in this monograph can be applied to a wide class of chaotic systems rather than a specific chaotic system. For example, in Chap. 4 a systematic method is proposed for stabilizing discrete-time chaotic systems and in

[3] Anticontrol of chaos is also known as chaotification. In this book, they are synonymous.

[4] H. Zhang and D. Liu, *Fuzzy Modeling and Fuzzy Control*. Birkhäuser, Boston, 2006.

Chap. 5 a method based on nonlinear geometric control theory is proposed which provides a systematic procedure to synchronize two or more chaotic systems.

The Content of This Book

The whole book involves nine chapters. As indicated by the title of the book, the main content of the book is composed of three parts: suppressing chaos, synchronizing chaos, and generating chaos. To make the book self contained, additional materials are added to provide readers with a brief review of the history of chaos control and some necessary mathematical preliminaries on dynamical systems.

In Chap. 1, we briefly review the history of chaos theory and chaos control. We first review the history of chaos by following the important events in the development of chaos theory. We start the review from the last decade of the 19th century to the 1980s in the 20th century. The work of many distinguished scientists, such as Poincaré, Birkhoff, van de Pol, Littlewood, Andoronov, Lorenz, Smale, Kolmogorov, Arnol'd, Feigenbaum, Li, Yorke, and May, is summarized. After that, we review the development of chaos control from three different aspects, i.e., from the points of view of suppression, synchronization, and chaotification. For each aspect, not only the main methods are introduced but also the ideas behind those methods are mentioned. Some representative methods are introduced, such as the Ott–Grebogi–Yorke (OGY) method and its extensions, the entrainment and migration method, the time-delay feedback method, and some state feedback methods. Chaos synchronization is introduced according to different synchronization patterns, such as complete synchronization, phase synchronization, lag synchronization, and generalized synchronization. Chaotification was proposed by Chen in 1997 and has attracted a lot of attention since then. Methods for chaotification will be reviewed, including the state feedback method, the state delay feedback method, the impulsive control method, and the Smale horseshoe method.

In Chap. 2, necessary mathematical background materials on nonlinear dynamics and chaos are introduced. Dynamical system theory is a powerful tool for chaos study. For the completeness of the book we provide a brief introduction to nonlinear dynamical systems. This chapter is rather difficult for readers with engineering background since many mathematical concepts, definitions, and theorems are involved. The content of this chapter includes two parts. Some concepts and definitions about nonlinear ordinary equations and dynamical systems are introduced first, such as the concepts of flow, fixed point, equilibrium state, invariant set, attractor, stable (unstable) manifold, Floquet index, Lyapunov exponents, and Smale horseshoe. We also state some important theorems, such as the theorem about existence and uniqueness of solutions, the Hartman-Grobman theorems, and the Lyapunov stability theorems. Some concepts and theorems about retarded functional differential equations (RFDEs) are introduced next, such as the definitions of solutions and the initial problem, existence and uniqueness of solutions, and stability of solutions. After that, some stability criteria for RFDEs are introduced.

In Chap. 3, the entrainment and migration control of chaos is introduced. It is usually the case that several attractors coexist in a complex dynamical system. These attractors can be stable or unstable. When a key parameter is changed the attractor may be altered either in appearance or in spatial position, or in both. One of the goals of chaos control is to steer the system's trajectory to the expected state. The background of entrainment and migration control is based on two facts: a multi-attractor chaotic system is sensitive to both initial conditions and parameters and each stable attractor has its basin of attraction. After introducing the basics of entrainment and migration control, a modified entrainment and migration control method: the open-plus-closed-loop (OPCL) control method, is introduced. The OPCL method is an effective method for polynomial chaotic systems when the order of the polynomial is less than two. To overcome this restriction, based on the OPCL method, we propose an improved method, the open-plus-nonlinear-closed-loop (OPNCL) control method. The OPNCL method can be applied to polynomial chaotic systems with arbitrary order and greatly improves the control accuracy. Finally, the OPNCL method is employed for controlling continuous-time polynomial chaotic systems and discrete-time polynomial chaotic systems. By simulating the Chua circuit, the logistic map, and the Hénon map, we validate the effectiveness of the OPNCL method.

The parameters of a chaotic system play an important role, whose variation will lead to completely different dynamics. Sometimes, we want to design a controller which is optimal in a certain sense. However, it is a difficult task to solve the HJB equation when one designs an optimal controller directly along the conventional route. To solve the aforementioned problems, in Chap. 4, we focus on two kinds of methods of suppressing chaos: the adaptive control method and the inverse optimal control method. We develop two new methods of parametric adaptive control for a class of discrete-time chaotic systems and a class of continuous-time chaotic systems with multiple parameters, respectively. The systems are assumed to be linearly parameterized in the adaptive control algorithm. Then, systems with nonlinear distributed parameters and uncertain noise are considered. We apply the inverse optimal control method to stabilize a new four-dimensional chaotic system. The merit of this approach is that it does not need to solve the complicated HJB equation.

In Chap. 5, the synchronization problems of both continuous-time systems and discrete-time chaotic systems are studied. After introducing some necessary preliminaries, a method is proposed to make the single output signal of the response system synchronized with that of the drive system with a scalar controller. The method is based on nonlinear geometric control theory and an exact linearization technique. Then, this method is generalized to the case of multiple output signals. An adaptive method is also proposed to synchronize two different continuous-time chaotic systems. Finally, for problems of synchronizing discrete-time chaotic systems with parametric perturbations, an adaptive controller is designed. This method is based on the theory of exact linearization of discrete-time systems. The methods developed in this chapter have a systematic procedure and can be used with a rather wide class of chaotic systems.

In Chap. 6, we study how to synchronize two identical or different chaotic systems by impulsive control methods. Impulsive control is an efficient method to deal with dynamical systems which cannot be controlled by continuous control. Additionally, in the synchronization process, the response system receives information from the drive system only at discrete time instants, which drastically reduces the amount of synchronization information transmitted from the drive system to the response system and makes this method more efficient in a great number of real-life applications. We first study the complete synchronization of a class of chaotic systems and, after that, we develop synchronization methods for the unified systems with channel time delay in the sense of practical stability. Then, robust synchronization schemes are studied for the chaotic systems with parametric uncertainty and parametric mismatch. Our goal is to develop practical impulsive control methods for different synchronization schemes.

In engineering applications, time delays always exist, and parameters of the system are inevitably perturbed by external disturbances. Moreover, the values of delays and parameters are often unknown in advance. In some special cases, the structure or parameters of the drive system are even unknown in advance. In Chap. 7, we study how to synchronize chaotic systems when time delay exists and the synchronized systems have different structures. Synchronization methods are developed for a class of delayed chaotic systems when the drive system and the response system have the same structures but different parameters. After that, the problem of synchronizing different chaotic systems is studied. Some concrete examples are presented to show how to design the controller. Based on that, a more general case, synchronizing two different delayed chaotic neural networks with unknown parameters, is studied.

In recent years, fuzzy logic systems have received much attention from control theorists as a powerful tool for nonlinear control. A motivation for using fuzzy systems and fuzzy control stems in part from the fact that they are particularly suitable in industrial processes when the physical systems or qualitative criteria are too complex to model and they have provided an efficient and effective way in the control of complex uncertain nonlinear or ill-defined systems. Among various kinds of fuzzy control or system methods, the T–S fuzzy system is widely accepted as a powerful tool for fuzzy control. Chap. 8 focuses on the modeling and synchronizing of chaotic and hyperchaotic systems. We first introduce fuzzy modeling methods for some classical chaotic systems via the T–S fuzzy models. Next, we model some hyperchaotic systems with T–S fuzzy models and then, based on these fuzzy models, we develop an H_∞ synchronization method for two different hyperchaotic systems. Finally, the problem of synchronizing a class of time-delay chaotic systems based on the T–S fuzzy model is considered.

As a reverse process of suppressing or eliminating chaotic behaviors in order to reduce the complexity of an individual system or a coupled system, chaotification aims at creating or enhancing the system's complexity for some special applications. In recent years, many conventional control methods and special techniques were applied to the chaotification of discrete-time dynamical systems or continuous-time dynamical systems. However, most of them are based on the assumption that the analytical representations of the nonlinear dynamical systems to be chaotified are known exactly. For an unknown or uncertain nonlinear dynamical system, the above methods are ineffective. Chap. 9 is devoted to the chaotification of dynamical systems which are originally nonchaotic. We develop a simple nonlinear state feedback controller to chaotify the discrete-time FHM with uncertain parameters. After that, we use an impulsive and nonlinear feedback control method to chaotify the continuous-time FHM. We believe that if a nonchaotic system can be approximated by a fuzzy model with adequate accuracy, then the chaotifying controller of the fuzzy model will also make the nonchaotic system chaotic. Finally, we chaotify two classes of continuous-time linear systems via a sampled data control approach.

Acknowledgments

The authors would like to acknowledge the help and encouragement they have received during the course of writing this book. A great deal of the materials presented in this book is based on the research that we conducted with several colleagues and former students, including Y. H. Xie, D. S. Yang, Z. S. Wang, W. Huang, J. Wang, Y. B. Quan, Z. B. Liu, and H. X. Guan. We wish to acknowledge especially Dr. T. D. Ma and Dr. Y. Zhao for their hard work on this book. The authors also wish to thank Prof. S. S. Wiggins, Prof. A. Medio, Prof. M. Lines, Prof. M. W. Hirsch, Prof. S. Smale, Prof. R. L. Devaney, and Prof. Y. A. Kuznetsov for their excellent books on the theory of nonlinear dynamics and chaos. We are very grateful to the National Natural Science Foundation of China (Grant Nos. 60521003, 60534010, 60572070, 60621001, 60728307, 60774048, and 60804006), the Program for Innovative Research Team in Universities (Grant No. IRT0421), the Program for Changjiang Scholars, the National High Technology Research and Development Program of China (863 Program) (Grant No. 2006AA04Z183), and the 111 Project of the Ministry of Education of China (B08015), which provided necessary financial support for writing this book.

Shenyang, China
Beijing, China
Chicago, USA
February 2009

Huaguang Zhang
Derong Liu
Zhiliang Wang

Contents

Chapter 1
Overview

Abstract In this chapter, we will provide a brief introduction to the history and development of chaos and chaos control. We begin our introduction with chaos theory. Some important events are narrated in chronological order. After that, we turn our attention to the field of chaos control. We present this subject in three parts; that is, suppressing chaos, synchronizing chaos, and chaotification. We will mention some main approaches used in research and point out open problems in each part.

1.1 The Origin and Development of Chaos Theory

The word 'chaos' generally refers to a phenomenon that is disordered and irregular. In modern scientific terminology, however, 'chaos' refers to a pseudo-random phenomenon generated in a deterministic system. Henri Poincaré is acknowledged as the first person to glimpse the possibility of chaos. When he studied the stability properties of the solar system at the end of the 19th century, Poincaré found that even in the case of three masses moving under Newton's law of attraction they could still exhibit very complicated behavior [56]. This kind of motion depends sensitively on the initial conditions, thereby rendering long-term prediction impossible [15]. Birkhoff developed geometric methods created by Poincaré and found many different types of long-term limiting behaviors, such as *ω-limit set* and *α-limit set*. The term 'dynamical systems' comes from the work [3] but chaos has always been in the background. During the first half of the 20th century, nonlinear oscillators were mostly studied due to their vital role in the development of such technologies as radio, radar, phase-locked loops, and lasers. Success in technology stimulates the invention of new mathematical tools. The pioneers in this area include van der Pol, Andronov, Littlewood, Cartwright, Levinson, and Smale. Especially, in the 1950s, a type of complexity unknown previously was revealed by Cartwright, Littlewood, and Levinson to show that a certain forced nonlinear oscillator had an infinite number of different periods. Smale extended the result of Cartwright, Littlewood, and Levinson in a general framework and illustrated the phenomenon with his 'horseshoe' mapping [64]. Meanwhile, in a separate development, Poincaré's geometric methods were extended to yield a much deeper understanding of celestial mechanics. One of the most important mathematical achievements, the KAM theorem, named by taking the first letters of Kolmogorov, Arnol'd, and Moser, was proved in 1963. This theorem tells us that a Hamiltonian system will still be Hamiltonian when it is subjected to tiny perturbations, which gives an answer that the solar system is stable in some degree. This theorem also implies that in energy preservation systems complicated behavior would still arise.

Prior to Smale, E. Lorenz, an American meteorologist, made an important contribution in 1963 when he used a computer to study a group of ordinary differential equations. These equations were reduced from the partial differential equations which described the turbulent motion of the atmosphere. Lorenz found that a small change in initial conditions led to very different outcomes in a relatively short time; this property is called *sensitive dependence on initial conditions*. This is really a great discovery since almost all traditional scientists at that time believed that two trajectories emitted from close initial points would always be close when time evolves. Laplace's famous assertion is an extreme reflection of this idea: if there is an omnipotent spirit who can distinguish any tiny difference of initial conditions and can discriminate all the forces in the universe, then he can know all the history and all the future about everything in the universe. From the uncertainty principle in quantum mechanics we know that one cannot exactly determine the position and velocity of a particle at the same time. This means that if a system has the property of sensitive dependence on initial conditions, then one cannot predict its long-term behavior since we cannot avoid the error of measurement on the initial conditions. Lorenz used the phrase 'butterfly effect' to refer to the phenomenon of sensitive dependence on initial conditions. He explained that the phenomenon means that a butterfly flapping its wings in Australia today could affect the weather in the United States a month later. It was not until the 1970s that Lorenz's work became known to the more theoretical mathematical community.

The 1970s were the boom years for chaos. In 1971 Ruelle and Takens proposed a new theory for the onset of turbulence in fluids, based on abstract considerations about strange attractors [58]. The word 'chaos' was first introduced by T. Y. Li and J. Yorke in 1975 [37] to designate systems that have aperiodic behavior more complicated than *equilibrium*, *periodic*, or *quasiperodic* motion. But, in fact, their work is a special case of the theorem obtained by Sharkovskii in 1964 [61], which, because of political reasons, was not known by western mathematicians for a long time. In 1976, May showed how chaos arises in iterated mappings in population dynamics, and wrote an influential review article that stressed the pedagogical importance of studying simple nonlinear systems, to counterbalance the often misleading linear intuition fostered by traditional education [41]. Next came the most surprising discovery of all, from the physicist Feigenbaum. He discovered that there are certain universal laws governing the transition from regular to chaotic behavior; roughly speaking, completely different systems can go chaotic in the same way. His work established a link between chaos and phase transitions, and enticed a generation of physicists to the study of dynamics.

Entering into the 1980s, computers become a powerful tool, used to help researchers visualize the complicated structures of strange attractors, calculate characteristic indices of the chaotic systems, and provide evidence required by proofs [19, 69]. Mandelbrot constructed the theory of *fractal geometry* at the end of the 1970s, and drew the first picture of a Mandelbrot set [39]. The theory of fractal geometry generalizes the notion of dimension from integers to real numbers and has become a powerful tool for characterizing the complicated structures of strange attractors. From the middle of the 1980s, more and more researchers have paid attention to how to control chaos, including suppression, synchronization, and chaotification.

1.2 Control of Chaos

There are many practical reasons for controlling chaos. In some situations, such as in a distributed artificial intelligence system, chaotic behavior is undesirable and will degrade the performance of the entire system. Naturally, chaos should be reduced as much as possible, or totally suppressed.

However, recent research has shown that chaos may be useful under certain circumstances. In the following, we will consider three examples where chaos is needed.

Since a chaotic attractor is usually embedded in a dense set of unstable limit cycles, if any of these limit cycles can be stabilized, it may be desirable to stabilize one that characterizes a certain maximal system performance. This character can be used to design a multi-purpose system.

Fluid mixing is another typical example in which chaos is not only useful but also very important. When two fluids are to be thoroughly mixed and the required energy is to be minimized, one will hope that the particle motion of the fluids is strongly chaotic, since otherwise it is hard to obtain rigorous mixing properties, due to the possibility of invariant 2-tori in the flow.

The sensitivity of chaotic systems to small perturbations can be used to steer system trajectories to a desired target quickly. NASA scientists, utilizing the sensitivity of the three-body problem of celestial mechanics to small perturbations, have used small amounts of residual hydrazine fuel to send the spacecraft ISEE-3/IEC more than 50 million miles across the solar system, which achieved the first scientific cometary encounter [62]. This would not have been possible in a nonchaotic system since a large control effort would be required to quickly direct the system trajectory to reach distant targets.

For a long time chaotic systems were thought of as unpredictable and uncontrollable: unpredictable, because a small disturbance in the initial conditions or parameters of a chaotic system produces changes of the original motion that grow exponentially; uncontrollable, because small disturbances usually lead to other chaotic states but not to any stable and regular motion [17]. However, the requirement for controlling chaos, which is either the demand of industrial applications or the academic interest, becomes stronger and stronger than ever before. The first paper on controlling chaos was published in 1989 [25], but it is the paper by Ott, Grebogi, and Yorke [46] that attracted more attention and their method is named the OGY method. In 1990, Pecora and Carroll opened another important application field, synchronization of chaos [49]. Several years later, Chen and Lai proposed a method for generating chaos in a nonchaotic discrete-time system [10]. From then on, a new research aspect, anticontrol of chaos, or chaotification, came into being. Up to now, numerous control methods have been proposed, developed, and tested. Many experiments have demonstrated that chaotic physical systems respond quite well to those newly developed, conventional or novel, simple or sophisticated, control strategies. Applications of chaos control have been proposed in such diverse areas of research as biology, medicine, physiology, epidemiology, chemical engineering, laser physics, electric power systems, fluid mechanics, aerodynamics, electronics, communications, and so on. Chaos control has become a process that manages the dynamics of a nonlinear system on a wider scale, with the hope that more benefits may thus be derived.

1.2.1 Suppression of Chaos

In the beginning, the goal of controlling chaos was to eliminate the harmful chaotic motion. A remarkable feature which distinguishes chaotic systems from other nonlinear systems is that usually a chaotic system has at least one chaotic attractor.[1] Thus, the key of the problem is how to utilize the attractor in the procedure of controller design. The OGY method is based on such an idea: for a neighborhood of a point located on the chaotic attractor there is always a periodic orbit, which may be stable or unstable, passing through the neighborhood. Thus, if one tunes the parameters of the chaotic system slightly such that the trajectory is displaced on the periodic orbit, then, from that time on, the trajectory will always be along the periodic orbit. Because of the limitation of the OGY method, that it is only adapted to the low-dimensional case in which the equilibrium points of the considered chaotic system are saddles and the observed state variables are not contaminated with noise, some modified OGY methods were proposed again [16, 18, 20, 27, 76]. In nature, the OGY method is a local method. One may wait for a long time before the trajectory goes into the neighborhood in which the controller will be turned on. However, because of the *ergodicity* of the chaotic motion on the attractor, there is always an instant when the trajectory will be in the neighborhood. To shorten the waiting time, a targeting method was proposed which can direct the trajectory of the chaotic system into the neighborhood in a very short time [62, 63]. But sometimes it is difficult

[1] In some literature, a strange attractor has a different meaning to a chaotic attractor. The former emphasizes the geometrical structure of the attractor and sometimes has a fractional dimension. The latter emphasizes the sensitivity to initial conditions. Two trajectories starting from different but close initial points will be divergent exponentially. Therefore, a system having a strange attractor may not be a chaotic one while a chaotic attractor may have integer dimension [47].

to obtain the parameters of a chaotic system. Even if the parameters can be obtained, it is difficult to adjust them as control inputs. So, the continuous feedback control method is considered. There is a problem which should be answered when continuous feedback is used: will the topological structure of the chaotic attractor be destroyed by the continuous feedback? At present, we can answer the question as follows: if all of the invariant sets of a chaotic system are *hyperbolic invariant sets*, then the small perturbation will not destroy the topological structure of its state space [80]. We know that many chaotic systems have hyperbolic invariant sets. Therefore, at least for these systems, the continuous feedback control method is feasible. The studies over the last decade show that almost all the methods that do well for conventional nonlinear control problems can be used to suppress chaos (see [8] and references therein).

One main goal of controlling chaos is to steer the trajectory of a chaotic system to a periodic one. So, the first step of the control procedure is to identify these periodic orbits.

Suppose that the mathematical model is known. For discrete-time systems, the periodic orbits can be obtained by calculation according to its definition, no matter whether stable or unstable. For continuous-time systems, the problem can be changed to the case of discrete time by the Poincaré section method. Thus, the key problem is how to identify the periodic orbits of discrete-time systems. In practice, one wants to use computers to find the periodic orbits of the discrete-time systems. The iterations of a discrete-time system can only get stable periodic orbits. For unstable periodic orbits (UPOs) the iteration method loses its effectiveness because even if a point is located on a UPO at present, at the next instant of the evolution it will deviate from the UPO due to small errors caused either by calculation or by measurement.

Four methods are known which work well for identifying UPOs. The first method is called the word-lifting technique which was proposed by Hao [22]. Using this method an analytical expression of periodic orbits can be obtained. The second one was proposed in [45] in which both symbol dynamics and graph theory are used to determine a periodic orbit. The main idea of this method is the construction of a directed graph which represents the structure of the state space for the dynamical system under investigation. The third one is to transform the original chaotic system by a homeomorphism (the term will be introduced in the following chapter). Without changing the topological structure of the state space, this transformation only changes the stabilities of the orbits. Thus, UPOs become stable periodic orbits of the new system. Once these stable periodic orbits are identified by some numerical algorithms [13, 35, 52, 53], we can get the UPOs of the original chaotic system by inverse transformation. The fourth method is very ingenious, in which the retarded state variables are taken as inputs [54]. Considering the fact that a chaotic trajectory can approach any periodic orbit, if we take an expected period as the delayed time and design a stabilizing controller, then the trajectory is just on the expected periodic orbit when the controlled chaotic system is stabilized. Only the structure of the system and period of orbit are needed in the delayed feedback method. The delayed feedback method is a two-in-one method. The procedure of identifying the UPOs is combined with the procedure of stabilization. The periodic orbit will be obtained provided that the controlled system is stabilized. This merit makes the method employed extensively [4, 5, 6, 7, 23, 33, 34, 66]. However, with deeper study a fatal problem is found with the delayed feedback method; that is, the delayed feedback method cannot stabilize the UPOs which have real eigenvalues of odd number. This is called the *odd number limitation*. Fortunately, there are some modified delayed feedback methods which can go round the obstacle of the odd number limitation [43, 44, 70].

If the mathematical model is unknown and only some measurable signals can be obtained, then, at first, one needs to utilize the technique of space reconstruction to get a model. Thus, the four methods discussed above can be used again [9, 65].

A merit of the strategy of feedback control is that the energy of input is small and will not destroy the structure of the state space and, when the control goal is reached, the controller can be turned off. However, the strategy requires the continuous monitoring of the state, which is difficult in some cases. For example, the extremely high temperature in plasma will make it very difficult to precisely measure some physical quantities in the plasma [24]. Non-feedback control methods for suppressing chaos can overcome this difficulty. There are many different kinds of non-feedback control methods. Some of them do not require the understanding of the chaotic system dynamics, such as the method with a periodic signal to suppress chaos [42] or the method with a noise signal to suppress chaos [55]. When implementing a control strategy, there are fewer limitations required by non-feedback control methods compared with that of feedback control methods. But, the cost is that with the non-feedback control strategy one can only achieve some general control goals, such as eliminating chaos, realizing periodic motion, etc. Some non-feedback control strategies can steer the trajectories of a chaotic system to some predesigned orbits, which, especially, may not be the inherent orbits of the original chaotic system [12, 29, 30, 31]. But, in that situation, one must have an exact understanding about the dynamics of the original system. When a non-feedback control method is used, the control signal would not reduce to zero even if the control goal has been realized, which will produce new problems in practice.

1.2.2 Synchronization of Chaos

The common meaning of synchronization is that the phases of two or more oscillators change according to some patterns. C. Huygens was the first person who found the phenomenon of synchronization of coupled pendulums. Actually, synchronization phenomena are abundant in science, nature, engineering, and social life. Systems as diverse as clocks, singing crickets, cardiac pacemakers, firing neurons, and applauding audiences exhibit a tendency to operate in synchrony [50]. These phenomena are universal and can be understood within a common framework based on modern nonlinear dynamics. Existing studies of suppressing chaos show that a periodic input signal can make a chaotic system periodic with the same frequency as that of the input signal. This situation can be looked at as a pattern of synchronization in which a periodic signal synchronizes with a chaotic signal (it can be thought of as a periodic motion with period of infinite length). Furthermore, if the input signal is replaced with a chaotic signal, what will happen? Will there be the phenomenon of synchronization? Intuitively, even if the two chaotic signals are generated from the chaotic systems with the same structure, this synchronization would yet not happen because of the sensitivity of chaotic systems. However, Pecora and Carroll firstly found the synchronization of chaos in the experiment of circuits. They took the Lorenz system as the drive system, and replicated the same system as the response system. When they replaced the first state variable of the response system with the first state variable of the drive system, the synchronization happened. This synchronization scheme is called the P–C scheme. They found that the synchronization in the P–C scheme is not only globally asymptotically stable but also structurally stable. By a strict analysis they pointed out that two chaotic systems in the P–C scheme can get synchronized only if the *conditional Lyapunov exponents* of the response system are all negative.[2]

Although the problems considered by Ott and Pecora seemed different and independent from each other, soon researchers realized the relationship between them. In fact, many control methods developed in suppressing chaos can also be used in synchronizing chaos. In this section, rather than talking more about the synchronization methods, we pay more attention to the the patterns of synchronization that can be obtained among chaotic systems. For convenience, suppose that we have the following chaotic systems:[3]

$$\dot{x} = f(t,x), \;\; y = h(x), \;\; x \in \mathbb{R}^n, \;\; h\colon \mathbb{R}^n \to \mathbb{R}^m. \tag{1.1}$$

Taking (1.1) as the drive system, we construct the response system as

$$\dot{\hat{x}} = \hat{f}(t,\hat{x},y), \;\; \hat{y} = \hat{h}(\hat{x}), \;\; \hat{x} \in \mathbb{R}^p, \;\; \hat{h}\colon \mathbb{R}^p \to \mathbb{R}^q, \tag{1.2}$$

where x and $\hat{x}$ are state variables and y and $\hat{y}$ are outputs. Suppose that the following relation holds when (1.1) and (1.2) get synchronized:

$$\hat{y} = \varphi(y), \;\; \varphi\colon \mathbb{R}^m \to \mathbb{R}^q, \tag{1.3}$$

where φ is continuous but not necessarily differentiable [26]. $(y,\hat{y})$ is called the manifold of synchronization.

Considering the dynamics of drive–response systems and the structure of the manifold of synchronization, we give the following categories of synchronization of chaos.

---

[2] It should be pointed out that this condition is not necessary for synchronization of chaos. Some chaotic systems cannot get synchronized as in P–C schemes, but they indeed can get synchronized [21, 48].

[3] We take a continuous-time system as an example. The same illustration can also be done for discrete-time systems.

(i) Complete synchronization. In this case, $\hat{x}(t) = x(t)$; that is, $m = q$ and φ is an identical mapping. If $\hat{f} = f$, the relation is called identically complete synchronization. If $\hat{f} \neq f$, the relation is called nonidentically complete synchronization. The manifold of complete synchronization is a diagonal hyperplane in the phase space composed of $x - \hat{x}$.

(ii) Generalized synchronization [59]. If the state of the response system is uniquely determined by the state of the drive system, that is, (1.3) holds with φ a certain mapping, then the drive system and the response system are thought of as in the pattern of generalized synchronization. No conditions are imposed on m and q. In this case, the manifold of synchronization is no longer a diagonal hyperplane of the combined phase space $x - \hat{x}$.[4]

(iii) Lag synchronization [57]. In this case, $\hat{x}(t) \approx x(t-\tau)$, where τ is a small positive number.

(iv) Anticipating synchronization [72]. Analogous to lag synchronization, when anticipating synchronization happens, the state variables between the drive system and the response system have the following relationship:

$$\hat{x}(t) = x(t+\tau), \quad \tau > 0.$$

(v) Phase synchronization. Let ϕ_1 and ϕ_2 be the phases of system (1.1) and system (1.2), respectively. The phase synchronization means that for rational numbers m and n, there is a small positive number ε such that

$$|m\phi_1 - n\phi_2| < \varepsilon.$$

The key in phase synchronization is the determination of phase ϕ_1 and phase ϕ_2 [51]. There is no unified definition in high-dimensional chaotic systems. There are three common methods for the low-dimensional chaotic systems [24]:

a. Choose a state variable, such as $x(t)$, and find the instants, $t_1, t_2, \ldots, t_n, \ldots$, at which the state variable gets its local maxima. Then, the phase of a chaotic trajectory can be defined as

$$\varphi_M = 2\pi \frac{t - t_n}{t_{n+1} - t_n} + 2\pi n, \quad t_{n+1} > t > t_n. \tag{1.4}$$

b. If the flow of chaotic oscillators has a proper rotation around a certain reference point, the phase can be defined in a more intuitive and straightforward manner. Project the trajectory on to a two-dimensional plane, for example the x–y plane, with the origin at the reference point. The phase φ_P can simply be defined by

---

[4] A numerical method to verify whether a stable generalized synchronization happens or not is to construct an auxiliary system with the same structure as (1.2). Provide the same input signals derived from the drive system to the response system and the auxiliary system. Choose the initial point of the auxiliary system in the neighborhood of the initial point of the response system. If a stable generalized synchronization happens, then the trajectories of the auxiliary system and the response system will converge to each other. Otherwise, they will be different [1].

$$\tan \varphi_P = y/x. \tag{1.5}$$

c. Using Hilbert transformation. Choose a scalar $s(t)$ and define a complex variable as

$$\xi(t) = s(t) + \mathrm{i}\hat{s}(t) = A(t)\mathrm{e}^{\mathrm{i}\varphi_H(t)}, \tag{1.6}$$

$$\hat{s}(t) = \pi^{-1} P \int_{-\infty}^{\infty} \frac{s(\tau)}{t-\tau} \mathrm{d}\tau,$$

where $P\int_{-\infty}^{\infty}$ means the Cauchy principal value of the integral.

Remark 1.1. The phenomena of lag synchronization and generalized synchronization need to be considered when a synchronization system is designed. First, the perturbations caused by the noise of the environment and measurement cannot be ignored. Therefore, the drive system and the response system are almost never identical. Second, no matter how fast the signal transmission is in the channel, time is always needed. Thus, at the port of the response system, the input signal is not at the same time instant as that of the drive system. In fact, the drive system and the response system are as follows:

$$\begin{cases} \dfrac{\mathrm{d}x(\bar{t})}{\mathrm{d}\bar{t}} = f(\bar{t}, x(\bar{t})), \ \bar{t} = t - \tau, \\ \dfrac{\mathrm{d}\hat{x}(t)}{\mathrm{d}t} = \hat{f}(t, \hat{x}(t), x(\bar{t})). \end{cases}$$

□

Remark 1.2. The phenomenon of phase synchronization is completely different from the preceding cases. In general, when synchronization occurs, the conditional Lyapunov exponents of the response system are all negative. At that time, the response system is a nonchaotic system with a chaotic output. When phase synchronization takes place, since the relation is on the phases of the synchronized systems, the conditional Lyapunov exponents may be positive. □

The synchronizations mentioned above are all schemes of unidirectional couples. There is much more content to be studied if the coupling scheme is bidirectional and the number of coupled systems is greater than two. No matter which scheme of synchronization, the above five kinds of phenomena of synchronization are still encountered the most.

1.2.3 Control and Synchronization of Spatiotemporal Chaos

All of the chaotic systems mentioned above are described by ordinary differential equations or difference equations. This means that for these chaotic systems, the variation rate of the state variables with respect to spatial position is ignored. We call these chaotic systems temporal chaotic systems. But, sometimes, a chaotic system is not isotropic and must be described with partial differential equations. For these chaotic systems, the synchronization can take place either in time or in space position. Since the state of a partial differential equation belongs to a function space whose dimension is infinite, a spatiotemporal chaotic system can provide more patterns than those in a temporal chaotic system. Thus, the considered problems of synchronizing two or more spatiotemporal chaotic systems are much more complicated than those considered for temporal chaotic systems. At present, research on control and synchronization of spatiotemporal chaos mainly follows two directions [24, 32, 79]:

(i) The studies of coupled map lattices. Each lattice is a discrete mapping and evolves with time. The adjacent lattices are joined with specified boundary conditions. There are two reasons for studying coupled map lattices. First, it can be obtained by discretizing a partial differential equation. Therefore, many important characteristics that belong to partial differential equations can be found in the systems of coupled map lattices. Second, mathematical tools required by coupled map lattices are much simpler than what is needed when one directly studies partial differential equations.
(ii) The studies of the underlying partial differential equations. Many of the important properties of a partial differential equation cannot be presented by a coupled map lattice. Therefore, it is necessary to study the original partial differential equations.

There is a marked difference between temporal chaotic systems and spatiotemporal chaotic systems when one wants to control them. A temporal chaotic system can be controlled only by altering its vector fields through a feedback controller. But even for a coupled map lattice this method is impractical since lots of controllers would be required. A feasible and effective method is to alter the state of some lattices and diffuse the alteration through coupling relations so that the control goal can be achieved [24].

Nowadays, more and more researchers have realized that the study of spatiotemporal chaos can provide instructive inspiration and necessary ways for solving the problem of turbulence.

1.3 Anticontrol of Chaos

Since lots of applications of chaos are found in many areas, a new notion, *anticontrol of chaos* (or *chaotification*), was proposed [10]. Later, the concept and method of synthesis of chaos were proposed [71]. Anticontrol of chaos became a new direction in the study of chaos control. It seems that based on the well-developed theory of chaos there should be no difficulties in chaotification. However, it is not so. At the beginning, the difficulties mainly exist in the following aspects:

(i) How to define chaos? The first definition of chaos was proposed by Li and Yorke in 1975 for the mappings on intervals of one-dimensional space [37].

Later, Marotto generalized this definition to the differential mappings on high-dimensional space [40]. Another definition in common use was proposed by Devaney from the topological point of view [14]. Based on Devaney's definition, Wiggins proposed a more concise one which has been widely used [28, 75]. However, these definitions are all for discrete-time systems. For continuous-time systems, there is no unified criterion for what is chaos. A practical method is to compute the *Lyapunov exponents* of the controlled system. If there is a positive Lyapunov exponent, then it is believed that the controlled system becomes chaotic [2]. But this method is not the ideal one since, in a great degree, the computation of Lyapunov exponents depends on the numerical algorithm as well as the precision of computation.

(ii) How to design a chaotifying controller? If the Lyapunov exponents are used as the criterion, how does one design a chaotifying controller? We may chaotify some systems by tuning certain parameters. But there are also many systems which are not sensitive to the variation of parameters and, therefore, would not be chaotified no matter how the parameters are tuned. Furthermore, for some complicated systems, such as a coupled map lattice or a partial differential equation, we cannot determine what method can really make them chaotified because of the limited understanding about their dynamics.

(iii) How to utilize the existing theoretical results? Some theoretical results, such as the Melnikov method and the Shilnikov method [75], although useful in theoretical analysis, are difficult to apply in practice.

A nice controller of chaotification should not only be easy to implement but also have a solid mathematical foundation. From 1996, Chen and Wang developed a series of feedback methods which can make a wide class of nonchaotic systems chaotified. By the inspiration of their works, many chaotifying methods were proposed [11].

According to the criteria of chaos judgement, the existing methods of chaotification can be categorized as follows:

(i) Boundedness + positive Lyapunov exponents + Marotto theorem. Since the motion of chaos is in a bounded region and is sensitively dependent on initial conditions, the key in the procedure of chaotification is to make the controlled system have a positive Lyapunov exponent while its trajectory remains in a bounded region. For discrete-time systems, by using, modulo operation, Chen and Lai first proposed a chaotifying method which can guarantee that all the Lyapunov exponents are positive. Later, Chen and Wang improved the result by replacing the modulo operation by a small amplitude controller. Moreover, by using the Marotto theorem, Chen and Lai provided a proof that the chaos generated by their methods satisfies the definition of Li–Yorke [10].

(ii) Introducing chaotic dynamics. This class includes three main methods:

   a. Using continuous feedback control. First, transform some well-known chaotic systems, such as the Lorenz system, the Rössler system, etc., into a normal format, and then extract their nonlinear terms as one part of the controller.

Another part of the controller is designed to stabilize the original system. These two parts constitute a total chaotification controller [73].

b. Using impulsive control. The system is excited by impulses at discrete instants. The impulsive signal is composed of two parts. One part is to eliminate the original motion, the other part is to make the system chaotic. Usually, this part is composed of chaotic signals [78].
c. Introducing time delay. This method is based on the following idea: the introduction of time delay makes a finite-dimensional system become an infinite-dimensional system. Thus, it is possible to produce more complicated behavior within it, especially chaos. Along this route, Wang studied further the problem of chaotification of continuous-time systems and he proposed a method which makes the outputs of the controlled system approximate the outputs of a chaotic system [74].

(iii) Proving the existence of a Smale horseshoe. In the literature [68], a piecewise controller was designed to make a Duffing oscillator generate a Smale horseshoe. This method is difficult to generalize to other systems. Recently, Yang et al. proposed a computer auxiliary proving method to show the existence of a Smale horseshoe. When the method is applied, Poincaré mapping needs to be constructed first. The merit of the method is that it makes the Melnikov method better operated [77].
(iv) Switching mechanism. Lü et al. proposed a method in 2002, which can generate a chaotic system with a complex attractor. The method is simple and effective, but no rigorous proof is given to explain why chaos occurs [38].

Over the past ten years, many important results have been obtained. But there are still a lot of problems remaining unsolved, such as generating hyperchaotic systems, chaotifying a partial differential equation, etc. Undoubtedly, these problems are great challenges for researchers.

1.4 Summary

In this chapter, we briefly introduced the history and presented the current status of chaos and chaos control. Due to the focus of the book, we do not list all the methods developed in chaos control. Our attention is to give an introduction to the development of theory so as to provide a rough description of chaos control for a new researcher in this area. Readers who are interested in how the synchronization of chaos is utilized in applications are referred to, e.g., [36] and [67].

References

1. Abarbanel HDI, Rulkov NF, Sushchik MM (1996) Generalized synchronization of chaos: the auxiliary system approach. Phys Rev E 53:4528–4535
2. Alligood KT, Sauer TD, Yorke JA (1997) *Chaos – An Introduction to Dynamical Aystems*. Springer, New York
3. Birkhoff GD (1966) *Dynamical Systems* (revised edtion). American Mathematical Society, Providence, RI
4. Bielawski S, Derozier D, Glorieux P (1993) Experimental characterization of unstable periodic orbits by controlling chaos. Phys Rev A 47:2492–2495
5. Bleich ME, Hochheiser D, Moloney JV, Socolar J (1997) Controlling extended systems with spatially filtered, time-delayed feedback. Phys Rev E 55:2119–2126
6. Bleich ME, Socolar J (2000) Delayed feedback control of a paced excitable oscillator. Int J Bifurc Chaos 10:603–609
7. Brandt ME, Chen G (2000) Time-delay feedback control of complex pathological rhythms in an atrioventricular conduction model. Int J Bifurc Chaos 10:2781–2784
8. Chen G, Dong X (1998) *From Chaos to Order: Methodologies, Perspectives and Applications*. World Scientific, Singapore
9. Chen G, Chen G, de Figueiredo RJP (1999) Feedback control of unknown chaotic dynamical systems based on time-series data. IEEE Trans Circuits Syst I 46:640–644
10. Chen G, Lai D (1996) Feedback control of Lyapunov exponents for discrete-time dynamical systems. Int J Bifurc Chaos 6:1341–1349
11. Chen G, Wang XF (2006) *Chaotification of Dynamical Systems: Theories, Methodologies, and Applications*. Shanghai Jiaotong University Press, Shanghai
12. Chen L (2001) An open-plus-closed-loop control for discrete chaos and hyperchaos. Phys Lett A 281:327–333
13. Davidchack RL, Lai Y (1999) Efficient algorithm for detecting unstable periodic orbits in chaotic systems. Phys Rev E 60:6172–6175
14. Devaney RL (1989) *An Introduction to Chaotic Dynamical Systems*, 2nd edn. Addison-Wesley, New York
15. Diacu F, Holmes P (1996) *Celestial Encounters: The Origins of Chaos and Stability*. Princeton University Press, Princeton, NJ
16. Ding M, Ott E, Grebogi C (1994) Controlling chaos in a temporally irregular environment. Phys D 74:386–394
17. Dyson F (1988) *Infinite in All Directions*. Harper and Row, New York
18. Epureanu BI, Dowell EH (2000) Optimal multi-dimensional OGY controller. Physica D 139:87–96
19. Enns R, McGuire G (1997) *Nonlinear Physics with Maple for Scientists and Engineers*. Birkhäuser, Boston
20. Grebogi C, Lai Y (1997) Controlling chaos in high dimensions. IEEE Trans Circuits Syst I 44:971–975
21. Gutiérrez JM, Iglesias A (1998) Synchronizing chaotic systems with positive conditional Lyapunov exponents by using convex combinations of the drive and response systems. Phys Lett A 239:174–180
22. Hao B (1989) *Elementary Symbolic Dynamics and Chaos in Dissipative Systems*. World Scientific, Singapore
23. Hikihara T, Kawagoshi T (1996) An experimental study on stabilization of unstable periodic motion in magneto-elastic chaos. Phys Lett A 211:29–36
24. Hu G, Xiao J, Zheng Z (2000) *Chaos Control*. Shanghai Scientific and Technological Education Publishing House, Shanghai
25. Hübler A (1989) Adaptive control of chaotic systems. Helv Phys Acta 62:343–347
26. Hunt BR, Ott E, Yorke JA (1997) Differentiable generalized synchronization of chaos. Phys Rev E 55:4029–4034

27. Iplikci S, Denizhan Y (2001) Control of chaotic systems using targeting by extended control regions method. Physica D 150:163–176
28. Ingraham R (1992) *A Survey of Nonlinear Dynamics ('Chaos Theory')*. World Scientific, Singapore
29. Jackson EA (1991) On the control of complex dynamic systems. Physica D 50:341–366
30. Jackson EA, Grosu I (1995) An open-plus-closed-loop (OPCL) control of complex dynamic systems. Physica D 85:1–9
31. Jackson EA, Hübler A (1990) Periodic entrainment of chaotic logistic map dynamics. Physica D 44:407–420
32. Kahan S, Montagne R (2001) Controlling spatio-temporal chaos. Physica A 290:87–91
33. Konishi K, Hirai M, Kokame H (1998) Decentralized delayed-feedback control of a coupled map model for open flow. Phys Rev E 58:3055–3059
34. Konishi K, Kokame H, Hirata K (1999) Coupled map car-following model and its delayed-feedback control. Phys Rev E 60:4000–4007
35. Lan Y, Cvitanović P (2004) Variational method for finding periodic orbits in a general flow. Phys Rev E 69:061217(1–10)
36. Larson LE, Liu JM, Tsimring LS (2006) *Digital Communications using Chaos and Nonlinear Dynamics*. Springer, New York
37. Li TY, Yorke JA (1975) Period three implies chaos. Am Math Mon 82:985–992
38. Lü J, Zhou TS, Chen G, Yang X (2002) Generating chaos with a switching piecewise-linear controller. Chaos 12:344–349
39. Mandelbrot B (1983) *The Fractal Geometry of Nature*. Freeman, Paris
40. Marotto FR (1978) Snap-back repellers imply chaos in $\mathbb{R}^n$. J Math Anal Appl 63:199–223
41. May RM (1976) Simple mathematical model with very complicated dynamics. Nature 261:459–467
42. Mirus KA, Sprott JC (1999) Controlling chaos in a high dimensional system with periodic parametric perturbations. Phys Lett A 254:275–278
43. Nakajima H (1997) On analytical properties of delayed feedback control of chaos. Phys Lett A 232:207–210
44. Nakajima H, Ueda Y (1998) Limitation of generalized delayed feedback control. Physica D 111:143–150
45. Osipanko G (2007) *Dynamical Systems, Graphs, and Algorithms*. Springer, Berlin
46. Ott E, Grebogi C, Yorke JA (1990) Controlling chaos. Phys Rev Lett 64:1196–1199
47. Ott E (2002) *Chaos in Dynamical Systems*. Cambridge University Press, London
48. Pan S, Yin F (1997) Using chaos to control chaotic systems. Phys Lett A 231:173–178
49. Pecora LM, Carroll TL (1990) Synchronization in chaotic systems. Phys Rev Lett 64:821–824
50. Pilovsky AS, Rosenblum MG, Kurths J (2001) *Synchronization: A Universal Concepts in Nonlinear Science*. Cambridge University Press, London
51. Pilovsky AS, Rosenblum MG, Osipov GV, Kurths J (1997) Phase synchronization of chaotic oscillators by external driving. Physica D 104:219–238
52. Pingel D, Schmelcher P, Diakonos FK, Biham O (2000) Theory and application of systematic detection of unstable periodic orbits in dynamical systems. Phys Rev E 62:2119–2134
53. Pingel D, Schmelcher P, Diakonos FK (2001) Detecting unstable periodic orbits in chaotic continuous-time dynamical systems. Phys Rev E 64:026214
54. Pyragas K (1992) Continuous control of chaos by self-controlling feedback. Phys Lett A 170:421–428
55. Ramesh M, Narayanan S (1999) Chaos control by nonfeedback methods in the presence of noise. Chaos Solitons Fractals 10:1473–1489
56. Robinson RC (2004) *An Introduction to Dynamical Systems: Continuous and Discrete*. Prentice Hall, New York
57. Rosenblum MG, Pikovsky AS, Kurths J (1997) From phase to lag synchronization in coupled chaotic oscillators. Phys Rev Lett 78:4193–4196
58. Ruelle D, Takens F (1971) On the nature of turbulence. Commun Math Phys 20:167–192
59. Rulkov NF, Sushchik MM, Tsimring LS, Abarband HDI (1995) Generalized synchronization of chaos in directionally coupled chaotic systems. Phys Rev E 51:980–994

60. Schiff SJ, Jerger K, Duong DH, Chang T, Spano ML, Ditto WL (1994) Controlling chaos in the brain. Nature 370:615–620
61. Sharkovskii AN (1964) Coexistence of cycles of a continuous map of a line into itself. J Ukr Math 16:61–71
62. Shinbort T, Grebogi C, Yorke JA, Ott E (1993) Using small perturbations to control chaos. Nature 363:411–417
63. Shinbort T, Ott E, Grebogi C, Yorke JA (1990) Using chaos to direct orbits to targets. Phys Rev Lett 65:3215–3218
64. Smale S (1967) Differentiable dynamical systems. Bull Amer Math Society 73:747–817
65. So P, Ott E, Sauer T, Gluckman BJ, Grebogi C, Schiff SJ (1997) Extracting unstable periodic orbits from chaotic time series data. Phys Rev E 55:5398–5417
66. Socolar JES, Sukow DW, Gauthier DJ (1994) Stabilizing unstable periodic orbits in fast dynamical systems. Phys Rev E 50:3245–3248
67. Stavroulakis P (2006) *Chaos Applications in Telecommunications*. CRC Press, New York
68. Tang KS, Man KF, Zhong GQ, Chen G (2001) Generating chaos via $x|x|$. IEEE Trans Circuits Syst I 48:636–641
69. Tucker W (2002) A rigorous ODE solver and Smale's 14th problem. Found Comput Math 2:53–117
70. Ushio T (1996) Limitation of delayed feedback control in nonlinear discrete-time systems. IEEE Trans Circuits Syst I 43:815–816
71. Vanecek A, Celikovsky S (1996) *Control Systems: From Linear Analysis to Synthesis of Chaos*. Prentice Hall, Upper Saddle River, NJ
72. Voss H (2000) Anticipating chaotic synchronization. Phys Rev E 61:5115–5119
73. Wang XF, Chen G (2003) Generating topologically conjugate chaotic systems via feedback control. IEEE Trans Circuits Syst I 50:812–817
74. Wang XF, Chen G, Yu X (2000) Anticontrol of chaos in continuous-time systems via time-delay feedback. Chaos 10:771–779
75. Wiggins S (2003) *Introduction to Applied Nonlinear Dynamical Systems and Chaos*, 2nd edn. Springer, New York
76. Yagasaki K, Uozumi T (1998) New approach for controlling chaotic dynamical systems. Phys Lett A 238:349–357
77. Yang XS, Li QD (2007) *Chaotic Systems and Chaotic Circuits*. Science Press, Beijing (in Chinese)
78. Zhang HG, Wang ZL, Liu D (2005) Chaotifying fuzzy hyperbolic model using impulsive and nonlinear feedback control approaches. Int J Bifurc Chaos 15:2603–2610
79. Zhao Y (2003) Introduction to some methods of chaos analysis and control for PDEs. In: Chen G (ed) *Chaos Control*. Springer, New York
80. Zhang Z (1997) The Principle of Differential Dynamical Systems. Science Press, Beijing

Chapter 2
Preliminaries of Nonlinear Dynamics and Chaos

Abstract This chapter provides a brief review of some concepts and tools related to the subject of the monograph – chaos suppression, chaos synchronization, and chaotification. After a quick review of the history of 'dynamical systems,' we provide a summary of important definitions and theorems, including equilibrium points, periodic orbits, quasiperiodic orbits, stable and unstable manifolds, attractors, chaotic attractors, Lyapunov stability, orbital stability, and symbolic dynamics, which are all from the theory of ordinary differential equations and ordinary difference equations. The results are summarized both for continuous time and for discrete time. Then, we present three examples for chaotic attractors including the logistic map, the Lorenz attractor, and the Smale horseshoe. At the end of the chapter, we provide some necessary definitions and theorems of functional differential equations (PDEs).

2.1 Introduction

Roughly speaking, a *dynamical system* consists of two ingredients: a rule or 'dynamics,' which is described by a set of equations (difference, differential, integral, functional, or abstract operator equations, or a combination of some of them) and specifies how a system evolves, and an initial condition or 'state' from which the system starts. A *nonlinear dynamical system* is a dynamical system described by a set of nonlinear equations; that is, the dynamical variables describing the properties of the system (for example, position, velocity, acceleration, pressure, etc.) appear in the equations in a nonlinear form. The most successful class of rules for describing natural phenomena are differential equations. All the major theories of physics are stated in terms of differential equations. This observation led the mathematician V. I. Arnol'd to comment, 'consequently, differential equations lie at the basis of scientific mathematical philosophy' [2]. This scientific philosophy began with the discovery of calculus by Newton and Leibniz and continues to the present day. The theory of *dynamical systems* grew out of the qualitative study of differential equations, which in turn began as an attempt to understand and predict the motions that surround us such as the orbits of the planets, the vibrations of a string, the ripples on the surface of a pond, and the forever evolving patterns of the weather. The first two hundred years of this scientific philosophy, from Newton and Euler to Hamilton and Maxwell, produced many stunning successes in formulating the 'rules of the world,' but only limited results in finding their solutions.

By the end of the 19th century, researchers had realized that many nonlinear differential equations did not have explicit solutions. Even the case of three masses moving under the laws of Newtonian attraction could exhibit very complicated behavior and its explicit solution was not possible to obtain (e.g., the motion of the sun, the earth, and the moon cannot be given explicitly in terms of known functions). Short-term solutions could be given by power series, but these were not useful in determining long-term behavior. The modern theory of nonlinear dynamical systems began with Poincaré at the end of the 19th century with fundamental questions concerning the stability and evolution of the solar system. Poincaré shifted the focus from finding explicit solutions to discovering geometric properties of solutions. He introduced many ideas in specific examples. In particular, he realized that a deterministic system in which the outside forces are not varying and are not random can exhibit behavior that is apparently random (i.e., chaotic). Poincaré's point of view was enthusiastically adopted and developed by G. D. Birkhoff. He found many different types of long-term limiting behavior. His work resulted in the book [4] from which the term 'dynamical systems' came. Other people, such as Lyapunov, Pontryagin, Andronov, Morser, Smale, Peixoto, Kolmogorov, Arnol'd, Sinai, Lorenz, May, Yorke, Feigenbaum, Ruelle, and Takens, all made important contributions to the theory of dynamical systems. The field of nonlinear dynamical systems and especially the study of chaotic systems has been hailed as one of the important breakthroughs in science in the 20th century. Today, nonlinear dynamical systems are used to describe a vast variety of scientific and engineering phenomena and have been applied to a broad spectrum of problems in physics, chemistry, mathematics, biology, medicine, economics, and various engineering disciplines.

This chapter is a brief review of some concepts and tools related to the subject of the monograph – chaos suppression, chaos synchronization, and chaotification. The goal of the chapter is to provide readers with some necessary background on nonlinear dynamical systems and chaos so as to ease the difficulty when they read subsequent chapters of this book. Readers interested in the complete theory of dynamical systems are recommended to refer to [1, 7, 11, 13, 17].

2.2 Background

Two types of models are extensively studied in the field of dynamical systems: the continuous-time model and the discrete-time model. Most continuous-time nonlinear dynamical systems are described by a differential equation of the form

$$\dot{x} = f(x,t;\mu), \tag{2.1}$$

with $x \in U \subset \mathbb{R}^n$, $t \in \mathbb{R}$, and $\mu \in V \subset \mathbb{R}^p$, where U and V are open sets in $\mathbb{R}^n$ and $\mathbb{R}^p$. Meanwhile, most discrete-time nonlinear dynamical systems are described by an equation of the form

$$x(k+1) = f(x(k),k;\mu), \quad k = 0,1,2,\dots. \tag{2.2}$$

We refer to (2.1) as a *vector field* or *ordinary differential equation* and to (2.2) as a *map* or *difference equation*. By a solution of (2.1) we mean a map, x, from some interval $I \subset \mathbb{R}$ into $\mathbb{R}^n$, which is denoted by

$$x\colon I \to \mathbb{R}^n,$$
$$t \mapsto x(t),$$

such that $x(t)$ satisfies (2.1), i.e.,

$$\dot{x}(t) = f(x(t),t;\mu).$$

The map x has the geometrical interpretation of a curve in $\mathbb{R}^n$, and (2.1) gives the tangent vector at each point of the curve, hence the reason for referring to (2.1) as a vector field. We will refer to the space of independent variables of (2.1) (i.e., $\mathbb{R}^n$) as the *phase space* or *state space*. One goal of the study of dynamical systems is to understand the geometry of solution curves in the phase space. It is useful to distinguish a solution curve which passes through a particular point in the phase space at a specific time, i.e., for a solution $x(t)$ with $x(t_0) = x_0$. We refer to this as specifying an *initial condition* or *initial value*. This is often included in the expression for a solution by $x(t,t_0,x_0)$. In some situations explicitly displaying the initial condition may be unimportant, in which case we will denote the solution merely as $x(t)$. In other situations the initial time may be always understood to be a specific value, say $t_0 = 0$; in this case we would denote the solution as $x(t,x_0)$. Similarly, it may be useful to explicitly display the parametric dependence of solutions. In this case we would write $x(t,t_0,x_0;\mu)$ or, if we were not interested in the initial condition, $x(t;\mu)$.

Ordinary differential equations that depend explicitly on time (i.e., $\dot{x} = f(x,t;\mu)$) are referred to as *nonautonomous* or *time-dependent* ordinary differential equations or vector fields, and ordinary differential equations that do not depend explicitly on time (i.e., $\dot{x} = f(x;\mu)$) are referred to as *autonomous* or *time-independent* ordinary differential equations or vector fields. The same terminology may be used in the same way for discrete-time systems. It should be noted that a nonautonomous vector field or map can always be made autonomous by redefining time as a new independent variable. This is done as follows. For a vector field $\dot{x} = f(x,t)$, by writing it as

$$\frac{\mathrm{d}x}{\mathrm{d}t} = \frac{f(x,t)}{1} \tag{2.3}$$

and using the chain rule, we can introduce a new independent variable s so that (2.3) becomes

$$\begin{cases} \dfrac{\mathrm{d}x}{\mathrm{d}s} \equiv \dot{x} = f(x,t), \\ \dfrac{\mathrm{d}t}{\mathrm{d}s} \equiv \dot{t} = 1. \end{cases} \tag{2.4}$$

If we define $y = (x,t)^{\mathrm{T}}$ and $\tilde{f}(y) = (f(x,t),1)^{\mathrm{T}}$, we see that (2.4) becomes

$$\frac{\mathrm{d}y}{\mathrm{d}s} = \tilde{f}(y), \quad y \in \mathbb{R}^{n+1}.$$

For the map $x(k+1) = f(x(k),k)$, if we define $y(k) = (x(k),k)^{\mathrm{T}}$ and $\tilde{f}(y) = (f(x(k),k),k+1)^{\mathrm{T}}$, we get the autonomous system under the new phase space

$$y(k+1) = \tilde{f}(y(k)), \quad y \in \mathbb{R}^{n+1}.$$

So, it is generally sufficient to consider autonomous systems

$$\dot{x} = f(x), \quad x \in \mathbb{R}^n, \tag{2.5}$$

and

$$x(k+1) = g(x(k)), \quad x \in \mathbb{R}^n. \tag{2.6}$$

2.3 Existence, Uniqueness, Flow, and Dynamical Systems

2.3.1 Existence and Uniqueness

Consider the autonomous vector field (2.5). Geometrically, $x(t)$ is a curve in $\mathbb{R}^n$ whose tangent vector $\dot{x}(t)$ exists for all t in its dominant field J and equals $f(x(t))$. For simplicity, we usually take initial time $t_0 = 0$. The main problem in differential equations is to find the solution for any *initial value problem*; that is, to determine the solution of the system that satisfies the initial condition $x(t_0) = x_0$ for each $x_0 \in \mathbb{R}^n$.

Unfortunately, nonlinear differential equations may have no solutions satisfying certain initial conditions.

Example 2.1 ([9]). Consider the following simple first-order differential equation:

$$\dot{x} = \begin{cases} 1, & \text{if } x < 0, \\ -1, & \text{if } x \geq 0. \end{cases}$$

This vector field on $\mathbb{R}$ points to the left when $x \geq 0$ and to the right if $x < 0$. Consequently, there is no solution that satisfies the initial condition $x(0) = 0$. Indeed, such a solution must initially decrease since $\dot{x}(0) = -1$, but, for all negative values of x, solutions must increase. This cannot happen. Note further that solutions are never defined for all time. For example, if $x_0 > 0$, then the solution through x_0 is given by $x(t) = x_0 - t$, but this solution is only valid for $-\infty < t < x_0$ for the same reason as above.

The problem in this example is that the vector field is not continuous at 0; whenever a vector field is discontinuous we face the possibility that nearby vectors may point in 'opposing' directions, thereby causing solutions to halt at these points. □

Example 2.2. Consider the following differential equation:

$$\dot{x} = 3x^{2/3}.$$

The identically zero function $u\colon \mathbb{R} \to \mathbb{R}$ given by $u(t) \equiv 0$ is clearly a solution with initial condition $u(0) = 0$. But $u_0(t) = t^3$ is also a solution satisfying this initial condition. Moreover, for any $\tau > 0$, the function given by

$$u_\tau(t) = \begin{cases} 0, & \text{if } t \leq \tau, \\ (t-\tau)^3, & \text{if } t > \tau \end{cases}$$

is also a solution satisfying the initial condition $u_\tau(0) = 0$. While the differential equation in this example is continuous at $x_0 = 0$, the problems arise because $3x^{2/3}$ is not differentiable at this point. □

From these two examples it is clear that, to ensure the existence and uniqueness of solutions, certain conditions must be imposed on the function f. In the first example, f is not continuous at the point 0, while, in the second example, f fails to be differentiable at 0. It turns out that the assumption that f is continuously differentiable is sufficient to guarantee both existence and uniqueness of the solution. In fact, we can furthermore guarantee the existence and uniqueness under a weaker condition, called the Lipschitz condition, on f. We now state several qualitative theorems about the solutions of system (2.5) [7].

Theorem 2.1 (Local Existence and Uniqueness). Let $U \subset \mathbb{R}^n$ be an open subset of real Euclidean space (or of a differentiable manifold M[1]), let $x_0 \in U$, and let $f\colon U \to \mathbb{R}^n$ be a (locally) Lipschitzian map, i.e.,

$$\|f(y) - f(x)\| \leq K\|x - y\|$$

for some $K < \infty$. Then, there are some constant $c > 0$ and a unique solution $x(\cdot, x_0)\colon (-c, c) \to U$ satisfying the differential equation described by (2.5) with initial condition $x(0) = x_0$. □

[1] Roughly speaking, a manifold is a set which locally has the structure of Euclidean space. In applications, manifolds are most often met as m-dimensional surfaces embedded in $\mathbb{R}^n$. If the surface has no singular points, i.e., the derivative of the function representing the surface has maximal rank, then by the implicit function theorem it can locally be represented as a graph. The surface is a C^r manifold if the (local) coordinate charts representing it are C^r.

The local existence theorem becomes global in all cases when we work on *compact* manifolds[2] M instead of open spaces like $\mathbb{R}^n$.

Theorem 2.2 (Global Existence). The differential equation $\dot{x} = f(x)$, $x \in M$, with M compact, and $f \in C^1$, has solution curves defined for all $t \in \mathbb{R}$. □

The local theorem can be extended to show that solutions depend in a 'nice' manner on initial conditions.

Theorem 2.3 (Dependence on Initial Value). Let $U \in \mathbb{R}^n$ be open and suppose that $f\colon U \to \mathbb{R}^n$ has a Lipschitz constant K. Let $y(t), z(t)$ be solutions of $\dot{x} = f(x)$ on the closed interval $[t_0, t_1]$. Then, for all $t \in [t_0, t_1]$,

$$\|y(t) - z(t)\| \leq \|y(t_0) - z(t_0)\| \mathrm{e}^{K(t-t_0)}.$$ □

2.3.2 *Flow and Dynamical Systems*

If $x(t)$ is a solution of (2.5), then $x(t+\tau)$ is also a solution for any $\tau \in \mathbb{R}$. So, it suffices to choose a fixed initial time, say, $t_0 = 0$, which is understood and therefore often omitted from the notation. If we denote by $\phi_t(x) = \phi(t,x)$ the state in $\mathbb{R}^m$ reached by the system at time t starting from x, then the totality of solutions of (2.5) can be represented by a one-parameter family of maps $\phi^t : U \to \mathbb{R}^m$ satisfying

$$\frac{\mathrm{d}}{\mathrm{d}t}[\phi(t,x)]\Big|_{t=\tau} = f[\phi(\tau,x)]$$

for all $x \in U$ and for all $\tau \in I$ for which the solution is defined. The family of maps $\phi_t(x) = \phi(t,x)$ is called the *flow* (or the flow map) generated by the vector field f. The set of points $\{\phi(t,x_0) : t \in I\}$ defines an *orbit* of (2.5), starting from a given point x_0. It is a solution curve in the state space, parameterized by time. The set $\{[t, \phi(t,x_0)] : t \in I\}$ is a *trajectory* of (2.5) and it evolves in the space of motions. However, in applications, the terms 'orbit' and 'trajectory' are often used as synonyms. A simple example of a trajectory in the space of motions $\mathbb{R} \times \mathbb{R}^2$ and the corresponding orbit in the state space $\mathbb{R}^2$ are given in Fig. 2.1. Clearly, the orbit is obtained by projecting the trajectory on to the state space. The flows generated by vector fields form a very important subset of a more general class of maps, characterized by the following definition.

Definition 2.1. A *flow* is a map $\phi : I \subset \mathbb{R} \times X \to X$ where X is a metric space, that is, a space endowed with a distance function, and ϕ has the following properties:

[2] A compact manifold is a manifold that is compact as a topological space, such as the circle (the only one-dimensional compact manifold) and the n-dimensional sphere and torus. For many problems in topology and geometry, it is convenient to study compact manifolds because of their 'nice' behavior. Among the properties making compact manifolds 'nice' are the facts that they can be covered finitely by many coordinate charts and that any continuous real-valued function is bounded on a compact manifold.

(i) $\phi(0,x) = x$ for every $x \in X$ (identity axiom);
(ii) $\phi(t+s,x) = \phi(s,\phi(t,x)) = \phi(t,\phi(s,x)) = \phi(s+t,x)$, that is, time-translated solutions remain solutions;
(iii) for fixed t, $\phi^t := \phi(t,\cdot)$ is a homeomorphism[3] of the phase space on X. □

Remark 2.1. A *distance* on a space X (or, a *metric* on X) is a function $d(\cdot,\cdot)\colon X \times X \to \mathbb{R}^+$ satisfying the following properties for all $x,y \in X$:

(i) $d(x,y) \geq 0$ and $d(x,y) = 0$ if and only if $x = y$;
(ii) $d(x,y) = d(y,x)$ (symmetry);
(iii) $d(x,y) \leq d(x,z) + d(z,y)$ (triangle inequality).

Notice that there also exist notions of distance which are perfectly meaningful but do not satisfy the definition above and therefore do not define a metric, for example the distance between a point and a set A

$$d(x,A) = \inf_{y \in A} d(x,y)$$

and the distance between two sets A and B

$$d(A,B) = \inf_{x \in A} \inf_{y \in B} d(x,y).$$ □

In the following, we give a formal definition of a 'dynamical system.'

Definition 2.2. A *dynamical system* is a triplet $\{T, X, \phi^t\}$ where T is a time set, X is a state space, and $\phi^t : X \to X$ is a flow parameterized by $t \in T$. □

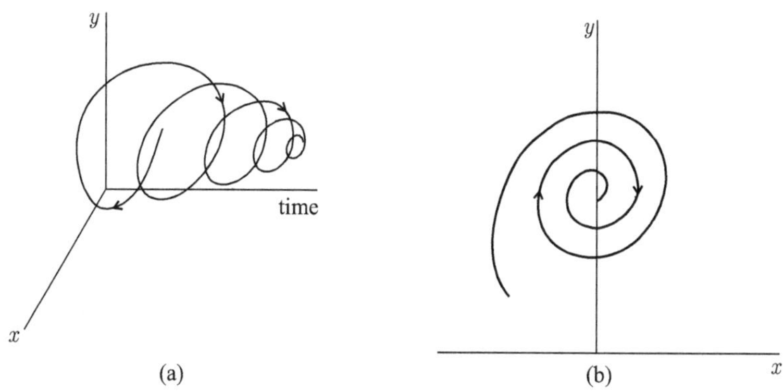

Fig. 2.1 A damped oscillator in $\mathbb{R}^2$: (a) space of motions; (b) state space

[3] $h\colon U \subset \mathbb{R}^n \to V \subset \mathbb{R}^m$ is said to be a C^r diffeomorphism if both h and h^{-1} are C^r. h is called a homeomorphism if $r = 0$.

2.4 Equilibrium, Periodic Orbit, Quasiperiodic Orbit, and Poincaré Map

2.4.1 Equilibrium of Continuous-Time Systems

Definition 2.3 (Equilibrium of Autonomous Systems). An *equilibrium solution* of (2.5) is a point $\bar{x} \in \mathbb{R}^n$ such that

$$f(\bar{x}) = 0,$$

i.e., a solution which does not change in time. Other terms often substituted for the term 'equilibrium solution' are 'fixed point,' 'stationary point,' 'rest point,' 'singularity,' 'critical point,' or 'steady state.' □

Remark 2.2. What about the notion of equilibria of nonautonomous vector fields? We should note that ideas developed for autonomous systems can lead to incorrect results for nonautonomous systems. For example, consider the following one-dimensional nonautonomous field:

$$\dot{x} = -x + t. \tag{2.7}$$

The solution through the point x_0 at $t = 0$ is given by

$$x(t) = t - 1 + \mathrm{e}^{-t}(x_0 + 1),$$

from which it is clear that all solutions asymptotically approach the solution $t - 1$ as $t \to \infty$. The frozen time or 'instantaneous' fixed points for (2.7) are given by

$$x = t.$$

At a fixed t, this is the unique point where the vector field is zero. However, $x = t$ is not a solution of (2.7). This is different from the case of an autonomous vector field where a fixed point is a solution of the vector field. □

Definition 2.4. Consider the following nonautonomous system:

$$\dot{x} = f(t, x), \tag{2.8}$$

where $f\colon [0,\infty) \times D \to \mathbb{R}^n$ is piecewise continuous in t and locally Lipschitz in x on $[0,\infty) \times D$, and $D \subset \mathbb{R}^n$ is a domain that contains the origin $x = 0$. The origin is an equilibrium point of (2.8) at $t = 0$ if

$$f(t, 0) = 0, \quad \forall t \geq 0.$$ □

For the discrete-time system (2.6), an equilibrium solution is the point $\bar{x} \in \mathbb{R}^n$ such that

$$\bar{x} = g(\bar{x}).$$

An equilibrium solution of a discrete-time system is usually called a fixed point.

2.4.2 Periodic and Quasiperiodic Orbits

Consider again the basic system of differential equation

$$\dot{x} = f(x) \tag{2.9}$$

and the derived flow ϕ. A solution $\phi(t,x^*)$ of system (2.9) through a point x^* is said to be *periodic* with period $T > 0$ if $\phi(T,x^*) = x^*$. The set $L_0 = \{\phi(t,x^*) : t \in [0,T)\}$ is a closed curve in the state space and is called a *periodic orbit* or *cycle*. T is called the period of the cycle and measures its time length. It should be emphasized that isolated periodic solutions are possible only for nonlinear differential equations. Moreover, a *limit cycle*[4] can be *structurally stable* in the sense that, if it exists for a given system of differential equations, it will persist under a slight perturbation of the system in the parameter space. On the contrary, linear systems of differential equations in $\mathbb{R}^m$ $(m \geq 2)$ may have a continuum of periodic solutions characterized by a pair of purely imaginary eigenvalues (the case of a *center*, which will be introduced later). But, these periodic solutions can be destroyed by arbitrarily small perturbations of the coefficients. In other words, these periodic solutions are not structurally stable.

For the discrete-time system of $x_{k+1} = G(x_k)$, an n-periodic orbit is defined as the set of points $L_0 = \{x_0, x_1, \ldots, x_{n-1}\}$ with $x_i \neq x_j$ $(i \neq j)$ such that

$$x_1 = G(x_0),\ x_2 = G(x_1), \ldots, x_{n-1} = G(x_{n-2}),\ x_0 = G(x_{n-1}).$$

It should be noted that each point in an n-periodic orbit is an *n-periodic point* since, for $k = 0, \ldots, n-1$,

$$x_k = G^n(x_k) \text{ and } G^j(x_k) \neq x_k \text{ for } 0 < j < n.$$

Periodic orbits of continuous-time systems and discrete-time systems are illustrated in Fig. 2.2.

To illustrate what quasiperiodic orbits are we will consider two examples, one for discrete time and one for continuous time.

Example 2.3. Consider the following unit-circle map:

$$M_c \colon S^1 \to S^1,\ z_{n+1} = M_c(z_n) = cz_n, \tag{2.10}$$

where $z_n = \mathrm{e}^{i2\pi\theta_n}$, $\theta_n \in \mathbb{R}$, α is a positive constant, and $c = \mathrm{e}^{i2\pi\alpha}$. The map (2.10) describes an anticlockwise jump of a particle on the unit circle S^1. The length of the circular arc between two adjacent jump points is α. If α is rational, that is, $\alpha = p/q$ with p and q integers, then any (initial) point on S^1 is a q-periodic point of the map M_c. If α is irrational, at each iteration a new point is added on the unit

---

[4] A limit cycle is an isolated periodic solution of an autonomous system. The points on the limit cycle constitute the *limit set*, which is the set of points in the state space that a trajectory repeatedly visits. A limit set is only defined for discrete-time or continuous-time autonomous systems.

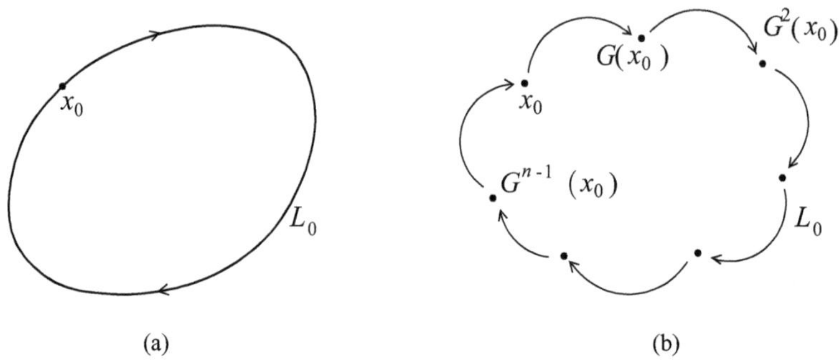

Fig. 2.2 Periodic orbits in: (a) a continuous-time system; (b) a discrete-time system

circle. For any jump point there exist other jump points arbitrarily close to it, but no point is revisited in finite time. This means that the map is topologically transitive (see Definition 2.12). However, points that are close to each other will remain close under the iteration. □

Example 2.4. The simplest example of quasiperiodic motion in continuous time is a system defined by a pair of oscillators of the form

$$\ddot{x} + \omega_1^2 x = 0, \;\; \ddot{y} + \omega_2^2 y = 0,$$

where $x, y \in \mathbb{R}$ and ω_1 and ω_2 are real constants. The above system can be rewritten in the form of first-order linear differential equations in $\mathbb{R}^4$:

$$\begin{cases} \dot{x}_1 = -\omega_1 x_2, \\ \dot{x}_2 = \omega_1 x_1, \\ \dot{y}_1 = -\omega_2 y_2, \\ \dot{y}_2 = \omega_2 y_1, \end{cases} \tag{2.11}$$

where $x_2 = x$ and $y_2 = y$. Transforming the variables x_1,x_2 and y_1,y_2 into polar coordinates, system (2.11) can be written as

$$\begin{cases} \dot{\theta}_1 = \omega_1, \\ \dot{r}_1 = 0, \\ \dot{\theta}_2 = \omega_2, \\ \dot{r}_2 = 0, \end{cases} \tag{2.12}$$

where θ_i and r_i $(i = 1, 2)$ denote the angle and the radius, respectively. We can see that the above equations describe a particle rotating on a two-dimensional torus for a given pair (r_1, r_2), $r_i > 0$ $(i = 1, 2)$ (see Fig. 2.3). There are two basic possibilities for the motion:

(i) ω_1/ω_2 is a rational number, in which case there exists a continuum of periodic orbits of period q;

(ii) ω_1/ω_2 is an irrational number, in which case the orbit starting from any initial point on the torus wanders on it, getting arbitrarily near any other point, without ever returning to that exact initial point. The flow generated by (2.12) is topologically transitive on the torus (see Fig. 2.4).

In both cases, points that are close to each other remain close under the action of the flow. □

A general definition of quasiperiodicity of an orbit as a function of time can be given as follows:

Definition 2.5 ([11], p. 128). A function $h\colon \mathbb{R} \to \mathbb{R}^m$ is called quasiperiodic if it can be written in the form of $h(t) = H(\omega_1 t, \omega_2 t, \ldots, \omega_m t)$, where H is periodic of period 2π in each of its arguments, and two or more of the m (positive) frequencies are incommensurable. □

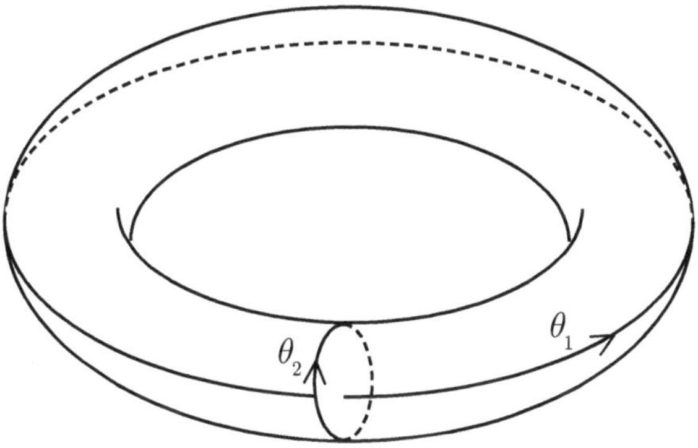

Fig. 2.3 The motion on two-dimensional torus

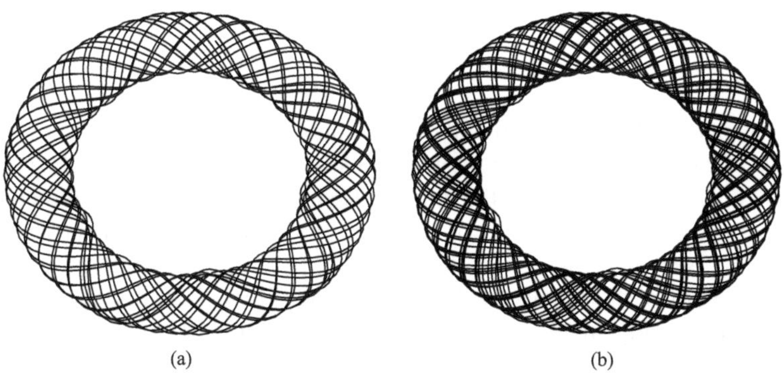

Fig. 2.4 The quasiperiodic motions with different evolution times

2.4.3 Poincaré Map

A *Poincaré map* is a classical technique for analyzing dynamical systems. It replaces the flow of an nth-order continuous-time system with an $(n-1)$st-order discrete-time system. The definition of the Poincaré map ensures that the limit sets of the discrete-time system correspond to the limit sets of the underlying flow. The Poincaré map's usefulness lies in the reduction of order and the fact that it bridges the gap between continuous-time and discrete-time systems.

The definitions of a Poincaré map are different for autonomous systems and nonautonomous systems. We present the two cases separately.

Case 1: In this case we consider an nth-order time-periodic nonautonomous system $\dot{x} = f(t,x)$, with period T. We can convert it into an $(n+1)$st-order autonomous system by appending an extra state $\theta := 2\pi t/T$. Then, the autonomous system is given by

$$\begin{cases} \dot{x} = f(x, \theta T/2\pi), & x(t_0) = x_0, \\ \dot{\theta} = 2\pi/T, & \theta(t_0) = 2\pi t_0/T. \end{cases} \tag{2.13}$$

Since f is time periodic with period T, system (2.13) is periodic in θ with period 2π. Hence, the planes $\theta = 0$ and $\theta = 2\pi$ may be identified and the state space transformed from the Euclidean space $\mathbb{R}^{n+1}$ to the cylindrical space $\mathbb{R}^n \times S^1$, where S^1 is the unit circle. The solution of (2.13) in the cylindrical state space is

$$\begin{pmatrix} x(t) \\ \theta(t) \end{pmatrix} = \begin{pmatrix} \phi_t(x_0, t_0) \\ 2\pi t/T \bmod 2\pi \end{pmatrix}, \tag{2.14}$$

where the modulo function restricts to $0 \leq \theta < 2\pi$. Consider the n-dimensional hyperplane $\Sigma \in \mathbb{R}^n \times S^1$ defined by

$$\Sigma := \{(x,\theta) \in \mathbb{R}^n \times S^1 : \theta = \theta_0\}.$$

Every T seconds, the trajectory of (2.14) intersects Σ (see Fig. 2.5). The resulting map $P_{\mathrm{N}} : \Sigma \to \Sigma$ $(\mathbb{R}^n \to \mathbb{R}^n)$ is defined by

$$P_{\mathrm{N}}(x) := \phi_{t_0+T}(x, t_0).$$

P_{N} is called the Poincaré map of the nonautonomous system.

Case 2: Consider an nth-order autonomous system with a limit cycle Γ shown in Fig. 2.6. Let x_0 be a point on the limit cycle and let Σ be an $(n-1)$-dimensional hyperplane transversal to Γ at x_0. The trajectory emanating from x_0 will hit Σ at x_0 in T seconds, where T is the minimum period of the limit cycle. Due to the continuity of ϕ_t with respect to the initial condition, the trajectory starting on Σ in a sufficiently small neighborhood of x_0 will, in approximately T seconds, intersect Σ in the vicinity of x_0. Hence, ϕ_t and Σ define a mapping P_{A} of some neighborhood $U \subset \Sigma$ of x_0 on to another neighborhood $V \subset \Sigma$ of x_0. P_{A} is a Poincaré map of the autonomous system.

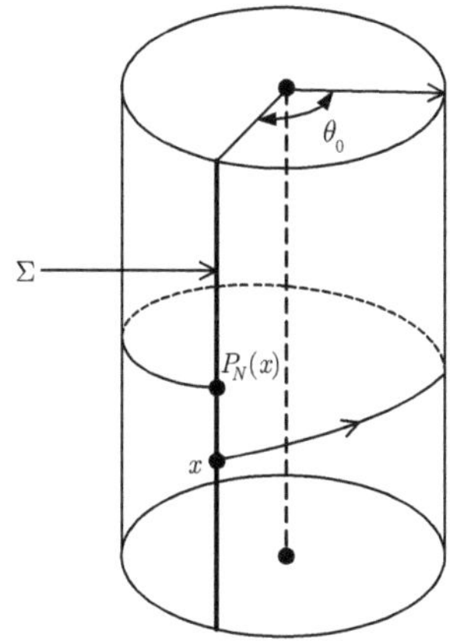

Fig. 2.5 The Poincaré map of a first-order nonautonomous system

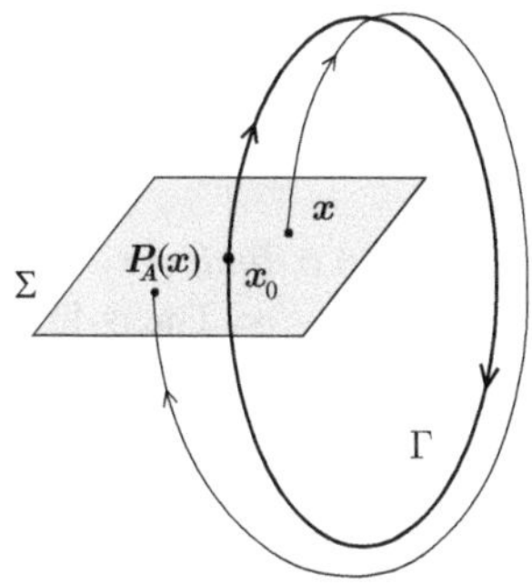

Fig. 2.6 The Poincaré map of a third-order autonomous system

Remark 2.3.

(i) P_A is defined locally, i.e., in a neighborhood of x_0. Unlike the nonautonomous case, it is not guaranteed that the trajectory emanating from any point on Σ will intersect Σ.
(ii) For a Euclidean state space, the point $P_A(x)$ is not the first point where $\phi_t(x)$ intersects Σ; $\phi_t(x)$ must pass through Σ at least once before returning to V. This is in contrast with the cylindrical state space in Fig. 2.5.
(iii) P_A is a diffeomorphism and is, therefore, invertible and differentiable [12]. □

2.5 Invariant and Attracting Sets

Definition 2.6 (Invariant Set). Let $S \subset \mathbb{R}^n$ be a set. Then,

(i) (Continuous time) S is said to be *invariant* under the vector field $\dot{x} = f(x)$ if for any $x_0 \in S$ we have $x(t,0,x_0) \in S$ for all $t \in \mathbb{R}$, where $x(0,0,x_0) = x_0$.

(ii) (Discrete time) S is said to be *invariant* under the map $x_{k+1} = g(x_k)$ if for any $x_0 \in S$ we have $g^n(x_0) \in S$ for all $n \in \mathbb{Z}$.

If we restrict ourselves to positive time (i.e., $t \geq 0$, and $n \geq 0$), then we refer to S as a *positively invariant set*, while, for negative time, as a *negatively invariant set*. □

The definition means that trajectories starting in the invariant set remain in the invariant set, for all of their future and all of their past.

Definition 2.7. An invariant set $S \subset \mathbb{R}^n$ is said to be a C^r $(r \geq 1)$ *invariant manifold* if S has the structure of a C^r differentiable manifold. Similarly, a positively (negatively) invariant set $S \subset \mathbb{R}^n$ is said to be a $C^r(r \geq 1)$ *positively (negatively) invariant manifold* if S has the structure of a C^r differentiable manifold. □

Definition 2.8. Let $\phi(t,x)$ be a flow on a metric space M. Then, a point $y \in M$ is called an *ω-limit point* of $x \in M$ for $\phi(t,x)$ if there exists an infinitely increasing sequence $\{t_i\}$ such that

$$\lim_{i \to \infty} d(\phi(t_i,x),y) = 0.$$

The set of all ω-limit points of x for $\phi(t,x)$ is called the *ω-limit set* and is denoted by $\omega(x)$. □

The definitions of *α-limit point* and *α-limit set* of a point $x \in M$ are obtained just by taking sequences t_i decreasing in i to $-\infty$. The α-limit set of x is denoted as $\alpha(x)$.

Definition 2.9. A point x_0 is called *nonwandering* if the following condition holds.
Flows: for any neighborhood U of x_0 and $T > 0$, there exists some $|t| > T$ such that

$$\phi(t,U) \cap U \neq \emptyset;$$

Maps: for any neighborhood U of x_0, there exists some $n \neq 0$ such that

$$g^n(U) \cap U \neq \emptyset.$$

The set of all nonwandering points of a flow or map is called the *nonwandering set* of that particular flow or map. □

Definition 2.10. A closed invariant set $A \subset \mathbb{R}^n$ is called an *attracting set* if there is some neighborhood U of A such that
Flows: for any $t \geq 0$, $\phi(t,U) \subset U$ and $\bigcap_{t>0} \phi(t,U) = A$;
Maps: for any $n \geq 0$, $g^n(U) \subset U$ and $\bigcap_{n>0} g^n(U) = A$. □

Definition 2.11. The *basin of attraction* of an attracting set A is given by
Flows: $\bigcup_{t \leq 0} \phi(t,U)$;
Maps: $\bigcup_{n \leq 0} g^n(U)$;
where U is any open set satisfying Definition 2.10. □

Definition 2.12. A closed invariant set A is said to be *topologically transitive* if, for any two open sets $U, V \subset A$,
Flows: there exists a $t \in \mathbb{R}$ such that $\phi(t,U) \cap V \neq \emptyset$;
Maps: there exists an $n \in \mathbb{Z}$ such that $g^n(U) \cap V \neq \emptyset$. □

Definition 2.13. An *attractor* is a topologically transitive attracting set. □

2.6 Continuous-Time Systems in the Plane

In this section and the next two sections we will discuss the types of equilibrium points of planar systems of continuous time and discrete time, respectively. In applications, we very often encounter linear systems described by two first-order differential equations (or a differential equation of second order), either because the underlying model is linear or because it is linearized around an equilibrium point. Systems in two-dimensional space are particularly easy to discuss in full detail and give rise to a number of interesting basic dynamic configurations. Moreover, in practice, it is very difficult or impossible to determine the exact values of the eigenvalues and eigenvectors for matrices of order greater than two. Thus, one can draw inspiration from the discussion about planar systems when studying high-dimensional systems.

The general form of a continuous-time planar system can be written as

$$\begin{pmatrix} \dot{x} \\ \dot{y} \end{pmatrix} = A \begin{pmatrix} x \\ y \end{pmatrix} = \begin{pmatrix} a_{11} & a_{12} \\ a_{21} & a_{22} \end{pmatrix} \begin{pmatrix} x \\ y \end{pmatrix}, \tag{2.15}$$

where $x, y \in \mathbb{R}$ and a_{ij} are real constants. If $\det(A) \neq 0$, the unique equilibrium, for which $\dot{x} = \dot{y} = 0$, is $x = y = 0$. The characteristic equation is

$$\lambda^2 - \mathrm{tr}(A)\lambda + \det(A) = 0,$$

and the eigenvalues are

$$\lambda_{1,2} = \frac{1}{2}\left(\mathrm{tr}(A) \pm \sqrt{\Delta}\right),$$

where $\Delta \equiv (\mathrm{tr}(A))^2 - 4\det(A)$ is called the *discriminant*. For system (2.15) the discriminant is

$$\Delta = (a_{11} - a_{22})^2 + 4a_{12}a_{21}.$$

The different types of dynamical behavior of (2.15) can be described in terms of the two eigenvalues of the matrix A, which in planar systems can be completely characterized by the trace and determinant of A. In the following we consider nondegenerate equilibria for which λ_1 and λ_2 are both nonzero, when there is no explicit claim. We distinguish behaviors according to the sign of the discriminant.

Case 1: $\Delta > 0$. Eigenvalues and eigenvectors are real. Solutions have the form

$$\begin{cases} x(t) = c_1 e^{\lambda_1 t} u_1^{(1)} + c_2 e^{\lambda_2 t} u_2^{(1)}, \\ y(t) = c_1 e^{\lambda_1 t} u_1^{(2)} + c_2 e^{\lambda_2 t} u_2^{(2)}, \end{cases}$$

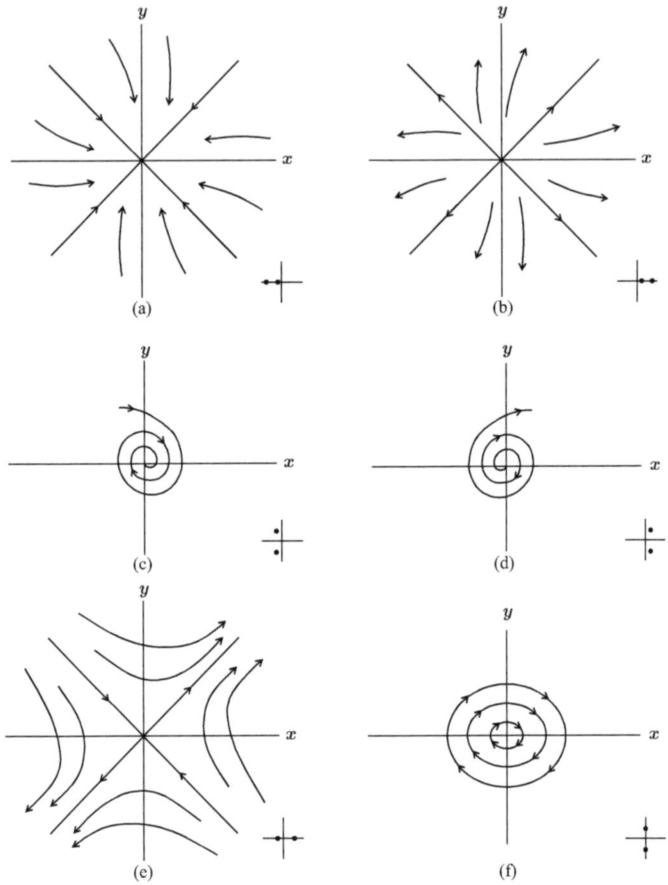

Fig. 2.7 Equilibrium types in the plane

where $u_1 = \left(u_1^{(1)}, u_1^{(2)}\right)^{\mathrm{T}}$ and $u_2 = \left(u_2^{(1)}, u_2^{(2)}\right)^{\mathrm{T}}$ are eigenvectors corresponding to the eigenvalues λ_1 and λ_2, respectively. We have three basic subcases corresponding to Fig. 2.7 (a), (b), and (e), respectively (eigenvalues are plotted in the complex plane).

(i) $\mathrm{tr}(A) < 0$, $\det(A) > 0$. In this case, eigenvalues and eigenvectors are real and both eigenvalues are negative (say, $0 > \lambda_1 > \lambda_2$). The two-dimensional

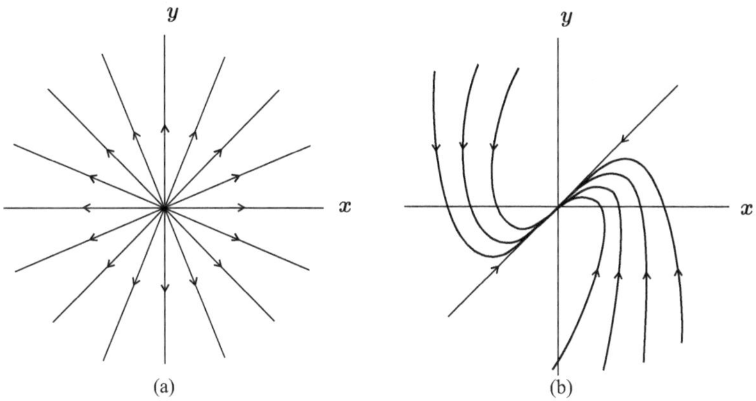

Fig. 2.8 Equilibrium types in the plane with repeated eigenvalue: (a) bicritical node; (b) Jordan node

state space coincides with the stable eigenspace.[5] The equilibrium is called a *stable node*, and the term 'node' refers to the characteristic shape of the ensemble of orbits around the equilibrium.

(ii) $\mathrm{tr}(A) > 0, \det(A) > 0$. In this case, eigenvalues and eigenvectors are real, both eigenvalues are positive (say, $\lambda_1 > \lambda_2 > 0$), and the state space coincides with the unstable eigenspace. The equilibrium is called an *unstable node*.

(iii) $\det(A) = 0$. In this case, $\Delta > 0$ independent of the sign of the trace of A. One eigenvalue is positive, and the other is negative (say, $\lambda_1 > 0 > \lambda_2$). There are, then, a one-dimensional stable eigenspace and a one-dimensional unstable eigenspace and the equilibrium is known as a *saddle point*.

Case 2: $\Delta < 0$. The eigenvalues and eigenvectors are complex conjugate pairs and we have

$$(\lambda_1, \lambda_2) = (\lambda, \bar{\lambda}) = \alpha \pm \mathrm{i}\beta$$

with

$$\alpha = \frac{1}{2}\mathrm{tr}(A), \quad \beta = \frac{1}{2}\sqrt{-\Delta}.$$

The solutions have the form

$$\begin{cases} x(t) = Ce^{\alpha t}\cos(\beta t + \phi), \\ y(t) = Ce^{\alpha t}\sin(\beta t + \phi), \end{cases}$$

and the motion is oscillatory. If $\alpha \neq 0$ there is no strict periodicity in the sense that there exists no τ such that $x(t) = x(t+\tau)$. However, a *conditional period* can be defined as the length of time between two successive maxima of a variable, which is equal to $2\pi/\beta$. The frequency is simply the number of oscillations per

[5] An eigenspace is spanned by eigenvectors. A stable eigenspace is spanned by the eigenvectors corresponding to negative eigenvalues, and an unstable eigenspace is spanned by the eigenvectors corresponding to positive eigenvalues.

time unit, that is, $\beta/2\pi$. The amplitude or size of the oscillations depends on the initial condition and $e^{\alpha t}$ (more on this point below). There are three subcases depending on the sign of $\mathrm{tr}(A)$ and therefore of $\mathrm{Re}(\lambda) = \alpha$; see the corresponding illustrations in Figs. 2.7 (c), (d), and (f), respectively.

(i) $\mathrm{tr}(A) < 0$, $\mathrm{Re}(\lambda) = \alpha < 0$. The oscillations are dampened and the system converges to the equilibrium. The equilibrium point is known as a *focus* or, sometimes, a *vortex*, due to the characteristic shape of the orbits around the equilibrium. In this case the focus or vortex is stable and the stable eigenspace coincides with the state space.

(ii) $\mathrm{tr}(A) > 0$, $\mathrm{Re}(\lambda) = \alpha > 0$. The amplitude of the oscillations gets larger with time and the system diverges from the equilibrium. The unstable eigenspace coincides with the state space and the equilibrium point is called an unstable focus or vortex.

(iii) $\mathrm{tr}(A) = 0$, $\mathrm{Re}(\lambda) = \alpha = 0$. In this special case we have a pair of purely imaginary eigenvalues. Orbits neither converge to, nor diverge from, the equilibrium point, but they oscillate regularly around it with a constant amplitude that depends only on initial conditions and the equilibrium point is called a *center*.

Case 3: $\Delta = 0$. The eigenvalues are real and equal to each other, $\lambda_1 = \lambda_2 = \lambda$. In this case, if $A \neq \lambda I$, only one eigenvector can be determined, say $u = \left(u^{(1)}, u^{(2)}\right)^{\mathrm{T}}$, defining a single straight line through the origin. We can write the general solution as

$$\begin{cases} x(t) = (c_1 u^{(1)} + c_2 v^{(1)})e^{\lambda t} + t c_2 u^{(1)} e^{\lambda t}, \\ y(t) = (c_1 u^{(2)} + c_2 v^{(2)})e^{\lambda t} + t c_2 u^{(2)} e^{\lambda t}, \end{cases}$$

with

$$\begin{cases} x(0) = c_1 u^{(1)} + c_2 v^{(1)}, \\ y(0) = c_1 u^{(2)} + c_2 v^{(2)}. \end{cases}$$

The equilibrium type is again a node, sometimes called a *Jordan node*. An example of this type is provided in Fig. 2.8 (b), where it is obvious that there is a single eigenvector. If $A = \lambda I$ the equilibrium is still a node, sometimes called a *bicritical node*. However, all half-lines from the origin are solutions, giving a star-shaped form (see Fig. 2.8 (a)).

Fig. 2.9 provides a very useful geometric representation in the $(\mathrm{tr}(A), \det(A))$ plane of the various cases discussed above. Quadrants III and IV of the plane correspond to saddle points, quadrant II to stable nodes and foci, and quadrant I to unstable nodes and foci. The parabola divides quadrants I and II into nodes and foci (the former below the parabola and the latter above it). The positive part of the $\det(A)$ axis corresponds to centers.

The analysis of systems with $n > 2$ variables can be developed along the same lines although geometric insight will fail when the dimension of the state space is larger than three. In order to give the reader a broad idea of common situations we

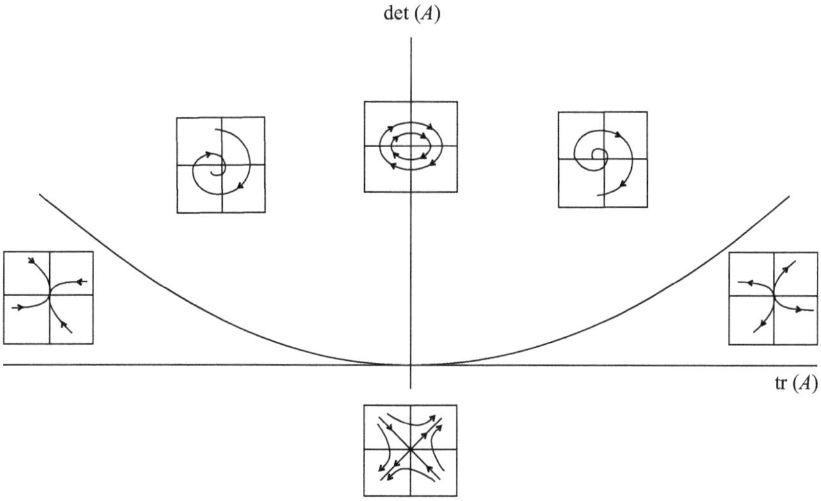

Fig. 2.9 Continuous-time dynamics in $\mathbb{R}^2$

depict sample orbits of three-dimensional systems in Fig. 2.10, which indicates the corresponding eigenvalues in the complex plane. The system in Fig. 2.10 (a) has two positive real eigenvalues and one negative real eigenvalue. The equilibrium point is an unstable saddle. In this case the plane associates with the positive real eigenvalues. All orbits eventually converge to the unstable eigenspace and are captured by the expanding dynamics. The only exceptions are those orbits initiating on the stable eigenspace (defined by the eigenvector associated with the negative eigenvalue) which converge to the plane at the equilibrium point. When A is an $m \times m$ matrix, we divide the eigenvectors (or, in the complex case, the vectors equal to the real and imaginary parts of them) into three groups, according to whether the corresponding eigenvalues have negative, positive, or zero real parts. Then, the subsets of the state space spanned (or generated) by each group of vectors are known as the stable, unstable, and center eigenspaces, respectively, and denoted by $E^{\mathrm{s}}, E^{\mathrm{u}}$, and E^{c}. Notice that the term saddle in $\mathbb{R}^m$ refers to all cases in which there exist some eigenvalues with positive and some with negative real parts. We use the term *saddle node* when eigenvalues are all real, *saddle focus* when some of the eigenvalues are complex. An example of the latter is presented in Fig. 2.10 (b). The two real vectors associated with the complex conjugate pair of eigenvalues, with negative real parts, span the stable eigenspace. Orbits approach asymptotically the unstable eigenspace defined by the eigenvector associated with the positive eigenvalue, along which the dynamics is explosive.

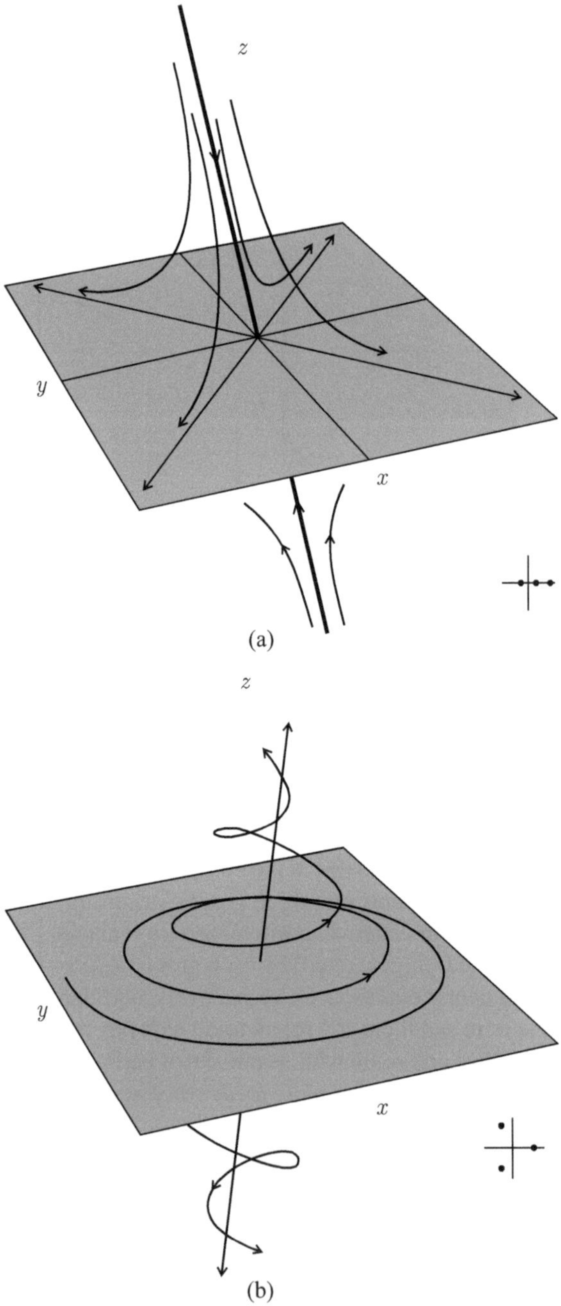

Fig. 2.10 Continuous-time dynamics in $\mathbb{R}^3$: (a) saddle node; (b) saddle focus

2.7 General Solutions of Discrete-Time Linear Systems

Now consider the following linear, discrete-time system described by

$$x(n+1) = Bx(n), \quad x \in \mathbb{R}^m. \tag{2.16}$$

It is easy to see that $x = 0$ is the only equilibrium solution. If κ_i is a real, distinct eigenvalue of the $m \times m$ matrix B and v_i is the corresponding real eigenvector so that $Bv_i = \kappa_i v_i$, it can be verified that

$$x(n) = \kappa_i^n v_i \tag{2.17}$$

is a solution of (2.16). Suppose that we have a pair of eigenvalues of B

$$(\kappa_j, \kappa_{j+1}) = (\kappa_j, \bar{\kappa_j}) = \sigma_j \pm \mathrm{i}\theta_j$$

with a corresponding pair of eigenvectors

$$(v_j, v_{j+1}) = (v_j, \bar{v_j}) = p_j \pm \mathrm{i}q_j,$$

where p_j and q_j are m-dimensional vectors. Then, the pair of functions

$$\begin{cases} x_j(n) = \dfrac{1}{2}\left(\kappa_j^n v_j + \bar{\kappa}_j^n \bar{v_j}\right) = r_j^n[p_j \cos(\omega_j n) - q_j \sin(\omega_j n)], \\ x_{j+1}(n) = -\dfrac{\mathrm{i}}{2}\left(\kappa_j^n v_j - \bar{\kappa}_j^n \bar{v_j}\right) = r_j^n[p_j \sin(\omega_j n) + q_j \cos(\omega_j n)] \end{cases} \tag{2.18}$$

are the jth and $(j+1)$st solutions of (2.16), respectively. Here, we have used the polar coordinate transformations

$$\begin{cases} \sigma_j = r_j \cos \omega_j, \\ \theta_j = r_j \sin \omega_j, \end{cases} \tag{2.19}$$

and a well-known result

$$(\cos\omega \pm \mathrm{i}\sin\omega)^n = \cos(\omega n) \pm \mathrm{i}\sin(\omega n).$$

From (2.19) we have $r_j = \sqrt{\sigma_j^2 + \theta_j^2}$. Then, r_j is simply the modulus of the complex eigenvalues. If $p_j^{(l)}$ and $q_j^{(l)}$ denote the lth elements of p_j and q_j, respectively, then in polar coordinates we have

$$\begin{cases} p_j^{(l)} = C_j^{(l)} \cos(\phi_j^{(l)}), \\ q_j^{(l)} = C_j^{(l)} \sin(\phi_j^{(l)}), \end{cases}$$

where $l = 1, 2, \ldots, m$. (2.18) can be rewritten as m-dimensional vectors whose lth elements have the form

$$\begin{cases} x_j^{(l)}(n) = C_j^{(l)} r_j^n \cos(\omega_j n + \phi_j^{(l)}), \\ x_{j+1}^{(l)}(n) = C_j^{(l)} r_j^n \sin(\omega_j n + \phi_j^{(l)}). \end{cases}$$

Assuming that we have m linearly independent solutions defined by (2.17) and (2.18), by the superposition principle the general solution of (2.16) can be written as a linear combination of the individual solutions, namely

$$x(n) = c_1 x_1(n) + c_2 x_2(n) + \cdots + c_m x_m(n),$$

where c_i are constants depending on the initial conditions.

When eigenvalues are repeated, the general solution becomes

$$x(n) = \sum_{j=1}^{h} \sum_{l=0}^{n_j-1} k_{jl} n^l \kappa_j^n,$$

where $n_j \geq 1$ is the multiplicity of the jth eigenvalue, $h \leq m$ is the number of distinct eigenvalues, and k_{jl} are independent vectors whose values depend on the m arbitrary initial conditions.

Inspection of (2.17) and (2.18) indicates that if the modulus of any of the eigenvalues is greater than 1, solutions tend to $+\infty$ or $-\infty$ as time goes to $+\infty$. On the contrary, if all eigenvalues have modulus smaller than 1, solutions converge asymptotically to the equilibrium point. Analogous to the situation of continuous time, we call the space spanned by the eigenvectors whose corresponding eigenvalues have modulus less than 1 (greater than 1) a stable eigenspace (unstable eigenspace) and call the space spanned by the eigenvectors whose corresponding eigenvalues have modulus equal to 1 a center eigenspace. We also use the same symbols, $E^{\mathrm{s}}, E^{\mathrm{u}}$, and E^{c}, to denote them.

2.8 Discrete-Time Systems in the Plane

The discrete-time autonomous system that is analogous to the continuous-time system (2.15) is described by

$$\begin{pmatrix} x_{n+1} \\ y_{n+1} \end{pmatrix} = B \begin{pmatrix} x_n \\ y_n \end{pmatrix} = \begin{pmatrix} b_{11} & b_{12} \\ b_{21} & b_{22} \end{pmatrix} \begin{pmatrix} x_n \\ y_n \end{pmatrix}. \tag{2.20}$$

We assume that the matrix $I - B$ is nonsingular. Thus, the origin is the unique equilibrium point of (2.20). The characteristic equation is analogous to the continuous case as well,

$$\kappa^2 - \mathrm{tr}(B)\kappa + \det(B) = 0,$$

and the eigenvalues are given by

$$\kappa_{1,2} = \frac{1}{2}\left(\mathrm{tr}(B) \pm \sqrt{\mathrm{tr}(B)^2 - 4\det(B)}\right).$$

We also assume that the equilibria of (2.20) are nondegenerate, i.e., $|\kappa_1|, |\kappa_2| \neq 1$. We will discuss the dynamics of the discrete-time system (2.20) for the following three cases.

Case 1: $\Delta > 0$. The eigenvalues are real and solutions take the form

$$\begin{cases} x(n) = c_1\kappa_1^n v_1^{(1)} + c_2\kappa_2^n v_2^{(1)}, \\ y(n) = c_1\kappa_1^n v_1^{(2)} + c_2\kappa_2^n v_2^{(2)}. \end{cases}$$

(i) If $|\kappa_1| < 1$ and $|\kappa_2| < 1$ the fixed point is a stable node. This means that solutions are sequences of points approaching the equilibrium as $n \to +\infty$. If $\kappa_1, \kappa_2 > 0$ the approach is monotonic; otherwise, there are improper os-

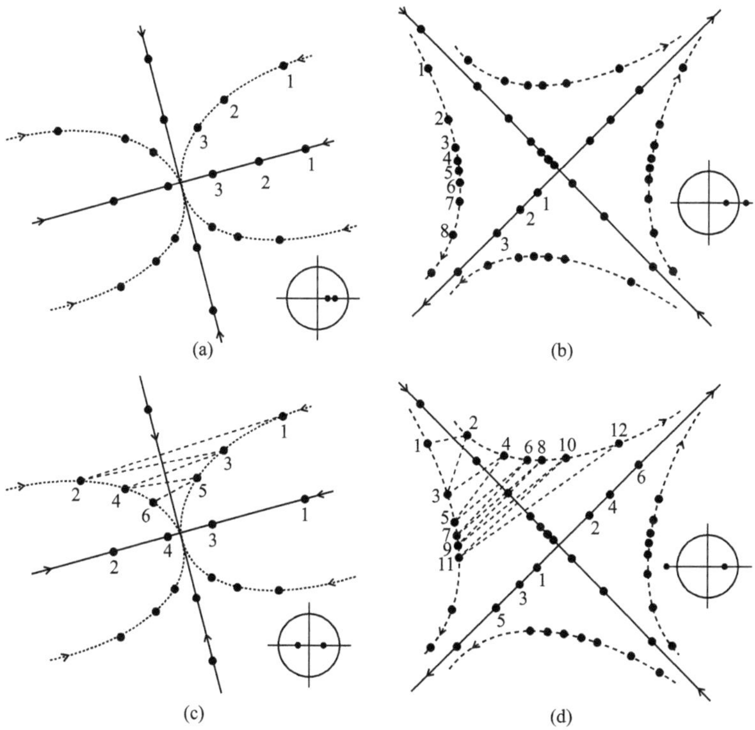

Fig. 2.11 Phase diagrams for real eigenvalues: (a) and (c) stable nodes; (b) and (d) saddle points

cillations[6] (see Figs. 2.11 (a) and (c), respectively). In this case, the stable eigenspace coincides with the state space.

(ii) If $|\kappa_1| > 1$ and $|\kappa_2| > 1$ the fixed point is an unstable node. In this case, solutions are sequences of points approaching equilibrium as $n \to -\infty$. If $\kappa_1, \kappa_2 > 0$ the approach is monotonic; otherwise, there are improper oscillations (as in Figs. 2.11 (a) and (c), respectively, but arrows point in the opposite direction and the time order of points is reversed). In this case, the unstable eigenspace coincides with the state space.

(iii) If $|\kappa_1| > 1$ and $|\kappa_2| < 1$ the fixed point is a saddle point. No sequences of points approach the equilibrium for $n \to \pm\infty$ except for those originating from points on the eigenvectors associated with κ_2. Again, if $\kappa_1,\ \kappa_2 > 0$ orbits move monotonically (see Fig. 2.11 (b)); otherwise they oscillate improperly (see Fig. 2.11 (d)). The stable and unstable eigenspaces are one dimensional.

Case 2: $\Delta < 0$. In this case, $\det(B) > 0$. Eigenvalues are a complex conjugate pair given by

$$(\kappa_1, \kappa_2) = (\kappa, \bar{\kappa}) = \sigma \pm i\theta$$

and solutions are sequences of points situated on a spiral whose amplitude increases or decreases in time according to the factor r^n, where

$$r = |\sigma \pm i\theta| = \sqrt{\sigma^2 + \theta^2} = \det(B)$$

is the modulus of the complex eigenvalue pair. Solutions are of the form

$$\begin{cases} x(n) = Cr^n \cos(\omega n + \phi), \\ y(n) = Cr^n \sin(\omega n + \phi). \end{cases}$$

(i) If $r < 1$ solutions converge to equilibrium and the equilibrium point is a stable focus (see Fig. 2.12 (a)).

(ii) If $r > 1$ solutions diverge and the equilibrium point is an unstable focus (as in Fig. 2.12 (a), but arrows point in the opposite direction and the time order of points is reversed).

(iii) If $r = 1$ the eigenvalues lie exactly on the unit circle, an exceptional case. There are two subcases which depend on the frequency of the oscillation $\omega/2\pi, \omega = \arccos[\mathrm{tr}(B)/2]$:

   a. $\omega/2\pi$ is rational and the orbit in the state space is a periodic sequence of points situated on a circle, the radius of which depends on initial conditions (see Fig. 2.12 (b));

   b. $\omega/2\pi$ is irrational and the sequence is nonperiodic or quasiperiodic, that is, starting from any point on the circle, orbits stay on the circle but no

[6] First-order, discrete-time equations (where the order is determined as the difference between the extreme time indices) can also have fluctuating behavior, called improper oscillations, owing to the fact that if their eigenvalue $\beta < 0$, β^n will be positive or negative according to whether n is even or odd. The term *improper* refers to the fact that in this case oscillations of variables have a 'kinky' form that does not properly describe the smoother ups and downs of real variables.

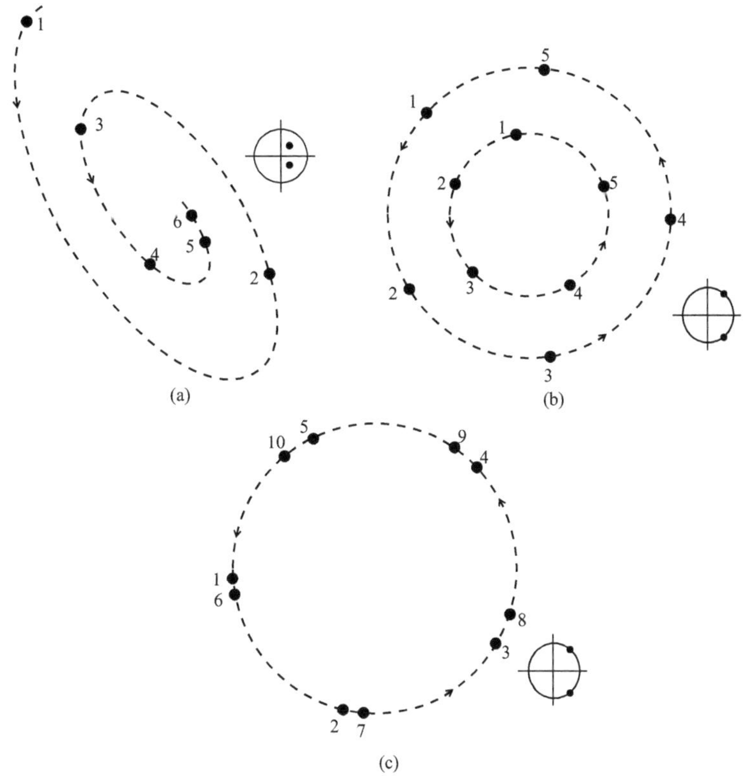

Fig. 2.12 Phase diagrams for complex eigenvalues: (a) a stable focus; (b) periodic cycles; (c) a quasiperiodic solution

sequence returns to the initial point in finite time. Therefore, solutions wander on the circle filling it up, without ever becoming periodic (see Fig. 2.12 (c)).

Case 3: $\Delta = 0$. There is a repeated real eigenvalue $\kappa_1 = \mathrm{tr}(B)/2$. The general form of solutions with a repeated eigenvalue κ is as follows:

$$\begin{cases} x(n) = (c_1 v^{(1)} + c_2 u^{(1)})\kappa^n + n c_2 v^{(1)} \kappa^n, \\ y(n) = (c_1 v^{(2)} + c_2 u^{(2)})\kappa^n + n c_2 v^{(2)} \kappa^n. \end{cases}$$

If $|\kappa| < 1, \lim_{n\to\infty} n\kappa^n = 0$. If the repeated eigenvalue is equal to 1 in absolute value, the equilibrium is unstable (with improper oscillations for $\kappa_1 = -1$). However, the divergence is linear not exponential.

The dynamics of the discrete case can be conveniently summarized by the diagram in Fig. 2.13. (For the sake of simplicity, we represent orbits as continuous rather than dotted curves.) If we replace the greater-than sign with the equal sign in conditions (i)–(iii) of Case 2, we obtain three lines intersecting each other in the (tr(B), det(B)) plane, defining a triangle. Points inside the triangle correspond to stable combinations of the trace and determinant of B.[7] The parabola defined by

$$\text{tr}(B)^2 = 4\det(B)$$

divides the plane into two regions corresponding to real eigenvalues (below the parabola) and complex eigenvalues (above the parabola). Combinations of trace and determinant above the parabola but in the triangle lead to stable foci, combinations below the parabola but in the triangle are stable nodes. All other combinations lead to unstable equilibria.

Fig. 2.14 is an example of a system in $\mathbb{R}^3$. There are a complex conjugate pair with modulus less than one, and one dominant real eigenvalue greater than one. The equilibrium point is a *saddle focus*.

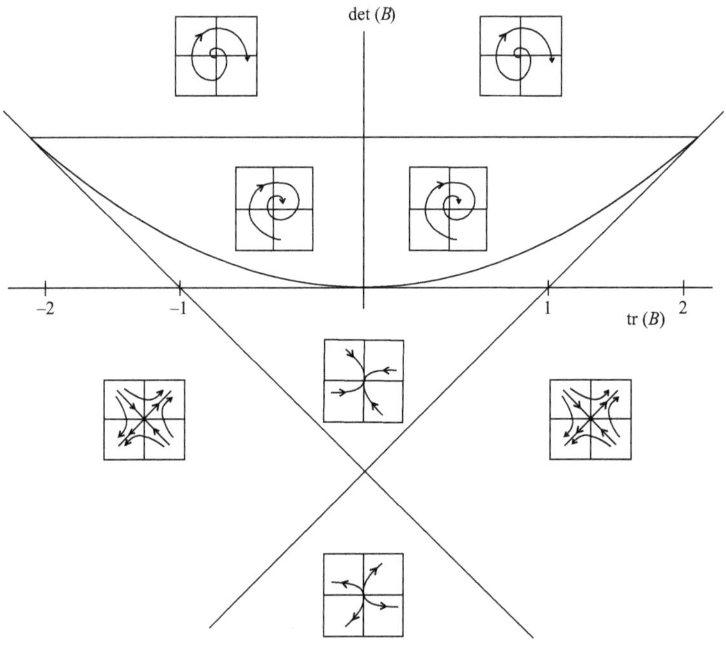

Fig. 2.13 Discrete-time dynamics in $\mathbb{R}^2$

[7] In Case 2, if $1+\text{tr}(B)+\det(B)=0$ while (ii) and (iii) hold, one eigenvalue is equal to -1; if $1-\text{tr}(B)+\det(B)=0$ while (i) and (iii) hold, one eigenvalue is equal to $+1$; and if $\det(B)=1$ while (i) and (ii) hold, the two eigenvalues are a complex conjugate pair with modulus equal to 1.

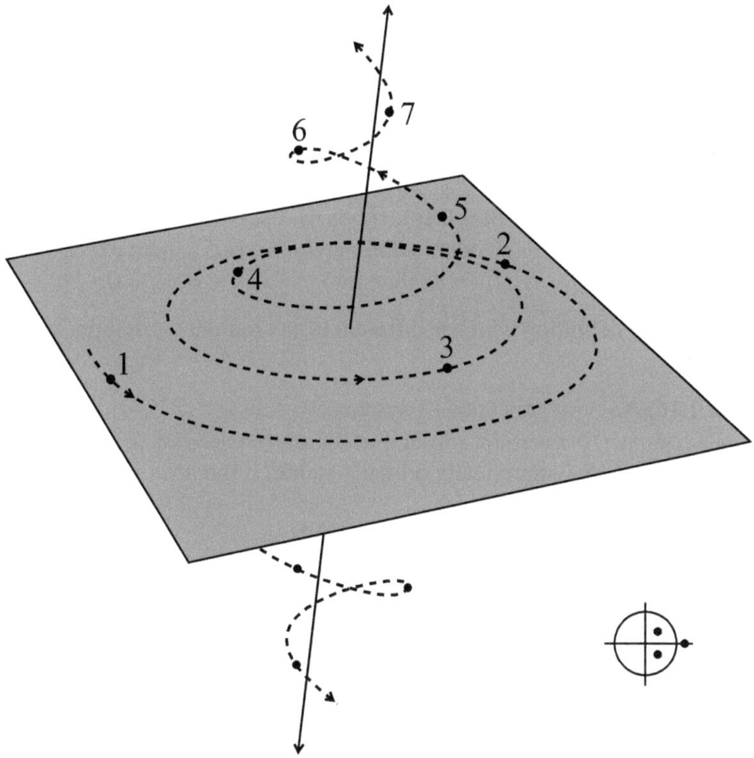

Fig. 2.14 A discrete-time dynamics in $\mathbb{R}^3$

2.9 Stabilities of Trajectories I: The Lyapunov First Method

Stability theory plays a central role in systems theory and engineering. The so-called Lyapunov first method or Lyapunov indirect method is used to study the stability of a system's trajectories through calculating the eigenvalues of a linearized system at the objective trajectories. This means that the Lyapunov first method is essentially a local method which can only be used in a neighborhood of the objective trajectory.

2.9.1 The Definition of Lyapunov Stability

Let $\bar{x}(t)$ be any solution of (2.5). Roughly speaking, $\bar{x}(t)$ is *stable* if solutions starting 'close' to $\bar{x}(t)$ at a given time remain close to $\bar{x}(t)$ for all later times. It is *asymptotically stable* if nearby solutions not only stay close, but also converge to $\bar{x}(t)$ as $t \to \infty$.

Definition 2.14 (Lyapunov Stability). $\bar{x}(t)$ is said to be stable (or Lyapunov stable) if, for any given $\varepsilon > 0$, there exists a $\delta = \delta(\varepsilon) > 0$ such that, for any other solution, $y(t)$, of (2.5) satisfying $\|\bar{x}(t_0) - y(t_0)\| < \delta$, we have $\|\bar{x}(t) - y(t)\| < \varepsilon$ for $t > t_0$, $t_0 \in \mathbb{R}$. □

A solution which is not stable is said to be unstable.

Definition 2.15 (Asymptotic Stability). $\bar{x}(t)$ is said to be asymptotically stable if it is Lyapunov stable and, for any other solution, $y(t)$, of (2.5), there exists a constant $b > 0$ such that, if $\|\bar{x}(t_0) - y(t_0)\| < b$, then $\lim_{t\to\infty} \|\bar{x}(t) - y(t)\| = 0$. □

A new stability definition which is different from Lyapunov's definitions is given as follows.

Definition 2.16. An orbit generated by system $\dot{x} = f(x)$ $(x \in \mathbb{R}^n)$, with initial condition x_0 on a compact, ϕ-invariant subset A of the state space (i.e., $\phi(A) \subset A$), is said to be orbitally stable (asymptotically orbitally stable) if the invariant set

$$\Gamma = \{\phi(t, x_0) : x_0 \in A,\ t \geq 0\}$$

(the forward orbit of x_0) is stable (asymptotically stable) according to Definition 2.14 (Definition 2.15). □

The analogous definitions of stability for autonomous dynamical systems in discrete time with the general form

$$x(n+1) = G(x(n)), \quad x \in \mathbb{R}^m \tag{2.21}$$

are as follows.

Definition 2.17. The equilibrium point $\bar{x}$ is *Lyapunov stable* (or, simply, stable) if, for every $\varepsilon > 0$, there exists $\delta(\varepsilon)$ such that

$$\|x_0 - \bar{x}\| < \delta(\varepsilon) \Rightarrow \|G^n(x_0) - \bar{x}\| < \varepsilon,\ \forall n > 0.$$ □

Definition 2.18. The equilibrium point $\bar{x}$ is *asymptotically stable* if

(i) it is stable and
(ii) $\exists b > 0$ such that

$$\|x_0 - \bar{x}\| < b \Rightarrow \lim_{n\to\infty} \|G^n(x_0) - \bar{x}\| = 0.$$ □

Property (ii) can be replaced by the following equivalent property:

(ii') there exists $b > 0$ and, for each $\varepsilon > 0$, there exists an integer $T = T(b, \varepsilon) > 0$ such that

$$\|x_0 - \bar{x}\| < b \Rightarrow \|G^n(x_0) - \bar{x}\| < \varepsilon,\ \forall n \geq T.$$

Figs. 2.15 and 2.16 are the visualization of Definitions 2.14 and 2.15. Notice that these two definitions imply that we have information on the infinite-time existence of solutions. This is obvious for equilibrium solutions but is not necessary for

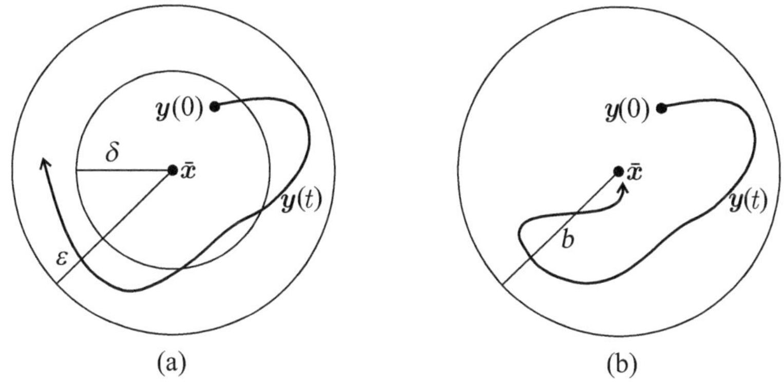

Fig. 2.15 (a) Lyapunov stability; (b) asymptotic stability

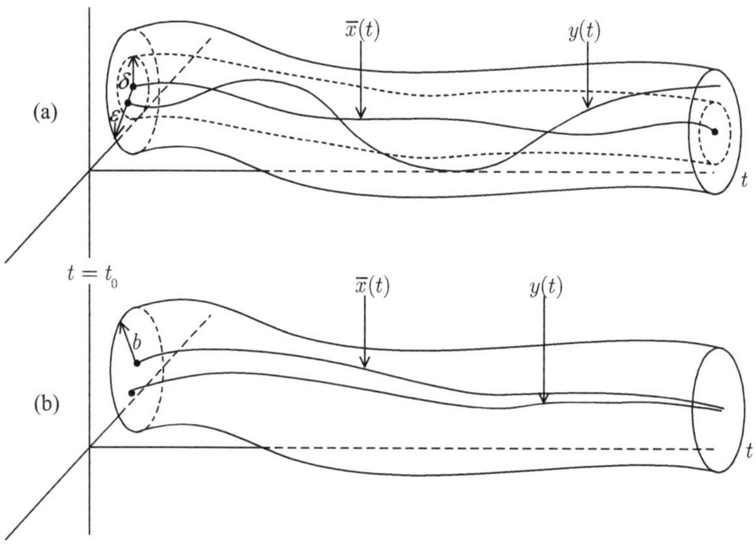

Fig. 2.16 (a) Lyapunov stability; (b) asymptotic stability

nearby solutions. Also, these definitions are for autonomous systems, since in the nonautonomous case it may be that δ and b depend explicitly on t_0.

In order to determine the stability of $\bar{x}(t)$ we must understand the nature of solutions near $\bar{x}(t)$. Let

$$x(t) = \bar{x}(t) + y. \tag{2.22}$$

Substituting (2.22) into (2.5) and performing Taylor expansion about $\bar{x}(t)$ gives

$$\dot{x} = \dot{\bar{x}}(t) + \dot{y} = f(\bar{x}(t)) + Df(\bar{x}(t))y + o(\|y\|^2), \tag{2.23}$$

where Df is the derivative of f called the Jacobian matrix of f, $\|\cdot\|$ denotes a norm on $\mathbb{R}^n$, and $o(\|y\|^2)$ denotes the higher-order infinitesimal term of $\|y\|^2$. Using the fact that $\dot{\bar{x}}(t) = f(\bar{x}(t))$, (2.23) becomes

$$\dot{y} = Df(\bar{x}(t))y + o(\|y\|^2). \tag{2.24}$$

Equation (2.24) describes the evolution of orbits near $\bar{x}(t)$. For stability questions we are concerned with the behavior of solutions arbitrarily close to $\bar{x}(t)$, so it seems reasonable that this question could be answered by studying the associated linear system

$$\dot{y} = Df(\bar{x}(t))y. \tag{2.25}$$

Usually it is difficult to determine the stability of $x(t)$ by (2.25) since there are no general analytical methods for finding the solution of linear ordinary differential equations with time-dependent coefficients. However, if $\bar{x}$ is an equilibrium solution, i.e., $\bar{x}(t) = \bar{x}$, then $Df(\bar{x}(t)) = Df(\bar{x})$ is a matrix with constant entries, and the solution of (2.25) through the point $y_0 \in \mathbb{R}^n$ at $t = 0$ can immediately be written as

$$y(t) = \mathrm{e}^{Df(\bar{x})t} y_0.$$

Thus, $y(t)$ is *asymptotically stable* if all eigenvalues of $Df(\bar{x})$ have negative real parts.

Theorem 2.4. Suppose that all of the eigenvalues of $Df(\bar{x})$ have negative real parts. Then, the equilibrium solution $x = \bar{x}$ of the nonlinear vector field (2.5) is asymptotically stable. □

Definition 2.19. Let $x = \bar{x}$ be a fixed point of $\dot{x} = f(x)$, $x \in \mathbb{R}^n$. Then, $\bar{x}$ is called a *hyperbolic fixed point* if none of the eigenvalues of $Df(\bar{x})$ has zero real part. □

The eigenvalues of the Jacobian matrix $Df(\bar{x})$ are also referred to as the eigenvalues of the fixed point $\bar{x}$.

Theorem 2.5 (Hartman–Grobman). If $\bar{x}$ is a hyperbolic fixed point of (2.5), then there is a homeomorphism h defined on some neighborhood N of $\bar{x}$ in $\mathbb{R}^n$, locally

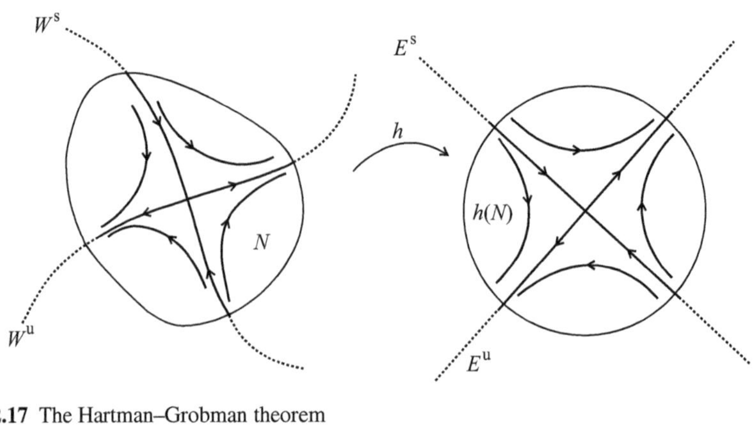

Fig. 2.17 The Hartman–Grobman theorem

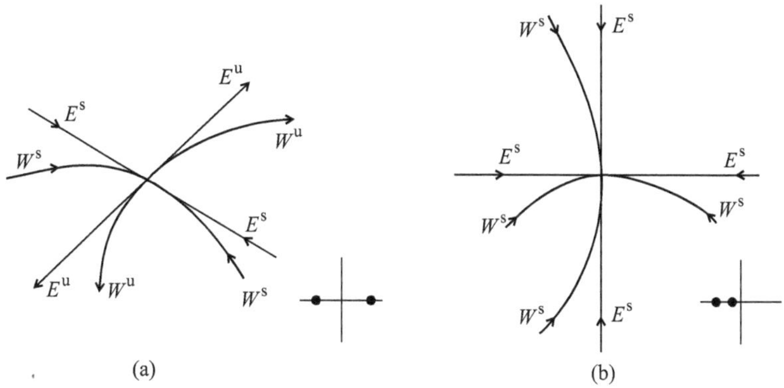

Fig. 2.18 Stable and unstable eigenspaces and manifolds in $\mathbb{R}^2$

mapping orbits of the nonlinear system (2.5) to those of the linear system (2.25). The map h preserves the sense of orbits and can also be chosen so as to preserve parameterization by time. □

If h is a homeomorphism, then from Theorem 2.5 we can deduce that asymptotic stability (or the lack of it) for the linear system (2.25) implies local asymptotic stability of the nonlinear system (2.5) (or the lack of it). However, homeomorphic equivalence does not preserve all the interesting geometric features of a dynamical system. For example, a linear system characterized by an asymptotically stable node is topologically conjugate to another linear system characterized by an asymptotically stable focus.

If the equilibrium point is not hyperbolic, that is to say, if there exists at least one eigenvalue with real part exactly equal to 0, the Hartman–Grobman theorem cannot be applied. The reason is that the linearized system is not sufficiently informative. In particular, the stability properties of the system depend on the higher-order terms of the expansion which have been ignored in the approximation (2.25).

In the above discussion of linear systems we emphasized the importance of certain invariant subspaces, i.e., the eigenspaces, defined by the eigenvectors of the Jacobian matrix. If the nonlinear system (2.5) has an isolated,[8] hyperbolic equilibrium $\bar{x}$, in the neighborhood of $\bar{x}$ there exist certain invariant surfaces, called stable and unstable manifolds, which are the nonlinear counterparts of the stable and unstable eigenspaces. Locally, these manifolds are continuous deformations, respectively, of the stable and unstable eigenspaces of the linear system (2.25) (because $\bar{x}$ is hyperbolic, there is no center eigenspace for (2.25)) and they are tangents to the eigenspaces of the linear system (2.25) at $\bar{x}$. We denote stable manifolds, unstable manifolds, and center manifolds by W^s, W^u, and W^c. Some simple examples of the phase diagrams of nonlinear systems and the corresponding linearized systems in $\mathbb{R}^2$ and $\mathbb{R}^3$ are provided in Figs. 2.17, 2.18, and 2.19.

[8] An equilibrium point is isolated if it has a surrounding neighborhood containing no other equilibrium point.

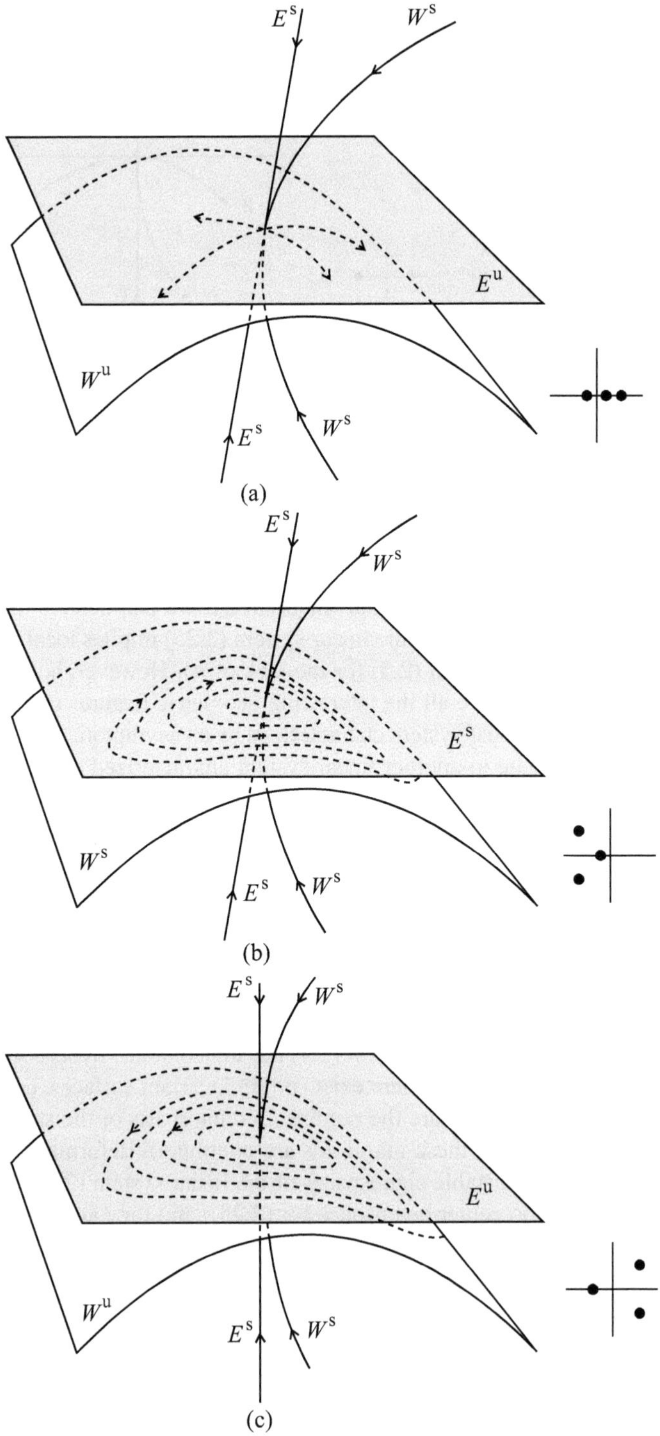

Fig. 2.19 Stable and unstable eigenspaces and manifolds in $\mathbb{R}^3$

The method of linear approximation can be applied in a perfectly analogous manner to nonlinear systems described by difference equations. Consider system (2.21), with a fixed point $\bar{x}$, and assume that G is differentiable. A local linear approximation of (2.21) near $\bar{x}$ is

$$\xi(n+1) = DG(\bar{x})\xi(n), \tag{2.26}$$

where $\xi = x - \bar{x}$ and $DG(\bar{x})$ is the Jacobian matrix of partial derivatives of G, evaluated at $\bar{x}$.

The discrete-time version of the Hartman–Grobman theorem for $x(n+1) = G(x(n))$ is perfectly analogous to that for flows except for the following important differences:

(i) For discrete-time systems, fixed points are hyperbolic if none of the eigenvalues of the Jacobian matrix, evaluated at the equilibrium, is equal to 1 in modulus.
(ii) The map h of the Hartman–Grobman theorem defining the local relationship between the nonlinear system (2.21) and the linearized system (2.26) is a diffeomorphism if the eigenvalues of $DG(\bar{x})$ satisfy a nonresonance condition. In the case of maps this condition requires that no eigenvalue κ_k of $DG(\bar{x})$ satisfies

$$\kappa_k = \prod_{i=1}^{m} \kappa_i^{c_i}$$

for any choice of $c_i \geq 0$ with $\sum_i c_i \geq 2$.

2.9.2 Floquet Theory

Local stability of a periodic solution of the system $\dot{x} = f(x)$ $(x \in \mathbb{R}^n)$ can be discussed in terms of eigenvalues of certain matrices. Suppose that the system has a periodic orbit $\Gamma = \{x^*(t) \colon t \in [0,T),\ x^*(t) = x^*(t+T)\}$. Define $\xi := x(t) - x^*(t)$. Linearizing $\dot{\xi}$ about $\xi = 0$, i.e., about the periodic orbit Γ, we obtain

$$\dot{\xi} = A(t)\xi, \tag{2.27}$$

where the matrix $A(t) := Df(x^*(t))$ has periodic coefficients of period T, so that $A(t) = A(t+T)$. Solutions of (2.27) take the general form of

$$B(t)\mathrm{e}^{\lambda t},$$

where the vector $B(t)$ is periodic in time with period T, $B(t) = B(t+T)$. Denote the fundamental matrix of (2.27) as $\Phi(t)$, that is, the $m \times m$ time-varying matrix whose m columns are solutions of (2.27). Thus, $\Phi(t)$ can be written as

$$\Phi(t) = Z(t)\mathrm{e}^{tR},$$

where $Z(t)$ is an $m \times m$, T-periodic matrix and R is a constant $m \times m$ matrix. Moreover, we can always set $\Phi(0) = Z(0) = I$, from which we get

$$\Phi(T) = \mathrm{e}^{TR}.$$

Therefore, the dynamics of orbits near the cycle Γ are determined by the eigenvalues $(\lambda_1, \dots, \lambda_m)$ of the matrix e^{TR} which are uniquely determined by (2.27).[9] The λs are called *characteristic (Floquet) multipliers* of (2.27), whereas the eigenvalues of R, $(\kappa_1, \dots, \kappa_m)$, are called *characteristic (Floquet) exponents.*

One of the roots (multipliers), say λ_1, is always equal to 1, so that one of the characteristic exponents, say κ_1, is always equal to 0, which implies that one of the solutions of (2.27) must have the form $B(t) = B(t+T)$. This can be verified by putting $B(t) = x^*(t)$ and differentiating it with respect to time. The presence of a characteristic multiplier equal to 1 (a characteristic exponent equal to 0) can be interpreted as that if, starting from a point on the periodic orbit Γ, the system is perturbed by a small displacement in the direction of the flow, it will remain on Γ. What happens for small, random displacements off Γ depends only on the remaining $m-1$ multipliers λ_j $(j = 2, \dots, m)$ (or the remaining κ_j exponents, $j = 2, \dots, m$), provided none of the other moduli is equal to 1 (provided none of them is equal to 0). In particular, we have

(i) If all the characteristic multipliers λ_j $(j = 2, \dots, m)$ satisfy the conditions $|\lambda_j| < 1$, then the periodic orbit is asymptotically (in fact, exponentially) orbitally stable.
(ii) If for at least one of the multipliers, say λ_k, $|\lambda_k| > 1$, then the periodic orbit is unstable.

2.10 Stabilities of Trajectories II: The Lyapunov Second Method

The so-called second or direct method of Lyapunov is one of the greatest landmarks in the theory of dynamical systems and has proved to be an immensely fruitful tool for analysis. The basic idea of the method is as follows. Suppose that there is a vector field in the plane with a fixed point $\bar{x}$, and we want to determine whether it is stable or not. Roughly speaking, according to our previous definitions of stability it would be sufficient to find a neighborhood U of $\bar{x}$ for which orbits starting in U remain in U for all positive time (for the moment we do not distinguish between stability and asymptotic stability). This condition would be satisfied if we could show that the vector field is either tangent to the boundary of U or pointing inward toward $\bar{x}$ (see Fig. 2.20). This situation should remain true even as we shrink U down to $\bar{x}$. Now, Lyapunov's method gives us a way of making this precise; we will show this for vector fields in the plane and then generalize our results to $\mathbb{R}^n$.

[9] The matrix e^{TR} itself is uniquely determined but for a similarity transformation, that is, we can substitute e^{TR} with $P^{-1}\mathrm{e}^{TR}P$ where P is a nonsingular $m \times m$ matrix. This transformation leaves eigenvalues unchanged.

Suppose that we have the vector field

$$\begin{cases} \dot{x} = f(x,y), \\ \dot{y} = g(x,y), \quad (x,y) \in \mathbb{R}^2, \end{cases} \tag{2.28}$$

which has a fixed point at $(\bar{x},\bar{y})$ (assume that it is stable). We want to show that in any neighborhood of $(\bar{x},\bar{y})$ the above situation holds. Let $V(x,y)$ be a scalar-valued function on $\mathbb{R}^2$, i.e., $V: \mathbb{R}^2 \to \mathbb{R}$ (and at least C^1), with $V(\bar{x},\bar{y}) = 0$, and such that the loci of points satisfying $V(x,y) = C =$ constant form closed curves for different values of C encircling $(\bar{x},\bar{y})$ with $V(x,y) > 0$ in a neighborhood of $(\bar{x},\bar{y})$ (see Fig. 2.21).

Recall that the gradient of V, ∇V, is a vector perpendicular to the tangent vector along each curve $V = C$ which points in the direction of increasing V. So, if the vector field were always to be either tangent or pointing inward for each of these curves surrounding $(\bar{x},\bar{y})$, we would have

$$\nabla V(x,y) \cdot (\dot{x},\dot{y}) \leq 0,$$

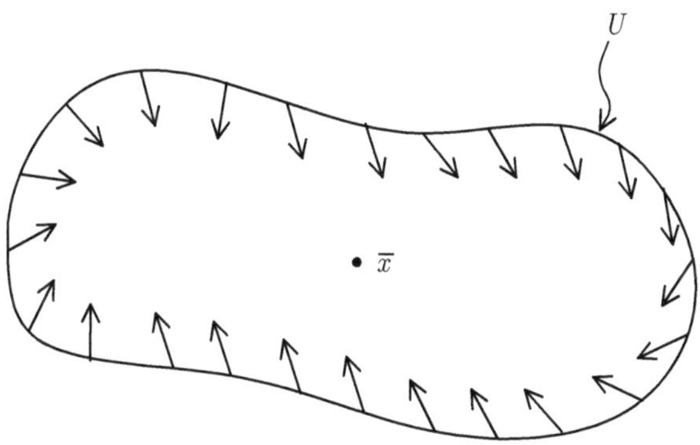

Fig. 2.20 The vector field on the boundary of U

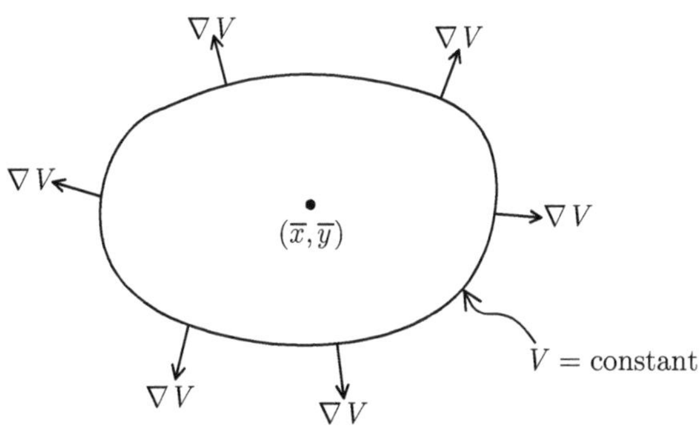

Fig. 2.21 Level set of V and ∇V denotes gradient vector of V at various points on the boundary

where the '·' represents the usual vector scalar product. (This is simply the derivative of V along orbits of (2.28), and is sometimes referred to as the *orbital derivative*.) We now state the general theorem which makes these ideas precise.

Theorem 2.6. Consider the following vector field:

$$\dot{x} = f(x), \quad x \in \mathbb{R}^n. \tag{2.29}$$

Let $\bar{x}$ be a fixed point of (2.29) and let $V: U \to \mathbb{R}$ be a C^1 function defined on some neighborhood U of $\bar{x}$ such that

(i) $V(\bar{x}) = 0$ and $V(x) > 0$ if $x \neq \bar{x}$.
(ii) $\dot{V} \leq 0$ in $U - \{\bar{x}\}$.

Then, $\bar{x}$ is stable. Moreover, if

(iii) $\dot{V} < 0$ in $U - \{\bar{x}\}$,

then $\bar{x}$ is asymptotically stable. □

We refer to V as a *Lyapunov function*. If U can be chosen to be all of $\mathbb{R}^n$, then $\bar{x}$ is said to be *globally asymptotically stable*, if (i) and (iii) hold.

Sometimes it is possible to prove asymptotic stability of a fixed point even when the Lyapunov function V in the relevant neighborhood of the point implies that $\dot{V} \leq 0$, but not necessarily $\dot{V} < 0$. For that case we have the following theorem.

Theorem 2.7 (Invariance Principle of LaSalle). Let $\bar{x} = 0$ be a fixed point of $\dot{x} = f(x)$ and V a Lyapunov function such that $\dot{V} \leq 0$ on some neighborhood N of $\bar{x} = 0$. If $x_0 \in N$ has its forward orbit, $\gamma^+(x_0) = \{\phi(t, x_0) : t \geq 0\}$, bounded with limit points in N, and M is the largest invariant subset of $E = \{x \in N\colon \dot{V}(x) = 0\}$, then

$$\phi(t, x_0) \to M \ \text{ as } \ t \to \infty. \qquad \square$$

According to Theorem 2.7, if a Lyapunov function $V(x)$ can be found such that $\dot{V}(x) \leq 0$ for $x \in N$, among the sets of points with forward orbits in N there exist sets of points defined by

$$V_k = \{x\colon V(x \leq k)\}$$

(k is a finite and positive scalar) which lie entirely in N. Since $\dot{V} \leq 0$, the sets V_k are invariant in the sense that no orbit starting in a V_k can ever move outside of it. If, in addition, it could be shown that the fixed point $\bar{x} = 0$ is the largest (or, for that matter, the only) invariant subset of E, Theorem 2.7 would guarantee its asymptotic stability.

The direct method can also be extended to discrete-time systems. We only state a result analogous to Theorem 2.6 in the following. A discrete-time version of the invariance principle of LaSalle will be introduced in Chap. 5.

Theorem 2.8. Consider the system described by the difference equation given in (2.21). Let $\bar{x} = 0$ again be an isolated equilibrium point at the origin. If there exists a C^1 function $V(x_n)\colon N \to \mathbb{R}$, defined on some neighborhood $N \subset \mathbb{R}^m$ of $\bar{x} = 0$, such that

(i) $V(0) = 0$,
(ii) $V(x) > 0$ in $N - \{0\}$,
(iii) $\Delta V(x_n) := V[G(x_n)] - V(x_n) \leq 0$ in $N - \{0\}$,

then $\bar{x} = 0$ is stable (in the sense of Lyapunov). Moreover, if

(iv) $\Delta V(x) < 0$ in $N - \{0\}$,

then $\bar{x} = 0$ is asymptotically stable. □

2.11 Chaotic Sets and Chaotic Attractors

More complicated invariant, attracting sets and attractors in structure than that of periodic or quasiperiodic sets are called chaotic. A dynamical system (discrete-time or continuous-time) is called chaotic if its typical orbits are aperiodic, bounded, and such that nearby orbits separate fast in time. Chaotic orbits never converge to a stable fixed or periodic point, but exhibit sustained instability, while remaining forever in a bounded region of the state space.

Definition 2.20. A flow ϕ (a continuous map G) on a metric space M is said to possess *sensitive dependence on initial conditions* on M if there exists a real number $\delta > 0$ such that for all $x \in M$ and for all $\varepsilon > 0$, there exist $y \in M$ ($y \neq x$) and $T > 0$ (an integer $n > 0$) such that $d(x,y) < \varepsilon$ and $d[\phi(T,x), \phi(T,y)] > \delta$ ($d[G^n(x), G^n(y)] > \delta$). □

Definition 2.21. A flow ϕ (a continuous map G) is said to be *chaotic* on a compact invariant set[10] A if:

(i) it is topologically transitive on A (Definition 2.12);
(ii) it has sensitive dependence on initial conditions on A. □

Remark 2.4. There is something that should be pointed out.

(i) Condition (i) of Definition 2.21 guarantees that the invariant set is single and indecomposable.
(ii) Condition (ii) of Definition 2.21 can be made sharper (and more restrictive) in two ways. First, the divergence of nearby points taking place at an exponential rate is required. This property can be made more precise by means of Lyapunov exponents which will be introduced later. Second, we may require that the divergence (exponential or otherwise) occurs for *each pair* $x, y \in A$. In this case, the flow ϕ or map G is called *expansive* on A.
(iii) The requirement that A is compact is necessary. Consider the following differential equation:

$$\dot{x} = ax, \ x \in \mathbb{R}, \ a > 0,$$

[10] Roughly speaking, a subset D of $\mathbb{R}^n$ is said to be compact if it can be covered by a finite collection of open sets $\{U_j\}_{j=1}^{l}$.

which is linear and its solution is $\phi(t,x) = x_0 e^{at}$. Therefore, the flow map ϕ is topologically transitive on the open, unbounded (and therefore noncompact) invariant sets $(-\infty, 0)$ and $(0, \infty)$. Also, for any two points $x_1, x_2 \in \mathbb{R}$ and $x_1 \neq x_2$ we have

$$|\phi(t,x_1) - \phi(t,x_2)| = e^{at}|x_1 - x_2|$$

and ϕ has sensitive dependence on initial conditions on $\mathbb{R}$. However, the orbits generated by ϕ are not chaotic.

(iv) This definition refers to a 'chaotic flow (or map) on a set A' or, for short, a 'chaotic set A.' It does not imply that all orbits of a chaotic flow (or map) on A are chaotic. In fact, there are many nonchaotic orbits on chaotic sets, in particular, many unstable periodic orbits. They are so important that some researchers add a third condition for chaos, that periodic orbits are dense on A [5]. This is an interesting property and it is automatically satisfied if the chaotic invariant set is hyperbolic [17].

(v) Two quite general results can be used to confirm the close relationship between chaos, as characterized in Definition 2.21, and dense periodic sets. The first result [3] states that for any continuous map on a metric space, transitivity and the presence of a dense set of periodic orbits imply sensitive dependence on initial conditions, that is, chaos. The second result [16] states that for any continuous map on an interval of R, transitivity alone implies the presence of a dense set of periodic orbits and, therefore, in view of the first result, it implies sensitive dependence on initial conditions, and therefore chaos.

(vi) There are several other different definitions of chaos based on orbits rather than sets. For example, in [1] (p. 196, Definition 5.2; p. 235, Definition 6.2; pp. 385–386, Definition 9.6), a chaotic set is defined as the ω-limit set of a chaotic orbit $G^n(x_0)$ which itself is contained in the ω-limit set. In this case, the presence of sensitive dependence on initial conditions (or a positive Lyapunov characteristic exponent) is not enough to characterize chaotic properties of orbits and additional conditions must be added to exclude unstable periodic or quasiperiodic orbits. □

2.12 Symbolic Dynamics and the Shift Map

Symbolic dynamics is a powerful tool for understanding the orbit structure of a large class of dynamical systems. In this section we only provide a brief introduction to this tool.

To establish the tool, three steps are needed. First, we define an auxiliary system characterized by a map, called a *shift map*, acting on a space of infinite sequences called the *symbol space*. Next, we prove some properties of the shift map. Finally, we establish a certain equivalence relation between a map we want to study and the shift map, and show that the relationship preserves the properties in question.

We begin by defining the symbol space and the shift map. Let S be a collection of symbols. In a physical interpretation, the elements of S could be anything, for example letters of an alphabet or discrete readings of some measuring device for the observation of a given dynamical system. To make ideas more clear, we assume here that S consists of only two symbols; let them be 0 and 1. Then, we have $S = \{0,1\}$. Next, we want to construct the space of all possible bi-infinite sequences of 0 and 1, defined as

$$\Sigma_2 := \cdots S \times S \times S \times \cdots .$$

A point in Σ_2, s, is therefore represented as a *bi-infinity-tuple* of elements of S, that is, $s \in \Sigma_2$ means

$$s = \{\dots s_{-n} \dots s_{-1} s_0 s_1 \dots s_n \dots\},$$

where $\forall i$, $s_i \in S$ (i.e., $s_i = 0$ or 1). For example, $s = \{\dots 00010100111 \dots\}$.

We can define a distance function $\bar{d}$ in the space Σ_2

$$\bar{d}(s,\bar{s}) = \sum_{i=-\infty}^{+\infty} \frac{d(s_i,\bar{s_i})}{2^{|i|}}, \tag{2.30}$$

where d is the discrete distance in $S = \{0,1\}$, that is

$$d(s_i,\bar{s_i}) = \begin{cases} 0 \text{ if } & s_i = \bar{s_i}; \\ 1 \text{ if } & s_i \neq \bar{s_i}. \end{cases}$$

This means that two points of Σ_2 are close to each other if their central elements are close, i.e., if the elements whose indexes have small absolute values are close. Notice that, from the definition of $\bar{d}(s_i,\bar{s_i})$, the infinite sum on the right-hand side of (2.30) is less than 3, and, therefore, converges.

Next, we define the *shift map* on Σ_2 as

$$T \colon \Sigma_2 \to \Sigma_2, \;\; T(s) = s' \text{ and } s'_i = s_{i+1}.$$

The map T maps each entry of a sequence from one place to the left. Similarly, the *one-sided shift map* T_+ can be defined on the space of one-sided infinite sequences, Σ_{2+}, that is, $s \in \Sigma_{2+}$, where $s = \{s_0 s_1 \dots s_n \dots\}$. In this case, we have

$$T_+ \colon \Sigma_{2+} \to \Sigma_{2+}, \;\; T_+(s) = s' \text{ and } s'_i = s_{i+1},$$

so that

$$T_+(s_0 s_1 s_2 \dots) = (s'_0 s'_1 s'_2 \dots) = (s_1 s_2 s_3 \dots).$$

It is obvious that the T_+ map shifts a one-sided sequence from one place to the left and drops its first element. Although maps T and T_+ have very similar properties, T is invertible whereas T_+ is not. The distance on Σ_{2+} is essentially the same as (2.30) with the difference that the infinite sum will now run from zero to ∞. The map T_+ can be used to prove chaotic properties of certain noninvertible, one-dimensional maps frequently employed in applications.

Theorem 2.9. The shift map T_+ on Σ_{2+} is chaotic according to Definition 2.21. □

Remark 2.5. The shift map T_+ on Σ_{2+} has a property that is stronger than *topological transitivity* called *topological mixing*. In general, we say that a map G is topologically mixing on a set A if for any two open subsets U and V of A there exists a positive integer N_0 such that $G^n(U) \cap V \neq \emptyset$ for all $n \geq N_0$. If a map G is topologically mixing, then for any integer n the map G^n is topologically transitive. □

The importance of the fact that the shift map is chaotic in a precise sense lies in that the chaotic properties of invariant sets of certain one- and two-dimensional maps and three-dimensional flows may sometimes be proved by showing that the dynamics on these sets are *topologically conjugate* to that of a shift map on a symbol space. This indirect argument is often the only available strategy for investigating nonlinear maps (or flows). We have encountered the concept of topological conjugacy in the Hartman–Grobman theorem (Theorem 2.5, which we called homeomorphic equivalence) between a nonlinear map (or flow) and its linearization in a neighborhood of a fixed point. We now provide some formal definitions.

Definition 2.22. Let X and Y be topological spaces, and let $f\colon X \to X$ and $g\colon Y \to Y$ be continuous functions. We say that f is *topologically semiconjugate* to g if there exists a continuous surjection[11] $h\colon Y \to X$ such that $f \circ h = h \circ g$. If h is a homeomorphism, then we say that f and g are topologically conjugate, and we call h a topological conjugation between f and g.

Similarly, a flow ϕ on X is topologically semiconjugate to a flow ψ on Y if there is a continuous surjection $h\colon Y \to X$ such that $\phi(h(y),t) = h(\psi(y,t))$ for each $y \in Y$, $t \in \mathbb{R}$. If h is a homeomorphism, then ϕ and ψ are topologically conjugate. □

Remark 2.6. Topological conjugation defines an *equivalence relation* in the space of all continuous surjections of a topological space to itself, by declaring f and g to be related if they are topologically conjugate. This equivalence relationship is very useful in the theory of dynamical systems, since each class contains all functions which share the same dynamics from the topological viewpoint. In fact, orbits of g are mapped to homeomorphic orbits of f through the conjugation. Writing $g = h^{-1} \circ f \circ h$ makes this fact evident: $g^n = h^{-1} \circ f^n \circ h$. Roughly speaking, topological conjugation is a 'change of coordinates' in the topological sense. □

However, the analogous definition for flows is somewhat restrictive. In fact, we require the maps $\phi(\cdot,t)$ and $\psi(\cdot,t)$ to be *topologically conjugate* for each t, which requires more than simply that orbits of ϕ be mapped to orbits of ψ homeomorphically. This motivates the definition of *topological equivalence*, which also partitions the set of all flows in X into classes of flows sharing the same dynamics, again from the topological viewpoint.

Definition 2.23. We say that two flows ψ and ϕ of a compact manifold M are *topologically equivalent* if there is an homeomorphism $h\colon Y \to X$, mapping orbits of

[11] A function $f\colon X \to Y$ is a surjection if, for every $y \in Y$, there is an $x \in X$ such that $f(x) = y$.

ψ to orbits of ϕ homeomorphically, and preserving orientation of the orbits. This means that

(i) $\{h(\psi(y,t)) : t \in \mathbb{R}\} = \{\phi(h(y),t) : t \in \mathbb{R}\}$ for each $y \in Y$;
(ii) for each $y \in Y$, there is $\delta > 0$ such that, if $0 < |s| < t < \delta$, and if s satisfies $\phi(h(y),s) = h(\psi(y,t))$, then $s > 0$. □

2.13 Lyapunov Exponent

Although sensitive dependence on initial conditions can be verified in some cases, it is not easy to verify for many systems. The *Lyapunov exponent* is a generalization of the eigenvalues at an equilibrium point, and it is used as a measure of exponential divergence of orbits. Suppose that $\phi(t,x_0)$ and $\phi(t,y_0)$ are solutions of an autonomous vector field $\dot{x} = f(x)$ starting from x_0 and y_0, respectively. By using the linear approximation for fixed t, we get

$$\phi(t,y_0) - \phi(t,x_0) \approx D_x\phi(t,x_0)(y_0 - x_0).$$

For any curve starting from initial condition x_s, letting

$$v(t) := \frac{\partial}{\partial s}\phi(t,x_s)\bigg|_{s=0} = D_x\phi(t,x_0)\frac{\partial x_s}{\partial s} = D_x\phi(t,x_0)v_0$$

and $v_0 := \dfrac{\partial x_s}{\partial s}$, then $v(t)$ satisfies the *first variation equation*

$$\frac{\mathrm{d}}{\mathrm{d}t}v(t) = Df_{(\phi(t,x_0))}v(t).$$

If $v_0 = y_0 - x_0$, then $v(t)$ would give the infinitesimal displacement at time t.

The growth rate of $\|v(t)\|$ is a number χ such that

$$\|v(t)\| \approx Ce^{\chi t},$$

where C is a constant. Taking the logarithm, we have

$$\frac{\ln\|v(t)\|}{t} \approx \frac{\ln(C)}{t} + \chi;$$

therefore,

$$\chi = \lim_{t\to\infty}\frac{\ln\|v(t)\|}{t}.$$

Definition 2.24. Let $v(t)$ be the solution of the first variation equation, starting from x_0 with $v(0) = v_0$. The *Lyapunov exponent* for initial condition x_0 and initial infinitesimal displacement v_0 is defined as

$$\chi(x_0, v_0) = \lim_{t \to \infty} \frac{\ln \|v(t)\|}{t},$$

whenever this limit exists. □

Remark 2.7. For most initial conditions x_0 for which the forward orbit is bounded, the Lyapunov exponents exist for all vectors v. In n-dimensional state space, there are at most n distinct values for $\chi(x_0, v)$ as v varies. If we count multiplicities, then there are exactly n values,

$$\chi_1(x_0) = \chi(x_0, v_1),\ \chi_2(x_0) = \chi_2(x_0, v_2), \ldots, \chi_n(x_0) = \chi(x_0, v_n).$$

We can order these so that

$$\chi_1(x_0) \geq \chi_2(x_0) \geq \cdots \geq \chi_n(x_0).$$ □

Several results on Lyapunov exponents are listed as follows. For detailed proofs please refer to [13].

Theorem 2.10. Assume that x_0 is a *fixed point* of the differential equation $\dot{x} = f(x)$. Then, the Lyapunov exponents at the fixed point are the real parts of the eigenvalues of the fixed point. □

Theorem 2.11. Let x_0 be an initial condition such that $\phi(t, x_0)$ is bounded and $\omega(x_0)$ does not contain any fixed points. Then,

$$\chi(x_0, f(x_0)) = 0.$$ □

Remark 2.8. The above theorem means that there is no growth or decay in the direction of the vector field, $v = f(x_0)$. □

Theorem 2.12. Let x_0 be an initial condition on a periodic orbit of period T. Then, the principal $(n-1)$ Lyapunov exponents are given by $(\ln|\lambda_j|)/T$, where λ_j are the *characteristic multipliers* of the periodic orbit and the eigenvalues of the Poincaré map. □

Theorem 2.13. Assume that $\phi(t, x_0)$ and $\phi(t, y_0)$ are two orbits for the same differential equation, which are bounded and converge exponentially (i.e., there are constants $a > 0$ and $C \geq 1$ such that $\|\phi(t, x_0) - \phi(t, y_0)\| \leq Ce^{-at}$ for $t \geq 0$). Then, the Lyapunov exponents for x_0 and y_0 are the same. So, if the limits defining the Lyapunov exponents exist for one of the points, they exist for the other point. The vectors which give the various Lyapunov exponents can be different at the two points. □

Theorem 2.14. Consider the system of $\dot{x} = f(x)$ in $\mathbb{R}^n$. Assume that x_0 is a point such that the Lyapunov exponents $\chi_1(x_0), \ldots, \chi_n(x_0)$ exist.

(i) Then, the sum of Lyapunov exponents is the limit of the average of the divergence along the trajectory,

$$\sum_{j=1}^{n} \chi_j(x_0) = \lim_{T\to\infty} \frac{1}{T} \int_0^{\mathrm{T}} \nabla \cdot f_{\phi(t,x_0)} \mathrm{d}t.$$

(ii) In particular, if the system has constant divergence δ, then the sum of the Lyapunov exponents at any point must equal δ.
(iii) In the three-dimensional case, assume that the divergence is a constant δ and that x_0 is a point for which the positive orbit is bounded and $\omega(x_0)$ does not contain any fixed points. If $\chi_1(x_0)$ is a nonzero Lyapunov exponent at x_0, then the other two Lyapunov exponents are 0 and $\delta - \chi_1$. □

The definition of the Lyapunov exponent for discrete-time systems is similar to the case of continuous time.

Definition 2.25. The Lyapunov exponent for the maps $x_{n+1} = G(x_n)$ is defined by

$$\chi(x_0, w) = \lim_{n\to\infty} \ln \frac{\|D^n G(x_0) w\|}{\|w\|},$$

where

$$D^n G(x_0) = DG(x_0) DG(x_1) \cdots G(x_{n-1})$$

and w is a vector in the tangent space at x_0. □

2.14 Examples

In this section, we will explore how chaos appears by investigating some examples.

2.14.1 Tent Map and Logistic Map

The tent map has the form of

$$G_\Lambda(y) = \begin{cases} 2y, & \text{if } 0 \le y \le \frac{1}{2}, \\ 2(1-y), & \text{if } \frac{1}{2} < y \le 1, \end{cases} \tag{2.31}$$

which is shown in Fig. 2.22.

Proposition 2.1. The tent map (2.31) is chaotic on $[0,1]$.

Proof. Consider that the graph of the nth iteration of G_Λ consists of 2^n linear pieces, each with slope $\pm 2^n$. Each of these linear pieces of the graph is defined on a subinterval of $[0,1]$ of length 2^{-n}. Then, for any open subinterval J of $[0,1]$, we can find a subinterval K of J of length 2^{-n}, such that the image of K under G_Λ^n covers the entire interval $[0,1]$. Therefore, G_Λ is topologically transitive on $[0,1]$. This fact, and the discussion in point (v) of Remark 2.4 [16], proves the proposition. □

Remark 2.9. From the geometry of the iterated map G_Λ^m, it appears that the graph of G_Λ^m on J intersects the bisector and therefore G_Λ^m has a fixed point in J. This proves that periodic points are dense in $[0,1]$. Also, for any $x \in J$ there exists a $y \in J$ such that $|G_\Lambda^n(x) - G_\Lambda^n(y)| \geq 1/2$ and, therefore, G_Λ has sensitive dependence on initial conditions. □

This result can be used to show that the logistic map

$$G_4\colon [0,1] \to [0,1], \ \ G_4(x) = 4x(1-x)$$

(see Fig. 2.23) is also chaotic. Consider the map $h(y) = \sin^2(\pi y/2)$. The map h is continuous and, restricted to $[0,1]$, is also one-to-one and onto. Its inverse is continuous and h is thus a homeomorphism. Consider now the diagram

$$\begin{array}{ccc} [0,1] & \xrightarrow{\ G_\Lambda\ } & [0,1] \\ {\scriptstyle h(y)}\downarrow & & \downarrow{\scriptstyle h(y)} \\ [0,1] & \xrightarrow[\ G_4\ ]{} & [0,1] \end{array}$$

where G_Λ is the tent map. Recalling the trigonometric relations:

(i) $\sin^2(\theta) + \cos^2(\theta) = 1$;
(ii) $4\sin^2(\theta)\cos^2(\theta) = \sin^2(2\theta)$;
(iii) $\sin(\pi - \theta) = \sin(\theta)$;

we can see that the diagram is commutative. Hence, the map G_4 is topologically conjugate to G_Λ and, therefore, its dynamics on $[0,1]$ is chaotic.

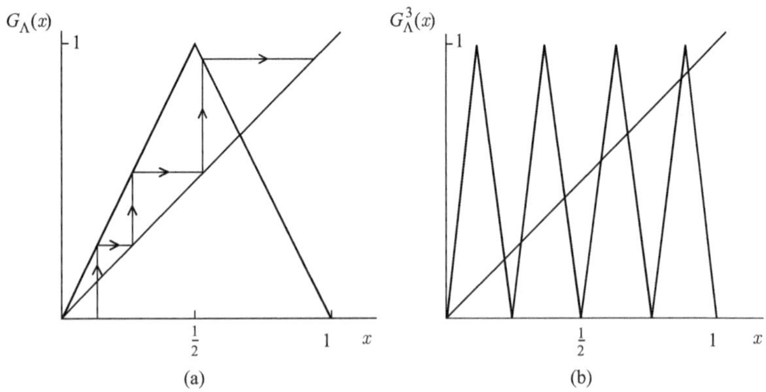

Fig. 2.22 The tent map: (a) G_Λ; (b) G_Λ^3

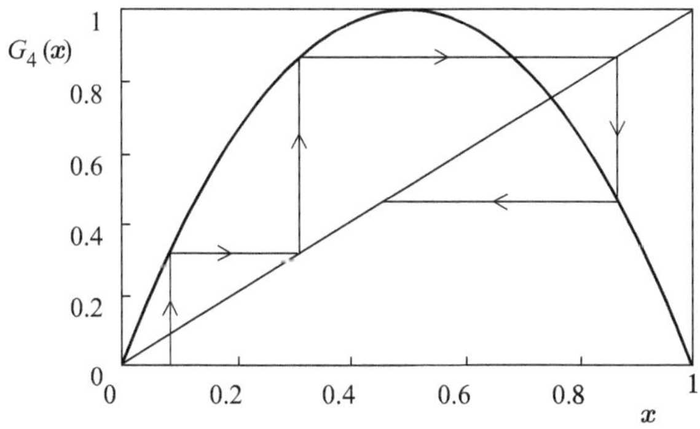

Fig. 2.23 The logistic map G_4

Both the tent map and the logistic map are all from an interval of $\mathbb{R}$ to itself. There are also many other one-dimensional mappings presenting chaotic dynamics. In fact, for one-dimensional mappings from $\mathbb{R}$ to itself we have the following so-called Li–Yorke theorem [13], famous for the phrase 'period three implying chaos,' and Sarkovskii's theorem [13], the generalization of Li–Yorke's result. Since many references include the proof of these two theorems, we will only state the content of the theorems.

Theorem 2.15 (Li–Yorke [13]). Assume that f is a continuous function from $\mathbb{R}$ to itself.

(i) If f has a period-3 point, then it has points of all periods.
(ii) Assume that there is a point x_0 such that either

a. $f^3(x_0) \leq x_0 < f(x_0) < f^2(x_0)$ or
b. $f^3(x_0) \geq x_0 > f(x_0) > f^2(x_0)$.

Then, f has points of all periods. □

This theorem was obtained by Li and Yorke in 1975 and soon after it was shown that Li–Yorke theorem is a special case of Sharkovskii's theorem. Before introducing Sharkovskii's theorem, we first define a new order for natural numbers as follows:

$$3 \rhd 5 \rhd 7 \cdots \rhd 2\cdot 3 \rhd 2\cdot 5 \rhd 2\cdot 7 \rhd \cdots \rhd 2^k\cdot 3 \rhd 2^k\cdot 5 \rhd 2^k\cdot 7 \rhd \cdots \rhd 2^k \rhd 2^{k-1} \rhd \cdots \rhd 2^2 \rhd 2 \rhd 1.$$

Theorem 2.16 (Sharkovskii's theorem [13]). Let f be a continuous function from $\mathbb{R}$ to itself. Suppose that f has period-n points and $n \rhd k$. Then, f has period-k points. □

2.14.2 Smale Horseshoe

The Smale horseshoe is the prototypical map possessing a chaotic invariant set. Therefore, a thorough understanding of the Smale horseshoe is absolutely essential for understanding what is meant by the term 'chaos' as it is applied to the dynamics of specific physical systems [10].

Consider the geometrical construction in Fig. 2.24. Take a square S on the plane (Fig. 2.24 (a)). Contract it in the horizontal direction and expand it in the vertical direction (Fig. 2.24 (b)). Fold it in the middle (Fig. 2.24 (c)) and place it so that it intersects the original square S along two vertical strips (Fig. 2.24 (d)). This procedure defines a map $f\colon \mathbb{R}^2 \to \mathbb{R}^2$. The image $f(S)$ of the square S under this transformation resembles a horseshoe. That is why it is called a *horseshoe map*. The exact shape of the image $f(S)$ is irrelevant; however, for simplicity we assume that both the contraction and expansion are linear and that the vertical strips in the intersection are rectangle. The map f can be made invertible and smooth together with its inverse. The inverse map f^{-1} transforms the horseshoe $f(S)$ back into the square S through stages (d)–(a). This inverse transformation maps the dotted square S shown in Fig. 2.24 (d) into the dotted horizontal horseshoe in Fig. 2.24 (a), which is assumed to intersect the original square S along two horizontal rectangles.

Denote the vertical strips in the intersection $S \cap f(S)$ by V_1 and V_2,

$$S \cap f(S) = V_1 \cup V_2$$

(see Fig. 2.25 (a)). Now make the most important step: perform the *second iteration* of the map f. Under this iteration, the vertical strips V_1 and V_2 will be transformed

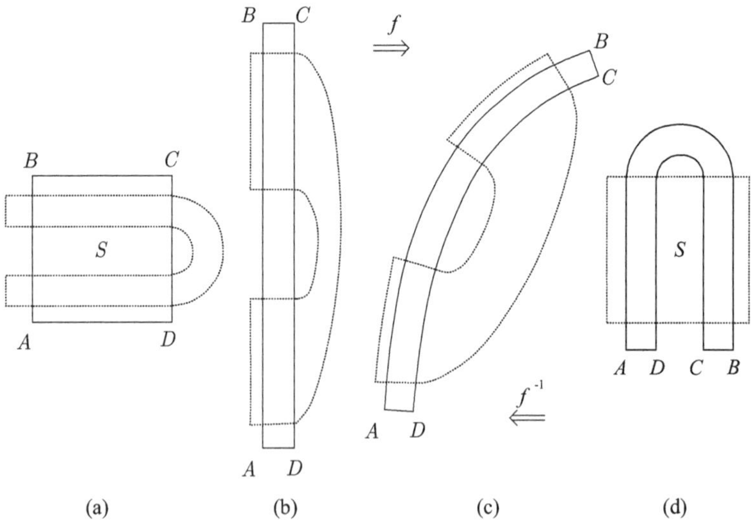

Fig. 2.24 Construction of the horseshoe map

into two 'thin horseshoes' that intersect the square S along four narrow vertical strips: V_{11}, V_{21}, V_{22}, and V_{12} (see Fig. 2.25 (b)). We write this as

$$S \cap f(S) \cap f^2(S) = V_{11} \cup V_{21} \cup V_{22} \cup V_{12}.$$

Similarly,

$$S \cap f^{-1}(S) = H_1 \cup H_2,$$

where H_1 and H_2 are the horizontal strips shown in Fig. 2.25 (c), and

$$S \cap f^{-1}(S) \cap f^{-2}(S) = H_{11} \cup H_{12} \cup H_{22} \cup H_{21},$$

with four narrow horizontal strips H_{ij} (Fig. 2.25 (d)). Notice that $f(H_i) = V_i, i = 1,2$, as well as $f^2(H_{ij}) = V_{ij}$, $i, j = 1.2$ (see Fig. 2.26).

Iterating the map f further, we obtain 2^k vertical strips in the intersection $S \cap f^k(S)$, $k \in \mathbb{N}$. Similarly, iteration of f^{-1} gives 2^k horizontal strips in the intersection $S \cap f^{-k}(S)$, $k \in \mathbb{N}$.

Most points leave the square S under iterations of f or f^{-1}. We consider all remaining points in the square under all iterations of f and f^{-1}:

$$\Gamma = \{x \in S\colon\ f^k(x) \in S,\ \forall k \in \mathbb{Z}\}.$$

Clearly, if the set Γ is nonempty, it is an *invariant set* of the discrete-time dynamical system defined by f. This set can be alternatively presented as an infinite intersection,

$$\Gamma = \cdots \cap f^{-k}(S) \cap \cdots \cap f^{-1}(S) \cap S \cap f(S) \cap \cdots f^k(S) \cap \cdots.$$

It is clear from this representation that the set Γ has a peculiar shape. Indeed, it should be located within

$$f^{-1}(S) \cap S \cap f(S),$$

which is formed by four small squares (see Fig. 2.27 (a)). Furthermore, it should be located inside

$$f^{-2}(S) \cap f^{-1}(S) \cap S \cap f(S) \cap f^2(S),$$

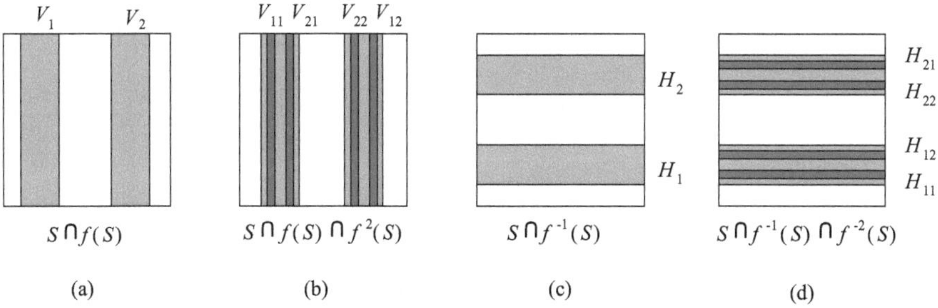

Fig. 2.25 Vertical and horizontal strips

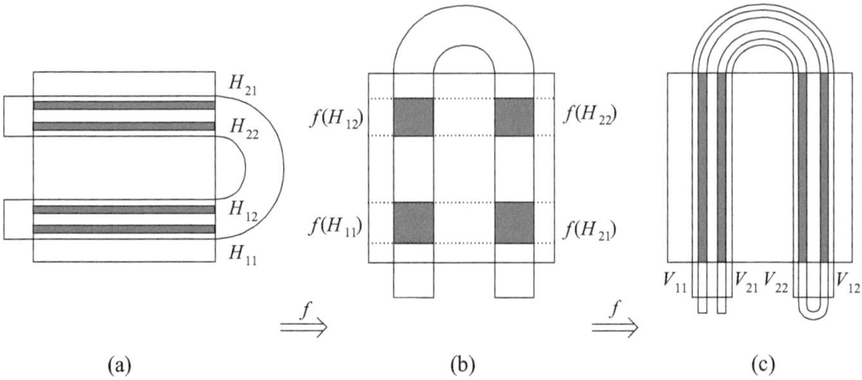

Fig. 2.26 Transformation $f^2(H_{ij}) = V_{ij}$, $i, j = 1, 2$

which is the union of sixteen smaller squares (see Fig. 2.27 (b)), and so forth. In the limit, we get a *Cantor set*. About the horseshoe map, we have the following lemma.

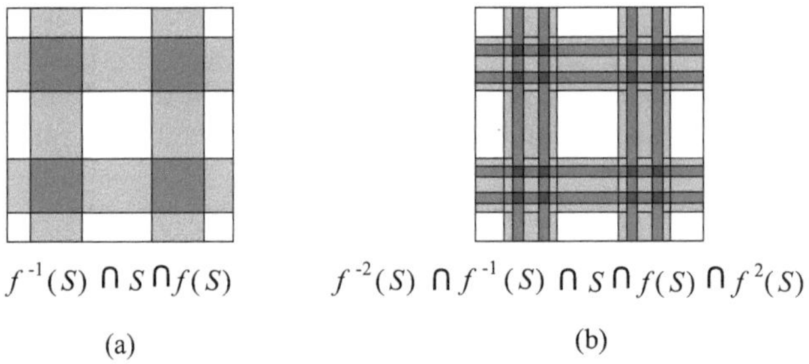

Fig. 2.27 Location of the invariant set

Lemma 2.1. There is a one-to-one correspondence $h\colon \Gamma \to \Sigma_2$ between points of Γ and all bi-infinite sequences of two symbols.

Proof. For any point $x \in \Gamma$, define a sequence of the two symbols $\{1,2\}$ by

$$\omega = \{\ldots, \omega_{-2}, \omega_{-1}, \omega_0, \omega_1, \omega_2, \ldots\}$$

by the formula

$$\omega_k = \begin{cases} 1, \text{ if } \; f^k(x) \in H_1, \\ 2, \text{ if } \; f^k(x) \in H_2, \end{cases} \tag{2.32}$$

for $k \in \mathbb{Z}$. Here, $f^0 = \text{id}$, the identity map. Clearly, this formula defines a map $h\colon \Gamma \to \Sigma_2$, which assigns a sequence to each point of the invariant set. To verify that this map is invertible, take a sequence $\omega \in \Sigma_2$, fix $m > 0$, and consider a set $R_m(\omega)$ of all points $x \in S$, not necessarily belonging to Γ, such that

$$f^k(x) \in H_{\omega_k},$$

for $-m \leq k \leq m-1$. For example, if $m=1$, the set R_1 is one of the four intersections $V_j \cap H_k$. In general, R_m belongs to the intersection of a vertical and a horizontal strip. These strips are getting thinner and thinner as $m \to \infty$, approaching in the limit a vertical and a horizontal segment, respectively. Such segments obviously intersect at a single point x with $h(x) = \omega$. Thus, $h\colon \Gamma \to \Sigma_2$ is a one-to-one map. It implies that Γ is nonempty. □

Remark 2.10. The map $h\colon \Gamma \to \Sigma_2$ is continuous together with its inverse (a homeomorphism) if we use the standard Euclidean metric in $S \subset \mathbb{R}^2$ and the metric given by (2.30) in Σ_2. □

Consider now a point $x \in \Gamma$ and its corresponding sequence $\omega = h(x)$, where h is the map previously constructed. Next, consider a point $y = f(x)$, that is, the image of x under the horseshoe map f. Since $y \in \Gamma$ by definition, there is a sequence that corresponds to y: $\theta = h(y)$. As one can easily see from (2.32), there is a simple relationship between these sequences ω and θ. Namely,

$$\theta_k = \omega_{k+1}, \quad k \in \mathbb{Z},$$

since

$$f^k(f(x)) = f^{k+1}(x).$$

In other words, the sequence θ can be obtained from the sequence ω by the *shift map* σ:

$$\theta = \sigma(\omega).$$

Therefore, the restriction of f to its invariant set $\Gamma \subset \mathbb{R}^2$ is *equivalent* to the shift map σ on the set of sequences Σ_2. This result can be formulated as the following lemma.

Lemma 2.2. $h(f(x)) = \sigma(h(x))$, for all $x \in \Gamma$. □

This lemma can be written as an even shorter one:

$$f|_\Gamma = h^{-1} \circ \sigma \circ h.$$

Combining Lemmas 2.1 and 2.2 with obvious properties of the shift dynamics on Σ_2, we get a theorem giving a rather complete description of the behavior of the horseshoe map.

Theorem 2.17. The horseshoe map f has a closed invariant set Γ that contains a countable set of periodic orbits of arbitrarily long period, and an uncountable set of nonperiodic orbits, among which there are orbits passing arbitrarily close to any point of Γ. □

Remark 2.11. The limit set Γ of a Smale horseshoe map is unstable and, therefore, not attracting. It follows that the existence of a Smale horseshoe does not imply the existence of a chaotic attractor. The existence of a Smale horseshoe does imply, however, that there is a region in state space that experiences sensitive dependence on initial conditions. Thus, even when there is no strange attractor in the flow, the dynamics of the system can appear chaotic until the steady state is reached. □

Remark 2.12. We can slightly perturb the constructed map f without qualitative changes to its dynamics. Clearly, Smale's construction is based on a sufficiently strong contraction or expansion, combined with a fold. Thus, a (smooth) perturbation $\tilde{f}$ will have similar vertical and horizontal strips, which are no longer rectangles but curvilinear regions. However, provided that the perturbation is sufficiently small, these strips will shrink to *curves* that deviate only slightly from vertical and horizontal lines. Thus, the construction can be carried through word for word, and the perturbed map $\tilde{f}$ will have an invariant set Γ on which the dynamics is completely described by the shift map σ on the sequence space Σ_2. This is an example of structurally stable behavior. □

2.14.3 The Lorenz System

Although plenty of numerical evidence of chaotic behavior arising from a variety of problems in different fields of applications has been provided, apart from the cases of one-dimensional, noninvertible maps, there are few rigorous proofs that specific mathematical models possess chaotic attractors as characterized in one or more of the above definitions, and those proofs are mostly restricted to artificial examples unlikely to arise in typical applications. We now turn to an example of a chaotic attractor derived from a celebrated model first discussed in 1963 by E. Lorenz to provide a mathematical description of atmospheric turbulence. Lorenz's investigation was enormously influential and stimulated a vast literature in the years that followed. Its extraordinary success was in part due to the fact that it showed how computer technology could be effectively used to study nonlinear dynamical systems. Lorenz's work provided strong numerical evidence that a low-dimensional system of differential equations with simple nonlinearities could generate extremely complicated orbits. The original Lorenz model is defined by the following three differential equations:

$$\begin{cases} \dot{x} = -\sigma x + \sigma y, \\ \dot{y} = -xz + rx - y, \\ \dot{z} = xy - bz, \end{cases} \tag{2.33}$$

where $x, y, z \in \mathbb{R}$ and $\sigma, r, b > 0$. System (2.33) is symmetrical under the transformation $(x,y,z) \to (-x,-y,z)$. The three equilibria are

$$\begin{aligned} &E_1: (0,0,0), \\ &E_2: (\sqrt{b(r-1)}, \sqrt{b(r-1)}, r-1), \\ &E_3: (-\sqrt{b(r-1)}, -\sqrt{b(r-1)}, r-1). \end{aligned}$$

In numerical analysis of the Lorenz model, the typical parameter configuration is $\sigma = 10$ and $b = 8/3$. When $r < r_H = 24.74$ there are two symmetric unstable periodic orbits with the Lorenz system. When $r > r_H$, the celebrated Lorenz attractor (the so-called 'butterfly') is observed numerically (see Fig. 2.28). The three Lyapunov exponents are $1.497, 0$, and -22.46, which imply that the Lorenz system has sensitive dependence on initial conditions.

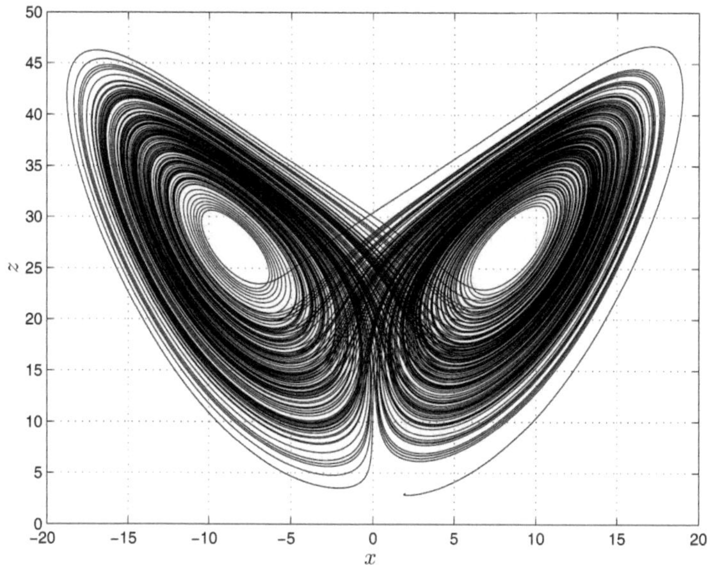

Fig. 2.28 The Lorenz attractor

Remark 2.13. When we refer to a computer to visualize the numerical solution of a chaotic system, an important issue arises. If we take into account the combined influence of round-off errors in numerical computations and the property of divergence of nearby trajectories for chaotic behavior, how can we trust numerical computations of trajectories to give us reliable results? (We note that the same problem arises in experimental measurements in which 'noise' plays the role of round-off errors.) If the system's behavior is chaotic, then even small numerical errors are amplified exponentially in time. Perhaps all of our results for chaotic systems are artifacts of the numerical computation procedure. Even if they are not artifacts, perhaps the numerical values of the properties depend critically on the computational procedures. If that is true, how general are our results? Although it is difficult to answer these questions once and for all, it is comforting to know that while it is true that the details of a particular trajectory do depend on the round-off errors in the numerical computation, the trajectory actually calculated does follow very closely some trajectory of the system. That is, the trajectory one calculates might not be exactly the one he thinks, but it is very close to one of the possible trajectories of the system. In more technical terms, we say that the computed trajectory shadows some possible trajectories of the system. (A proof of this shadowing property for chaotic systems is given in [6] and [14].) In general, we are most often interested in properties that are averaged over a trajectory; in many cases those average values are independent of the particular trajectory we follow. So, as long as we follow some possible trajectory for the system, we can have confidence that our results are a good characterization of the system's behavior. Recently, W. Tucker's work [15] strengthened the above discussion, in which he has shown, using a computer-assisted proof, that the Lorenz system not only has sensitive dependence on initial conditions, but also has a chaotic attractor. □

Although a full mathematical analysis of the observed attractor is still lacking, some of the attractor's properties have been established through a combination of numerical evidence and theoretical arguments. Before presenting the analysis we first consider the following three facts about system (2.33).

(i) The trace of the Jacobian matrix

$$\mathrm{tr}[Df(x,y,z)] = \frac{\partial \dot{x}}{\partial x} + \frac{\partial \dot{y}}{\partial y} + \frac{\partial \dot{z}}{\partial z} = -(b+\sigma+1) < 0$$

is constant and negative along orbits. Thus, any three-dimensional volume of initial conditions is contracted along orbits at a rate equal to

$$\gamma = -(b+\sigma+1) < 0;$$

that is to say, the system is *dissipative*.

(ii) It is possible to define a trapping region such that all orbits outside of it tend to it, and no orbits ever leave it. To see this, consider the function

$$V(x,y,z) = x^2 + y^2 + (z-r-\sigma)^2 = K^2(r+\sigma)^2 \tag{2.34}$$

defining a sphere with center at $(x=y=0; z=\sigma+r)$ and radius $K(\sigma+r)$. The time derivative of (2.34) along the solution of (2.33) is

$$\dot{V}(x,y,z) = -2\sigma x^2 - 2y^2 - 2b\left(z - \frac{r+\sigma}{2}\right)^2 + b\frac{(r+\sigma)^2}{2}.$$

$\dot{V} = 0$ defines an ellipsoid outside of which $\dot{V} < 0$. For a sufficiently large value of the radius (for sufficiently large K, given r and σ), the sphere (2.34) will contain all three fixed points and all orbits on the boundary of the sphere will point inward. Consequently, system (2.33) is 'trapped' inside the sphere.

(iii) Numerical evidence indicates that for $r \in (13.8, 14)$, there exist two symmetric *homoclinic orbits* (that is, orbits that connect a fixed point to itself) asymptotically approaching the origin for $t \to \pm\infty$, tangentially to the z axis; see the sketch in Fig. 2.29.

Keeping the three facts in mind, we can investigate the Lorenz system by constructing a geometric model which, under a certain hypothesis, provides a reasonable approximation of the dynamics of the original model with 'canonical' parameter values $\sigma = 10$, $b = 8/3$, and $r > r_H$.

We first consider a system of differential equations in $\mathbb{R}^3$ depending on a parameter μ with the following properties:

(i) for a certain value μ_h of the parameter, there exists a pair of symmetrical homoclinic orbits, asymptotically converging to the origin, and tangential to the positive z axis;

(ii) the origin is a saddle-point equilibrium and the dynamics in a neighborhood N of the equilibrium, for μ in a neighborhood of μ_h, is approximated by the system

$$\begin{cases} \dot{x} = \lambda_1 x, \\ \dot{y} = \lambda_2 y, \\ \dot{z} = \lambda_3 z, \end{cases}$$

where $\lambda_2 < \lambda_3 < 0 < \lambda_1$ and $-\lambda_3/\lambda_1 < 1$;

(iii) the system is invariant under the change of coordinates $(x,y,z) \to (-x,-y,z)$.

Under these conditions, for $(x,y,z) \in N$ and μ near μ_h, it is possible to construct a two-dimensional cross section Σ, such that the transversal intersections of orbits with Σ define a two-dimensional Poincaré map $P\colon \Sigma \to \Sigma$. For values of $|x|$ and $|\mu - \mu_h|$ sufficiently small, the dynamics of P can further be approximated by a one-dimensional, noninvertible map $G_\mu[-a,a] - \{0\} \to \mathbb{R}$ defined on an interval of the x axis, but not at $x = 0$.

A typical formulation of the map G_μ is

$$G_\mu(x) = \begin{cases} a\mu + cx^\delta, & \text{if } x > 0; \\ -a\mu - c|x|^\delta, & \text{if } x < 0; \end{cases}$$

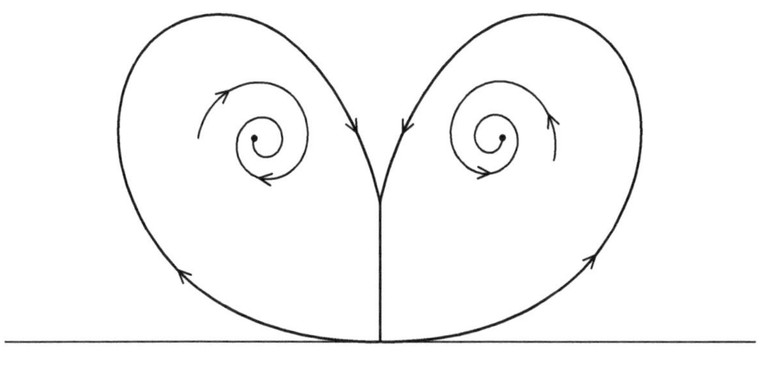

Fig. 2.29 Homoclinic orbits in the Lorenz model

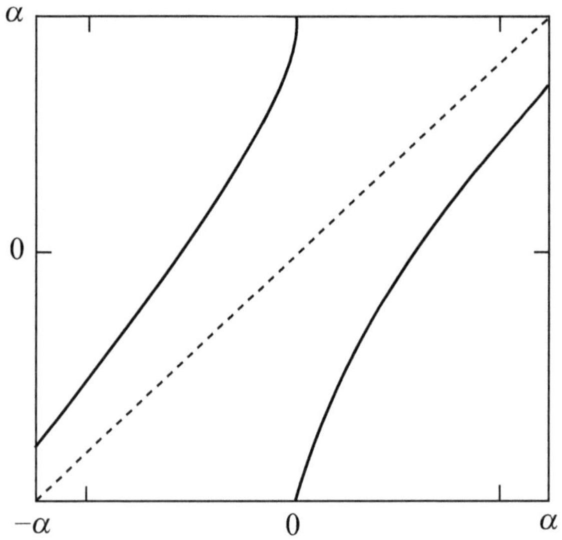

Fig. 2.30 One-dimensional map for the Lorenz model

where $a < 0$, $c > 0$, $\delta = -\lambda_3/\lambda_1$, and $0 < \delta < 1$.

Assuming that the one-dimensional approximation remains valid for values of the parameter μ outside the neighborhood of $\mu = \mu_h$ (that is, outside the neighborhood of the homoclinic orbit), values of $\mu > \mu_h$ can be chosen so that there exists a closed interval $[-\alpha, \alpha]$ with $\alpha > 0$ such that $G_\mu[-\alpha, \alpha]\backslash\{0\} \to [-\alpha, \alpha]$ and

$$\lim_{x \to 0^-} G_\mu(x) = \alpha > 0, \quad \lim_{x \to 0^+} G_\mu(x) = -\alpha < 0.$$

Then, $G'_\mu(x) > 1$ for $x \in [-\alpha, \alpha]$ $(x \neq 0)$, and $\lim_{x \to 0^\pm} G'_\mu(x) = +\infty$. The map G_μ on the interval is depicted in Fig. 2.30.

Because $G'_\mu(x) > 1$ for all $x \in [-\alpha, \alpha]$ $(x \neq 0)$, G_μ is a piecewise-expanding map and has therefore sensitive dependence on initial conditions. There are no fixed points or stable periodic points and most orbits on the interval $[-\alpha, 0) \cup (0, \alpha]$ are attracted to a chaotic invariant set. Although increasing μ beyond the homoclinic value μ_h leads to stable chaotic motion, if we take μ very large, the system reverts to simpler dynamical behavior and stable periodic orbits reappear.

Remark 2.14. The idea that some essential aspects of the dynamics of the original system (2.33) could be described by a one-dimensional map was first put forward by Lorenz himself. In order to ascertain whether the numerically observed attractor could be periodic rather than chaotic, he plotted successive maxima of the variable z along an orbit on the numerically observed attractor. In doing so, he discovered that the plot of z_{n+1} against z_n has a simple shape, as illustrated in Fig. 2.31. The points of the plot lie almost exactly on a curve whose form changes as the parameter r varies. Setting $\sigma = 10$, $b = 8/3$, and $r = 28$ (the traditional 'chaotic values'), we

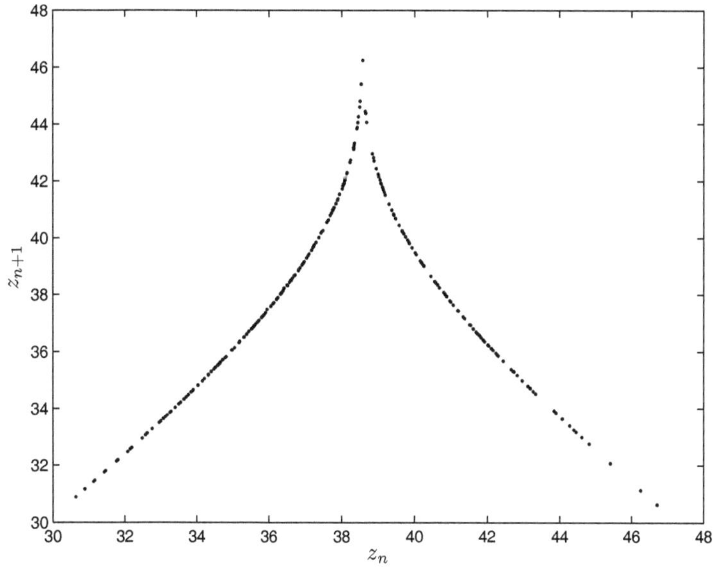

Fig. 2.31 Successive maxima of z for the Lorenz attractor

obtain a curve resembling a distorted tent map. It has slope everywhere greater than 1 in absolute value so, again, it approximates a piecewise-expanding map. For such a map there cannot be stable fixed points or stable periodic orbits and, for randomly chosen initial values, orbits converge to a chaotic attractor. □

2.15 Basics of Functional Differential Equations Theory

In many applications, the system under consideration is not governed by a principle of causality; that is, the future state of the system is dependent not only on the present state but also on the past states. The theory about these systems has been extensively developed, which formed the framework of the subject – *functional differential equations* (FDEs). In this section, we only provide some necessary notions and theorems about FDEs so that this book can be read smoothly.

Let $\mathbb{R} = (-\infty, \infty)$ and $\mathbb{R}^+ = [0, \infty)$. Let $C = C([-\tau, 0], \mathbb{R}^n)$ denote the space of continuous functions mapping the interval $[-\tau, 0]$ into $\mathbb{R}$. We designate the norm of an element ϕ in C by $\|\phi\|_\tau = \sup_{-\tau \leq \theta \leq 0} \|\phi(\theta)\|$. Each $x \in \mathbb{R}^n$ can also be looked as an element in C: $x(\theta) = x$, $\theta \in [-\tau, 0]$. If $\sigma \in \mathbb{R}$, $\alpha \geq 0$, and $x \in C[\sigma - \tau, \sigma + \alpha]$, then, for any $t \in [\sigma, \sigma + \alpha]$, we let $x_t \in C$ be defined by $x_t(\theta) = x(t + \theta)$, $-\tau \leq \theta \leq 0$. If D is a subset of $\mathbb{R} \times C$, f: $D \to \mathbb{R}^n$ is a given function, and '$\cdot$' represents the right-hand derivative, we say that the relation

$$\dot{x} = f(t, x_t) \tag{2.35}$$

is a retarded functional differential equation on D and will denote this equation by RFDE. We write RFDE(f) if we wish to emphasize that the equation is defined by f.

Definition 2.26. A function x is said to be a solution of (2.35) on $[\sigma - \tau, \sigma + \alpha]$ if there are $\sigma \in \mathbb{R}$ and $\alpha > 0$ such that $x \in C([\sigma - \tau, \sigma + \alpha], \mathbb{R}^n)$, $(t, x_t) \in D$, and $x(t)$ satisfies (2.35) for $t \in [\sigma, \sigma + \alpha)$. For given $\sigma \in \mathbb{R}$, $\phi \in C$, we say that $x(\sigma, \phi, f)$ is a solution of (2.35) with initial value ϕ at σ or simply a solution through (σ, ϕ) if there is an $\alpha > 0$ such that $x(\sigma, \phi, f)$ is a solution of (2.35) on $[\sigma - \tau, \sigma + \alpha]$ and $x_\sigma(\sigma, \phi, f) = \phi$. □

In the following we only state the main results; the proofs can be found in [8].

Theorem 2.18 (Existence). Suppose that Ω is an open subset in $\mathbb{R} \times C$ and $f^0 \in C(\Omega, \mathbb{R}^n)$. If $(\sigma, \phi) \in \Omega$, then there is a solution of the RFDE(f^0) passing through (σ, ϕ). More generally, if $W \subseteq \Omega$ is compact and $f^0 \in C(\Omega, \mathbb{R}^n)$ is given, then there is a neighborhood $V \subseteq \Omega$ of W such that $f^0 \in C^0(V, \mathbb{R}^n)$, there is a neighborhood $U \subseteq C^0(V, \mathbb{R}^n)$ of f^0, and an $\alpha > 0$ such that, for any $(\sigma, \phi) \in W$, $f \in U$, there is a solution $x(\sigma, \phi, f)$ of the RFDE(f) through (σ, ϕ) which exists on $[\sigma - \tau, \sigma + \alpha]$. □

Theorem 2.19 (Continuous Dependence). Suppose that $\Omega \subseteq \mathbb{R} \times C$ is open, $(\sigma^0, \phi^0) \in \sigma$, $f^0 \in C(\Omega, \mathbb{R}^n)$, and x_0 is a solution of the RFDE(f^0) through (σ^0, ϕ^0) which exists and is unique on $[\sigma^0 - \tau, b]$. Let $W^0 \subseteq \Omega$ be the compact set defined by

$$W^0 = \{(t, x_t^0) | t \in [\sigma^0, b]\}$$

and let V^0 be a neighborhood of W^0 on which f^0 is bounded. If (σ^k, ϕ^k, f^k), $k = 1, 2, \ldots$, satisfies $\sigma^k \to \sigma^0$, $\phi^k \to \phi^0$, and $\|f^k - f^0\|_{V^0} \to 0$ as $k \to \infty$, then there is a k^0 such that the RFDE(f^k) for $k \geq k^0$ is such that each solution $x^k = x^k(\sigma^k, \phi^k, f^k)$ through (σ^k, ϕ^k) exists on $[\sigma^k - \tau, b]$ and $x^k \to x^0$ uniformly on $[\sigma^0 - \tau, b]$. Since all x^k may not be defined on $[\sigma^0 - \tau, b]$, by $x^k \to x^0$ uniformly on $[\sigma^0 - r, b]$, we mean that, for any $\varepsilon > 0$, there is a $k_1(\varepsilon)$ such that $x^k(t)$, $k \geq k_1(\varepsilon)$, is defined on $[\sigma^0 - \tau + \varepsilon, b]$, and $x^k \to x^0$ uniformly on $[\sigma^0 - \tau + \varepsilon, b]$. □

Theorem 2.20. Suppose that Ω is an open set in $\mathbb{R} \times C$, $f \colon \Omega \to \mathbb{R}^n$ is continuous, and $f(t, \phi)$ is Lipschitzian in ϕ in each compact set in Ω, i.e.,

$$\|f(t, \phi) - f(t, \psi)\| \leq K \|\phi - \psi\|,$$

for arbitrary $\phi, \psi \in C$, where K is a constant. If $(\sigma, \phi) \in \Omega$, then there is a unique solution of (2.35) through (σ, ϕ). □

Definition 2.27. If, for all $(\sigma, \phi) \in \mathbb{R} \times C$, the solution of (2.35) through (σ, ϕ), $x(\sigma, \phi)(t)$, exists on $[\sigma - \tau, \infty)$, then we say that (2.35) has global solutions. □

Theorem 2.21. If there exist continuous functions $M, N\colon \mathbb{R} \to \mathbb{R}^+$ such that

$$\|f(t,\phi)\| \le M(t) + N(t)\|\phi\|, \quad (t,\phi) \in \mathbb{R} \times C,$$

then (2.35) has global solutions. □

In the following, we introduce the stability theory for (2.35). Analogous to what we have done with ordinary differential equations, the stability of a general solution $x(t)$ of (2.35) is equivalent to the stability of the zero solution of a new RFDE. Therefore, without loss of generality, we assume that $f(t,0) = 0$ for all $t \in \mathbb{R}$.

Definition 2.28.

(i) The solution $x = 0$ of (2.35) is said to be stable if, for any $\sigma \in \mathbb{R}$ and $\varepsilon > 0$, there exists a $\delta = \delta(\varepsilon, \sigma) > 0$ such that $\|\phi\|_\tau < \delta$ implies that $\|x(t,\sigma,\phi)\| < \varepsilon$ for $t \ge \sigma$.

(ii) The solution $x = 0$ of (2.35) is said to be uniformly stable if the δ in (i) is independent of σ.

(iii) The solution $x = 0$ of (2.35) is said to be attractive if, for any $\sigma \in \mathbb{R}$, there exists a $b = b(\sigma)$ such that $\|\phi\|_\tau \le b$ implies that $x(t,\sigma,\phi) \to 0$ $(t \to \infty)$. That is, for any $\varepsilon > 0$ and $\|\phi\|_\tau \le b$, there exists a $T(\sigma,\varepsilon,\phi)$ such that $\|x(t,\sigma,\phi)\| < \varepsilon$ whenever $t \ge \sigma + T(\sigma,\varepsilon,\phi)$. If $b = +\infty$, the solution $x = 0$ of (2.35) is said to be globally attractive.

(iv) The solution $x = 0$ of (2.35) is said to be asymptotically stable if it is both stable and attractive.

(v) The solution $x = 0$ of (2.35) is said to be uniformly attractive if in (iii) b is independent of σ and T only depends on ε.

(vi) The solution $x = 0$ of (2.35) is said to be uniformly asymptotically stable if it is both uniformly stable and uniformly attractive.

(vii) The solution $x = 0$ of (2.35) is said to be globally asymptotically stable if it is stable and globally attractive. □

Definition 2.29. The solution $x = 0$ of (2.35) is said to be exponentially stable if there exists a $\beta > 0$ and, for any $\varepsilon > 0$, there is a $\delta = \delta(\varepsilon) > 0$ such that, for any $\sigma \in \mathbb{R}$, $\|\phi\|_\tau < \delta$ implies that

$$\|x(t,\sigma,\phi)\| \le \varepsilon \exp[-\beta(t-\sigma)]$$

for $t \ge \sigma$. The solution $x = 0$ of (2.35) is said to be globally exponentially stable if there exist a $\beta > 0$ and an $\eta > 0$ such that, for any $(\sigma,\phi) \in \mathbb{R} \times C$, the following inequality holds:

$$\|x(t,\sigma,\phi)\| \le \eta \|\phi\|_\tau \exp[-\beta(t-\sigma)]$$

for $t \ge \sigma$. □

Definition 2.30. A solution $x(t,\sigma,\phi)$ of (2.35) is bounded if there is a $\beta(\sigma,\phi)$ such that $\|x(t,\sigma,\phi)\| < \beta(\sigma,\phi)$ for $t \ge \sigma - \tau$. The solution is uniformly bounded if, for any $\alpha > 0$, there is a $\beta = \beta(\alpha) > 0$ such that, for all $\sigma \in \mathbb{R}$, $\phi \in C$, and $\|\phi\| \le \alpha$, we have $\|x(t,\sigma,\phi)\| \le \beta(\alpha)$ for all $t \ge \sigma$. □

In the following, we introduce some sufficient conditions for the stability of the solution $x = 0$ of (2.35) which generalize the second method of Lyapunov for ordinary differential equations.

If the functional $V\colon \mathbb{R} \times C \to \mathbb{R}^+$ is continuous and $x(t, \sigma, \phi)$ is the solution of (2.35) through (σ, ϕ), we define

$$\dot{V}(t,\phi) = \overline{\lim_{h \to 0^+}} \frac{1}{h}[V(t+h, x_{t+h}(t,\phi)) - V(t,\phi)].$$

The function $\dot{V}(t,\phi)$ is the upper right-hand derivative of $V(t,\phi)$ along the solution of (2.35).

If the function $V\colon \mathbb{R} \times \mathbb{R} \to \mathbb{R}^+$ is continuous, we define

$$\dot{V}(t,\phi(0)) = \overline{\lim_{h \to 0^+}} \frac{1}{h}[V(t+h, x(t,\phi)(t+h)) - V(t,\phi(0))].$$

The function $\dot{V}(t,\phi(0))$ is the upper right-hand derivative of $V(t,x)$ along the solution of (2.35).

Sometimes we write $\dot{V}_{(2.35)}(t,\phi)$ and $\dot{V}_{(2.35)}(t,\phi(0))$ to emphasize the dependence on (2.35), respectively.

Theorem 2.22. Suppose that $f\colon \mathbb{R} \times C \to \mathbb{R}^n$ takes $\mathbb{R}\times$ (bounded set of C) into bounded sets of $\mathbb{R}^n$, and $u, v, w\colon \mathbb{R}^+ \to \mathbb{R}^+$ are continuous nondecreasing functions, $u(s)$ and $v(s)$ are positive for $s > 0$, and $u(0) = v(0) = 0$. If there is a continuous functional $V\colon \mathbb{R} \times C \to \mathbb{R}^+$ such that

$$u(\|\phi(0)\|) \le V(t,\phi) \le v(\|\phi\|_\tau),$$

$$\dot{V}_{(2.35)}(t,\phi) \le -w(\|\phi(0)\|),$$

then the solution $x = 0$ of (2.35) is uniformly stable. If $u(s) \to \infty$ as $s \to \infty$, the solutions of (2.35) are uniformly bounded. If $w(s) > 0$ for $s > 0$, then the solution $x = 0$ is uniformly asymptotically stable. □

Corollary 2.1. Suppose that $f\colon \mathbb{R} \times C \to \mathbb{R}^n$ takes $\mathbb{R}\times$ (bounded set of C) into bounded sets of $\mathbb{R}^n$, and $u, w\colon \mathbb{R}^+ \to \mathbb{R}^+$ are continuous nondecreasing functions, $u(s)$ and $w(s)$ are positive for $s > 0$, $u(0) = w(0) = 0$, and $u(s) \to \infty$ $(s \to \infty)$. If there is a continuous functional $V\colon \mathbb{R} \times C \to \mathbb{R}^+$ such that

$$u(\|\phi(0)\|) \le V(t,\phi), \; V(t,0) = 0,$$

$$\dot{V}_{(2.35)}(t,\phi) \le -w(\|\phi(0)\|),$$

then the solution $x = 0$ of (2.35) is globally asymptotically stable. □

Theorem 2.23. Suppose that $f\colon \mathbb{R} \times C \to \mathbb{R}^n$ takes $\mathbb{R}\times$ (bounded set of C) into bounded sets of $\mathbb{R}^n$, and $u, v, w\colon \mathbb{R}^+ \to \mathbb{R}^+$ are continuous nondecreasing functions, $u(s), v(s)$, and $w(s)$ are positive for $s > 0$, and $u(0) = v(0) = 0$. Suppose that $P(s)$ is a continuous and nondecreasing function satisfying $P(s) > s$ for $s > 0$. If there is a continuous function $V: \mathbb{R} \times \mathbb{R}^n \to \mathbb{R}^+$ such that

$$u(\|x\|) \leq V(t,\phi) \leq v(\|x\|)$$

and

$$\dot{V}_{(2.35)}(t,\phi(0)) \leq -w(\|\phi(0)\|) \text{ when } V(t+\theta,\phi(\theta)) < P(V(t,\phi(0))), \theta \in [-\tau,0],$$

then, the solution $x = 0$ of (2.35) is uniformly asymptotically stable. Moreover, if $u(s) \to \infty$ as $s \to \infty$, then the solution $x = 0$ of (2.35) is globally asymptotically stable. □

Theorem 2.24. Suppose that α, β, p, and μ are positive constants. If there is a continuous function $V: \mathbb{R} \times \mathbb{R}^n \to \mathbb{R}^+$ such that

$$\alpha\|x\|^p \leq V(t,x) \leq \beta\|x\|^p$$

and

$$\dot{V}_{(2.35)}(t,\phi(0)) \leq -\mu V(t,\phi(0)) \text{ when } \sup_{-\tau\leq\theta\leq 0} [e^{\mu\theta} V(t+\theta,x(t+\theta))] = V(t,x(t)),$$

then, the solution $x = 0$ of (2.35) is globally exponentially stable. □

2.16 Summary

One basic goal in studying dynamical systems is to explore how the trajectories of a system evolve as time proceeds. So, in this chapter, we began with the theorems on existence and uniqueness of solutions of ordinary differential equations. In the subsequent sections we focused on a special class of solutions: equilibrium solutions. Besides that notion, we introduced other definitions such as fixed point, periodic orbit, quasiperiodic orbit, ω-limit set, invariant set, etc., and, furthermore, the key concepts of the book, chaos and chaotic attractors, were introduced. Two powerful tools in studying chaotic systems, Lyapunov exponents and symbolic dynamics, were discussed briefly. The stability issue was discussed and a detailed category of the types of fixed point of planar systems for both continuous time and discrete time was provided. Three famous examples on chaos were carefully presented through which we wanted to give a concrete understanding about chaos. Finally, we provided some necessary preliminaries on retarded functional differential equations for the purpose of self-containment of the book.

References

1. Alligood KT, Sauer TD, Yorke JA (1997) *Chaos – An Introduction to Dynamical Systems*. Springer, New York
2. Arnold VI (1988) *Geometrical Methods in the Theory of Ordinary Differential Equations*, 2nd edn. Springer, New York
3. Banks J, Brooks J, Cairns G, Davis G, Stacey P (1992) On Devaney's definition of chaos. Am Math Mon 99:332–334
4. Birkhoff GD (1927) *Dynamical Systems*. American Mathematical Society, Providence, RI
5. Devaney RL (1989) *An Introduction to Chaotic Dynamical Systems*, 2nd edn. Addison-Wesley, New York
6. Grebogi C, Hammel SM, Yorke JA, Sauer T (1990) Shadowing of physical trajectories in chaotic dynamics: containment and refinement. Phys Rev Lett 65:1527–1530
7. Guckenheimer J, Holmes P (1983) *Nonlinear Oscillations, Dynamical Systems, and Bifurcations of Vector Fields*. Springer, New York
8. Hale J (1977) *Theory of Functional Differential Equations*. Springer, New York
9. Hirsch MW, Smale S (1974) *Differential Equations, Dynamical Aystems and Linear Algebra*. Academic Press, New York
10. Kuznetsov YA (1998) *Elements of Applied Bifurcation Theory*, 2nd edn. Springer, New York
11. Medio A, Lines M (2001) *Nonlinear Dynamics: A Primer*. Cambridge University Press, London
12. Parker TS, Chua LO (1989) *Practical Numerical Algorithms for Chaotic Systems*. Springer, New York
13. Robinson RC (2004) *An Introduction to Dynamical Systems: Continuous and Discrete*. Prentice Hall, Upper Saddle River, NJ
14. Sauer T, Grebogi C, Yorke JA (1997) How long do numerical chaotic solutions remain valid? Phys Rev Lett 79:59–62
15. Tucker W (2002) A rigorous ODE solver and Smale's 14th problem. Found Comput Math 2:53–117
16. Vellekoop M, Berglund R (1994) On intervals: transitivity → chaos. Am Math Mon 101:353–355
17. Wiggins S (2003) *Introduction to Applied Nonlinear Dynamical Systems and Chaos*, 2nd edn. Springer, New York

Chapter 3
Entrainment and Migration Control of Chaotic Systems

Abstract In this chapter, the entrainment and the migration methods of chaos control are discussed. The background of entrainment and migration control is based on two facts: a multi-attractor chaotic system is sensible to both initial values and parameters and each stable attractor has its basin of attraction. A common goal of chaos control is to steer the trajectory of a chaotic system to a periodic orbit which itself is a solution of the chaotic system. Sometimes, one wants to steer the trajectory to a predesigned orbit which is not the solution of the chaotic system. For some multiple attractor chaotic systems, one would like to direct the trajectory from one attractor to the others. To achieve these goals, entrainment and migration methods are developed. We first introduce these methods and their extension, the open-plus-closed-loop (OPCL) method. Based on the OPCL method, a new control scheme, the open-plus-nonlinear-closed-loop (OPNCL) method, is developed. In addition, we apply the method of OPNCL to a class of dynamical systems, both continuous time and discrete time.

3.1 Introduction

Usually, several attractors can coexist in a complex dynamical system. These attractors can be stable or unstable. When a key parameter is changed the attractor may be altered in either appearance or spatial position, or in both. The background of entrainment and migration control is based on two facts: a multi-attractor chaotic system is sensible to both initial values and parameters and each stable attractor has its basin of attraction. A common goal of chaos control is to steer the trajectory of a chaotic system to a periodic orbit which itself is a solution of the chaotic system. Sometimes, one wants to steer the trajectory to a predesigned orbit which is not the solution of the chaotic system. For some multiple attractor chaotic systems, one would like to direct the trajectory from one attractor to others. To achieve these goals, entrainment and migration methods are proposed. In this chapter, we will introduce these methods and their expansion, the OPCL method. Based on the OPCL method, a new control scheme, the OPNCL method, is proposed. Besides that, we will apply the method of OPNCL to a class of dynamical systems, both continuous time and discrete time.

3.2 Basics on Entrainment and Migration

3.2.1 Entrainment Control

Consider the following dynamical systems:

$$x_{n+1} = f(x_n) \tag{3.1}$$

or

$$\dot{x} = f(x), \tag{3.2}$$

where $x \in \mathbb{R}^m$ and $f\colon \mathbb{R}^m \to \mathbb{R}^m$ is smooth. Sometimes, it is difficult to measure and feed-back the state variables. Moreover, sometimes one wishes to steer the trajectory of systems (3.1) and (3.2) to a certain region in the phase space, or even to some predesigned orbits. For example, the target orbits can be given by

$$x_n = g_n \tag{3.3}$$

or

$$x(t) = g(t). \tag{3.4}$$

These target orbits may be fixed points, periodic orbits, nonperiodic orbits, or even chaotic orbits. We would like to achieve the specified target, for example, (3.3), by a control signal $\hat{S}_n$. That is to say, the following controlled discrete-time system:

$$x_{n+1} = f(x_n) + \hat{S}_n$$

takes (3.3) as its solution. Thus, we must have that

$$\begin{cases} x_{n+1} = f(x_n) + \hat{S}_n, \\ \hat{S}_n = g_{n+1} - f(g_n). \end{cases} \tag{3.5}$$

For the case of continuous time, we have analogous equations

$$\begin{cases} \dot{x} = f(x) + \hat{S}(t), \\ \hat{S}(t) = \dot{g}(t) - f(g(t)). \end{cases} \tag{3.6}$$

It should be noted that $\hat{S}_n$ and $\hat{S}(t)$ are independent of the state variables of systems (3.1) and (3.2). This control method belongs to an open-loop control scheme. Thus, the targets (3.3) and (3.4) are not arbitrary. They should be stable solutions of the controlled systems. We take (3.6) as an example. If $g(t)$ is a periodic orbit, its stability can be determined by the vector field

$$\delta \dot{x} = Df_x \delta x, \tag{3.7}$$

where $Df_x = (\partial f / \partial x)|_{x=g(t)}$ and $\delta x = x(t) - g(t)$. Because $g(t)$ and $f(x)$ are known, the conditions to hold for stability are that all Floquet characteristic multipliers are negative. Thus, we can select a certain region C, of the state space, so that the distance between any couple of points in this region is contracted by (3.7). This kind of region is called a convergence region. In this region, for any distance, Δx, it always holds that

$$\Delta x^{\mathrm{T}} Df_x \Delta x < 0. \tag{3.8}$$

Therefore, if $g(t)$ is in this region, it will be stable. For any bounded dissipative system this region always exists. In the discrete-time case, the variation equation given in (3.7) becomes

$$\delta x_{n+1} = Df_x \delta x_n, \quad \delta x_n = x_n - g_n.$$

The condition (3.8) becomes

$$|\Delta x^{\mathrm{T}} Df_x \Delta x| < \Delta x^{\mathrm{T}} \Delta x. \tag{3.9}$$

Once the target has been selected, the next step is to determine $BE(g)$, i.e., the *basin of entrainment* (the range of initial conditions of $x(t)$ that will satisfy the condition $\lim_{t \to \infty} \|x(t) - g(t)\| = 0$). This step is far more difficult than determining a stable target. A point in a contractive region must converge to the target orbit but, compared with the basin of entrainment, this contractive region is very small. In general, we cannot get an analytical expression for a basin of entrainment, but we can get a rough understanding of a basin of entrainment by means of numerical simulations or experiments.

Now, it is clear how the entrainment method is implemented. First, a target orbit $g(t)$ (or g_n) should be selected according to the requirements of the stability in practice. The basin of entrainment of the target orbit must cover a part of the attractor of the chaotic system to be controlled. Then, the controlled systems are as follows:

(i) for the discrete-time case, we have

$$x_{n+1} = f(x_n) + (g_{n+1} - f(g_n))S_n; \tag{3.10}$$

(ii) for the continuous-time case, we have

$$\dot{x} = f(x) + (\dot{g}(t) - f(g(t)))S(t). \tag{3.11}$$

Here, $S(t) = 0$ when the controller does not work and $S(t) = 1$ when the controller begins to work. The same definition is for S_n as well. The controller is turned on when the trajectory of the chaotic system goes into the overlapping region of the basin of entrainment and the chaotic attractor. Since these kinds of controllers were first proposed by Hübler [3], we will refer to them as Hübler open-loop controllers.

There are some distinguishable merits of the entrainment method. The selection of the target orbits is very flexible. The target orbits can be ones which are not solutions of the chaotic systems, which enhances the ability of control and, meanwhile, avoids real-time monitoring and feedback. Therefore, the controller would be simplified significantly. However, there are some problems with this method. On one hand, a total understanding about the original chaotic system is required when one wants to determine the control signals. This is not always achievable. On the other hand, the control signals usually have large amplitude and the original dynamics changes in a large range. It should be emphasized that both methods (3.10) and (3.11) cannot stabilize unstable periodic orbits. From (3.7) and (3.9) it is easy to see that the stability of the target orbits is not changed by the controller.

3.2.2 Attractor Selection by Migration Control

One interesting point with entrainment control is that $g(t)$ and g_n are not required to be periodic orbits. They may be a migration orbit from one point in the state space to another. Thus, we can use entrainment control to achieve an important control task, the migration of state.

One of the features of nonlinear systems is that there usually exist multiple attractors corresponding to some parametric conditions. Since each attractor has its own basin of attraction, different initial conditions will result in different final states. If one wants to migrate an orbit from a chaotic attractor to a periodic orbit, or to migrate a periodic orbit to another periodic orbit, then the migration orbit between different points in the state space is a reasonable selection.

Suppose that $g(t)$ is the selected orbit which is used to migrate $g(0) = x_S$ to $g(t \to \infty) = x_T$. We require that

$$\lim_{t \to \infty} \|x(t) - g(t)\| = 0.$$

The control equation that should be obeyed by the target is (3.11). Since the starting point of the migration orbit can be located at the chaotic attractor, the whole understanding about the distribution of the basin of entrainment is not required. The only requirement is the stability of $g(t)$. After the starting attractor and the terminal attractor are determined, we should select the starting point x_S and the terminal point x_T of the migration orbit. The basin of attraction may be quite big, which makes the selection of x_T very flexible. Since there is much freedom in the selection of $g(t)$, it is not difficult to guarantee local stability.

We take the Lorenz system as an example. The Lorenz system is of the following form:

$$\begin{cases} \dot{x}_1 = \sigma(x_2 - x_1), \\ \dot{x}_2 = \rho x_1 - x_1 x_3 - x_2, \\ \dot{x}_3 = x_1 x_2 - \beta x_3. \end{cases} \tag{3.12}$$

Let $g = (g_1(t), g_2(t), g_3(t))^{\mathrm{T}}$ be a migration orbit. Adding the Hübler open-loop control, (3.12) becomes

$$\begin{cases} \dot{u}_1 = \sigma(u_2 - u_1), \\ \dot{u}_2 = \rho u_1 - u_2 - u_1 u_3 - u_1 g_3 - g_1 u_3, \\ \dot{u}_3 = u_1 u_2 - \beta u_3 + u_1 g_2 + u_2 g_1, \end{cases} \tag{3.13}$$

where $u_i = x_i(t) - g_i(t),\ i = 1,2,3$. The stability of $g(t)$ can be guaranteed by requiring that there is a positive-definite function $V(u)$ such that $\dot{V}(u)$ is negative definite. One such function can be chosen as

$$V(u) = \frac{1}{2}(\gamma u_1^2 + u_2^2 + u_3^2),$$

where γ is a positive constant to be specified. If we set

$$a^2 + \beta b^2 = \gamma\sigma,\ 2a = \gamma\sigma + \rho - g_3(t),\ 2\beta b = g_2(t),\ \dot{\gamma} = 0, \tag{3.14}$$

then we have that

$$\begin{aligned} \dot{V}(u) &= (\gamma\sigma + \rho - g_3(t))u_1 u_2 - \gamma\sigma u_1^2 - u_2^2 - \beta u_3^2 + g_2(t) u_1 u_3 \\ &= -(a u_1 - u_2)^2 - \beta(b u_1 - u_3)^2. \end{aligned} \tag{3.15}$$

From (3.14) we get

$$\begin{cases} a(t) = 1 + \left[1 + g_3(t) - \rho - \dfrac{g_2^2(t)}{4\beta}\right]^{\frac{1}{2}}, \\ b(t) = \dfrac{g_2(t)}{2\beta}, \\ \gamma(t) = \dfrac{a^2(t) + \beta b^2(t)}{\sigma}. \end{cases}$$

In order to make (3.15) negative, $a(t)$ must be a real number, which leads to

$$g_3(t) > \rho - 1 + \frac{g_2^2(t)}{4\beta}. \tag{3.16}$$

Noting that $\dot{\gamma} = 0$, we get another restriction on $g(t)$:

$$a(t)\dot{g}_3(t) - g_2(t)\dot{g}_2(t)/2\beta = 0. \tag{3.17}$$

Thus, any orbit which satisfies conditions (3.16) and (3.17) must be stable. Furthermore, its basin of entrainment is the whole state space. Conditions (3.16) and

(3.17) are just sufficient conditions, not necessary ones. When $\sigma = 10$, $\beta = 8/3$, and $14 < \rho < 24$, the Lorenz system has two attractors, and they are

$$A^+ : \ (\sqrt{\beta(\rho-1)},\ \sqrt{\beta(\rho-1)},\ \rho-1),$$
$$A^- : \ (-\sqrt{\beta(\rho-1)},\ -\sqrt{\beta(\rho-1)},\ \rho-1).$$

Suppose that the Lorenz system starts from A^+. We want to migrate the trajectory of the Lorenz system to A^- along the migration orbit $g(t)$. We select the following migration orbit:

$$\begin{aligned} g_1(t) &= g_2(t) = [\beta(\rho-1)]^{1/2}(2\mathrm{e}^{-t/5}-1), \\ g_3(t) &= (\rho-1)\{0.05\sin[\pi(1-\mathrm{e}^{-t/5})]+1\}. \end{aligned} \tag{3.18}$$

The orbit (3.18) does not satisfy condition (3.17), but it is stable. The initial point is selected in the basin of attraction of A^+. The controller is turned on at $t = 0$. Therefore, the trajectory of the Lorenz system is entrained to the basin of attraction of A^-. Then, the controller is turned off. The trajectory goes to A^- under the domination of the dynamics of the Lorenz system. This procedure is shown in Fig. 3.1.

If the values of σ and β are fixed but the value of ρ varies in the range $[24.1, 24.7]$, a chaotic attractor appears besides the two fixed points. We take two steps to entrain the trajectory of the Lorenz system from the chaotic attractor to A^-.

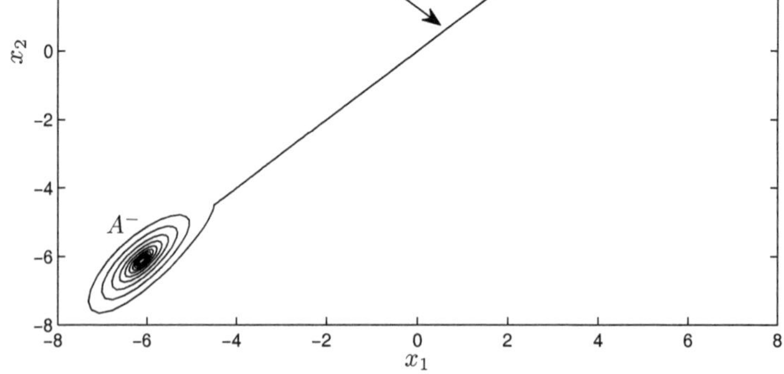

Fig. 3.1 Migrating the trajectory of the Lorenz system from A^+ to A^-

First, we choose a fixed point

$$g_1^* = g_2^* = [\beta(\rho - 1)]^{\frac{1}{2}}, \quad g_3^* = (\rho - 1) + 50$$

in the convergent region as the goal of entrainment. From (3.15) to (3.17) we know that this point is the unique fixed point of system (3.13). So, whenever the controller is turned on, the trajectory of the Lorenz system will be entrained to this point. Therefore, the difficulties encountered in the selection of the target orbit, which are caused by the complex structure of the basin of entrainment of each attractor, can be avoided. Second, we select the following migration orbit [2]:

$$g_1(t) = g_2(t) = [\beta(\rho - 1)]^{1/2}[2\mathrm{e}^{-(t-t_0)/5} - 1],$$
$$g_3(t) = (\rho - 1) + 50\mathrm{e}^{-(t-t_0)/5}, \quad t \geq t_0,$$

where t_0 is the instant at which the entrainment to g^* is finished and the migration from g^* to A^- is started. The simulation result is shown in Fig. 3.2.

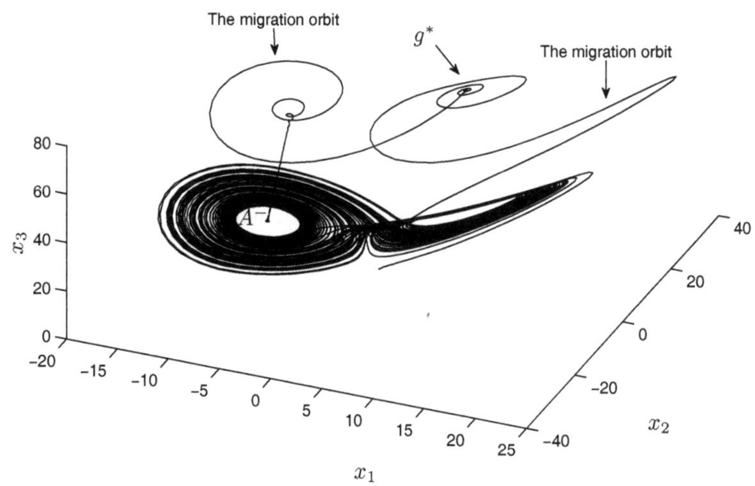

Fig. 3.2 Migrating the trajectory of the Lorenz system from the chaotic attractor to A^-

3.3 OPCL Control Scheme

Although Hübler's controller is successfully used in many applications, there are cases in which this method cannot be implemented. It was believed that all of the strange attractors had the property that these attractors intersected some convergent region in the phase space. Therefore, it was conjectured that this might be a general feature of strange attractors. This would have been nice (but not essential), since the initial state of a system on a strange attractor could then be selected arbitrarily close to $g(t)$, and hence certainly in its basin of entrainment. However, the Chua oscillator is a notable exception whose strange attractor can lie entirely in a nonconvergent region. Therefore, the Chua oscillator cannot be entrained near the strange attractor using the entrainment method (however, it can be entrained outside the attractor's region). Moreover, linear feedback can only stabilize the oscillator to one of the unstable periodic orbits. To overcome the limitations of both the open-loop and the closed-loop control methods, the OPCL method was proposed.

Theorem 3.1. Let $g(t)$ be an arbitrary smooth function, and $x(t)$ be a solution of

$$\dot{x} = F(x,t) + S(t)K(g,x,t) \quad (x \in \mathbb{R}^n), \tag{3.19}$$

where $S(t) = 0$ (if $t < t_0$) and $S(t) = 1$ (if $t \geq t_0$), and $F(x,t)$ is everywhere Lipschitzian. There exists a function $K(g,x,t)$, linear in $x(t)$, such that none of the basins of entrainment (associated with $g(t)$ and t_0), defined by

$$BE(g|t_0) = \{x(t_0) : \lim_{t\to\infty} \|x(t) - g(t)\| = 0\}, \tag{3.20}$$

are empty sets.

Proof. Consider a special kind of control, $K(g,t)$, which is the sum of Hübler's open-loop control and a particular form of a linear closed-loop control, namely

$$K(g,t) = \dot{g} - F(g,t) + C(g,t)(g(t) - x(t)), \tag{3.21}$$

where the matrix $C(g,t)$ is given by

$$C(g,t) = \frac{\partial F(g,t)}{\partial g} - A \tag{3.22}$$

and $A = (a_{ij})$ is any constant matrix whose eigenvalues all have negative real parts. The control (3.21) and (3.22) is referred to as the OPCL control.

Set $x(t) = g(t) + u(t)$, where $u(t_0)$ is arbitrarily small. Substitute $x(t)$ into (3.19) when $t \geq t_0$, and expand $F(g(t) + u(t),t)$ for small $u(t)$. This yields the following variational equation:

$$\frac{\mathrm{d}u}{\mathrm{d}t} = \frac{\partial F(g,t)}{\partial g} u(t) - C(g,t)u(t) = Au(t). \tag{3.23}$$

Since all eigenvalues of A have negative real parts, the asymptotic stability of (3.23) is established. Since this asymptotic stability also holds for the nonlinear system (3.19), $BE(g)$ is not an empty set. □

Remark 3.1. This is a remarkably general theorem, regardless of whether $g(t)$ is or is not a solution of $\dot{g} = F(g,t)$, and it does not restrict $g(t)$ to any region of the phase space. This latter point is particularly important for the possible application of this kind of control for the purpose of transferring a system from one attractor to another ('migration control'). What is not established by the theorem is the range of the basin of entrainment, $BE(g)$, in the phase space. It is to be expected that $BE(g)$ will depend on both the particular dynamical system and some characteristics of $g(t)$. Indeed, substituting $x(t) = g(t) + u(t)$ into (3.19) and expanding $F(g+u,t)$, we get

$$\frac{\mathrm{d}u}{\mathrm{d}t} = Au(t) + \sum_{k=2}^{\infty} \frac{1}{k!} \frac{\partial^k F(g,t)}{\partial g^k} u^k.$$

The above discussion can easily be seen from this formula. □

3.4 Global Control of a Class of Continuous-Time Polynomial Chaotic Systems

3.4.1 OPNCL Control Scheme

The OPCL control is very effective for systems whose right-hand sides of differential equations are mth-order polynomials, where $m \leq 2$. When $m > 2$, it is difficult to determine the basin of entrainment, which will be very complex with the increase of m. To overcome this difficulty, an improved OPCL method, called the OPNCL method, was proposed [5].

Consider the following mth-order polynomial systems:

$$\dot{x} = F(x,t) \quad (x \in \mathbb{R}^n) \tag{3.24}$$

or

$$\dot{x}_i = F_i(x,t) = \sum_{j_1+j_2+\cdots+j_n=0}^{m} a^{(i)}_{j_1 j_2 \cdots j_n}(t) x_1^{j_1} x_2^{j_2} \cdots x_n^{j_n} \quad (i = 1,2,\ldots,n). \tag{3.25}$$

Without loss of generality, we suppose that there is at least one nonzero parameter, for example, $a^{(i)}_{j_1 j_2 \cdots j_n}(t) \neq 0$ $(1 \leq i \leq n)$. For system (3.24) we add the following OPNCL controller:

$$K(g,x,t) = \dot{g} - F(g,t) + D(g,x,t)(g(t) - x(t)), \tag{3.26}$$

where $D(g,x,t)$ is a nonlinear function of $g(t)-x(t)$; its concrete form will be given later. Adding the control term $K(g,x,t)$ to system (3.24) through a switching function $S(t)$, we get

$$\dot{x} = F(x,t) + S(t)K(g,x,t), \tag{3.27}$$

where $K(g,x,t)$ is specified by (3.26). For an arbitrary target $g(t)$, we will prove that the basin of entrainment of (3.24) is not an empty set. Furthermore, the entrainment is global. Set $S(t)=1$ and let $x(t)=g(t)+u(t)$, where $u(t)\in\mathbb{R}^n$. Substituting $x(t)$ into (3.27), we get

$$\dot{g}+\dot{u} = F(g+u,t)+K(g,g+u,t).$$

Substituting (3.26) into the above equation, we get

$$\dot{u} = F(g+u,t)-F(g,t)-D(g,x,t)u = \sum_{k=1}^{m}\frac{1}{k!}\frac{\partial^k F(g,t)}{\partial g^k}u^k - D(g,x,t)u. \tag{3.28}$$

To calculate $\dfrac{\partial^k F(g,t)}{\partial g^k}u^k$, we note that $F=(F_1,F_2,\dots,F_n)^{\mathrm{T}}$ and $F_i=F_i(x_1,x_2,\dots,$ $x_n)$. Thus, we have

$$\begin{aligned}
&F_i(x_1+h_1,x_2+h_2,\dots,x_n+h_n)-F_i(x_1,x_2,\dots,x_n)\\
&=\frac{1}{1!}\sum_{j=1}^{n}\frac{\partial F_i}{\partial x_j}h_j+\frac{1}{2!}\sum_{j,k=1}^{n}\frac{\partial^2 F_i}{\partial x_j\partial x_k}h_jh_k+\frac{1}{3!}\sum_{j,k,l=1}^{n}\frac{\partial^3 F_i}{\partial x_j\partial x_k\partial x_l}h_jh_kh_l+\cdots.
\end{aligned}$$

Note that

$$\begin{aligned}
\sum_{j,k=1}^{n}\frac{\partial^2 F_i}{\partial x_j\partial x_k}h_jh_k &= \frac{\partial^2 F_i}{\partial x_1\partial x_1}h_1^2+\frac{\partial^2 F_i}{\partial x_1\partial x_2}h_1h_2+\cdots+\frac{\partial^2 F_i}{\partial x_1\partial x_n}h_1h_n\\
&\quad+\frac{\partial^2 F_i}{\partial x_2\partial x_1}h_2h_1+\frac{\partial^2 F_i}{\partial x_2\partial x_2}h_2^2+\cdots+\frac{\partial^2 F_i}{\partial x_2\partial x_n}h_2h_n\\
&\quad+\cdots\\
&\quad+\frac{\partial^2 F_i}{\partial x_n\partial x_1}h_nh_1+\frac{\partial^2 F_i}{\partial x_n\partial x_2}h_nh_2+\cdots+\frac{\partial^2 F_i}{\partial x_n\partial x_n}h_n^2\\
&=\Bigg(\frac{\partial^2 F_i}{\partial x_1\partial x_1}h_1+\frac{\partial^2 F_i}{\partial x_1\partial x_2}h_2+\cdots+\frac{\partial^2 F_i}{\partial x_1\partial x_n}h_n,\\
&\quad\frac{\partial^2 F_i}{\partial x_2\partial x_1}h_1+\frac{\partial^2 F_i}{\partial x_2\partial x_2}h_2+\cdots+\frac{\partial^2 F_i}{\partial x_2\partial x_n}h_n,\\
&\quad\dots,\frac{\partial^2 F_i}{\partial x_n\partial x_1}h_1+\frac{\partial^2 F_i}{\partial x_n\partial x_2}h_2+\cdots+\frac{\partial^2 F_i}{\partial x_n\partial x_n}h_n\Bigg)\begin{pmatrix}h_1\\h_2\\\vdots\\h_n\end{pmatrix}
\end{aligned}$$

$$= \left[\left(\frac{\partial^2 F_i}{\partial x_1 \partial x_1}, \frac{\partial^2 F_i}{\partial x_1 \partial x_2}, \dots, \frac{\partial^2 F_i}{\partial x_1 \partial x_n} \right) \begin{pmatrix} h_1 \\ h_2 \\ \vdots \\ h_n \end{pmatrix}, \right.$$

$$\left. \dots, \left(\frac{\partial^2 F_i}{\partial x_n \partial x_1}, \frac{\partial^2 F_i}{\partial x_n \partial x_2}, \dots, \frac{\partial^2 F_i}{\partial x_n \partial x_n} \right) \begin{pmatrix} h_1 \\ h_2 \\ \vdots \\ h_n \end{pmatrix} \right] \begin{pmatrix} h_1 \\ h_2 \\ \vdots \\ h_n \end{pmatrix}.$$

Analogously, we get the expressions of $\frac{\partial^3 F_i}{\partial x_j \partial x_k \partial x_l} h_j h_k h_l, \dots$.

To entrain the solution of system (3.24) to the target orbit $g(t)$, we require that the condition

$$\lim_{t \to +\infty} (x(t) - g(t)) = 0$$

holds. This will require that $\lim_{t\to+\infty} u(t) = 0$ holds. So, we set

$$D(g,x,t) = \sum_{k=1}^{m} \frac{1}{k!} \frac{\partial^k F(g,t)}{\partial g^k} (x(t) - g(t))^{k-1} - A, \tag{3.29}$$

where $A = (a_{ij})$ is an $n \times n$ constant matrix with all its eigenvalues having negative real parts. Substituting (3.29) into (3.28), we get a linear differential equation with respect to $u(t)$:

$$\dot{u}(t) = Au(t). \tag{3.30}$$

The zero solution of this system is asymptotically stable, which implies that the solution of (3.27) is asymptotically stable. According to (3.20), the definition of $BE(g)$, we know that $BE(g)$ is not empty. Since the stability of $u(t)$ is independent of initial conditions, $BE(g)$ is global, which means that, for any starting point x_0, the relation $\lim_{t\to+\infty} \|x(t) - g(t)\| = 0$ always holds.

By summarizing the above analysis, we obtain the following theorem.

Theorem 3.2 ([5]). Suppose that $g(t) \in C^r$ $(r \geq 1)$ and $x(t)$ is the solution of system (3.27). Then, there exists an OPNCL controller $K(g,x,t)$, which is nonlinear with respect to $x(t)$, such that the basin of entrainment (3.20) is not empty. Furthermore, by this control, the entrainment is global. □

In the rest of this section, we will study the ability of entrainment with respect to controllers of open-loop, linear open-loop, OPCL, and OPNCL.

Consider the controlled system (3.27). For convenience, we suppose that $x(t) = g(t) + u(t)$, where $g(t)$ is the target orbit, and set $S(t) = 1$.

(i) $K(g,x,t)$ is the Hübler open-loop action. In this case

$$K(g,x,t) = H(\dot{g},g,t) = \dot{g} - F(g,t).$$

Substituting it into (3.27), we have

$$\dot{u} = F(u+g,t) - F(g,t) = \sum_{k=1}^{m} \frac{1}{k!} \frac{\partial^k F(g,t)}{\partial g^k} u^k. \tag{3.31}$$

This equation is a complicated nonautonomous differential equation. In general, it is difficult to judge whether the zero solution of the above equation is stable. The basin of entrainment usually exists in the neighborhood of $u(t) = 0$ even if the basin of entrainment is not an empty set.

(ii) $K(g,x,t)$ is a linear feedback action. In this case

$$K(g,x,t) = -A(g(t) - x(t)),$$

where A is a constant matrix with all its eigenvalues having negative real parts. Substituting the above equation into (3.27), we have

$$\dot{u} = Au + F(u+g,t) - \dot{g}. \tag{3.32}$$

Expanding $F(u+g,t)$ with respect to $u+g$ at $u=0$, we obtain

$$F(u+g,t) = F(g,t) + \sum_{k=1}^{m} \frac{1}{k!} \frac{\partial^k F(g,t)}{\partial g^k} u^k.$$

If $g(t)$ is a solution of (3.24), then (3.32) can be changed to

$$\dot{u} = \left(A + \frac{\partial F(g,t)}{\partial g} \right) u + \sum_{k=2}^{m} \frac{1}{k!} \frac{\partial^k F(g,t)}{\partial g^k} u^k. \tag{3.33}$$

The discussion about $u = 0$ is analogous to (i). If $g(t)$ is not a solution of (3.24), we must have $\dot{g} - F(g,t) \to 0$ so that $u \to 0$ when $t \to \infty$. That is to say, $g(t)$ is restricted to asymptotically approximate a solution of (3.24). So, when linear feedback is used, the target orbit is usually selected as an unstable periodic orbit embedded in a chaotic attractor.

(iii) $K(g,x,t)$ is an OPCL action. In this case

$$K(g,x,t) = \dot{g} - F(g,t) + C(g,t)(g(t) - x(t)), \tag{3.34}$$

where $C(g,t)$ has the form

$$C(g,t) = \frac{\partial F(g,t)}{\partial g} - B$$

and $B = (b_{ij})_{n \times n}$ is a constant matrix with all its eigenvalues having negative real parts. Substituting (3.34) into (3.27), we have

$$\dot{u} = Bu + \sum_{k=2}^{m} \frac{1}{k!} \frac{\partial^k F(g,t)}{\partial g^k} u^k. \tag{3.35}$$

The judgement of stability of $u = 0$ for system (3.35) is easier than that of (3.32) because the linear part of (3.35) is a constant matrix. When $m = 2$, system (3.25) has the form

$$\dot{x}_i = a_i(t) + \sum_{j=1}^{n} b_{ij}(t)x_j + \sum_{j=1}^{n}\sum_{k=1}^{n} c_{jk}^{(i)}(t)x_j x_k \quad (i = 1,2,\ldots,n). \tag{3.36}$$

Adding OPCL action to (3.36) and transforming it into the equation about u, we get

$$\dot{u}_i = \sum_{j=1}^{n} b_{ij}(t)u_j + \sum_{j=1}^{n}\sum_{k=1}^{n} c_{jk}^{(i)}(t)u_j u_k \quad (i = 1,2,\ldots,n). \tag{3.37}$$

It is clear that (3.37) is independent of the target orbit $g(t)$.[1] The stability of $u = 0$ of (3.37) can be discussed by the Lyapunov direct method. However, when $m \geq 3$, the determination of the basin of entrainment is more complicated than the case of $m = 2$, because the right-hand side of (3.35) may include $g(t)$ as its parameters.

From the above discussion and Theorem 3.2 we can find the superiority of the OPNCL method. First, the OPNCL method is independent of the target orbit $g(t)$; second, the basin of entrainment of the OPNCL method is global; that is to say, the starting point may be arbitrarily selected in phase space.

3.4.2 Simulations

For validating the effectiveness of the OPNCL method, we take Chua's circuit as an example [1]. The diagram of Chua's circuit is very simple and is shown in Fig. 3.3.

Chua's circuit is composed of an inductor L, two capacitors C_1 and C_2, a linear resistor G, and a nonlinear resistor g. The system equations of Chua's circuit are as follows:

$$\begin{cases} C_1 \dfrac{\mathrm{d}V_{C_1}}{\mathrm{d}t} = G(V_{C_2} - V_{C_1}) - g(V_{C_1}), \\ C_2 \dfrac{\mathrm{d}V_{C_2}}{\mathrm{d}t} = G(V_{C_1} - V_{C_2}) + i_L, \\ L\dfrac{\mathrm{d}i_L}{\mathrm{d}t} = -V_{C_2}. \end{cases} \tag{3.38}$$

Here, V_{C_1} and V_{C_2} are voltages of C_1 and C_2, respectively, i_L is the current through the inductor L, and the nonlinear resistor g has the following feature:

$$g(V_{C_1}) = m_0 V_{C_1} + \frac{1}{2}(m_1 - m_0)(|V_{C_1} + 1| - |V_{C_1} - 1|),$$

[1] Both the Rössler system and the Lorenz system belong to this case whose linear parts are constants.

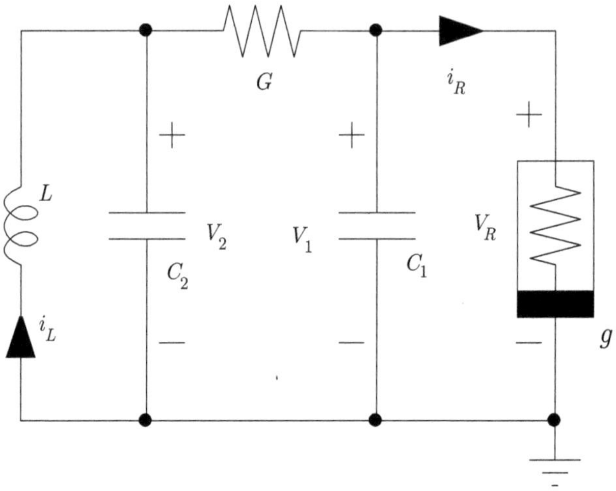

Fig. 3.3 The diagram of Chua's circuit

where $m_0, m_1 < 0$ are parameters. For the convenience of discussion, we implement a transformation

$$x_1 = V_{C_1},\ x_2 = V_{C_2},\ x_3 = \frac{i_L}{G},\ \tau = \frac{tG}{C_2},$$

and replace the piecewise-linear nonlinear function by a third-order polynomial

$$g(V_{C_1}) = a_0 V_{C_1}[2(1-G)V_{C_1}^2 - 1],$$

where a_0 is a suitable selected parameter. If we set

$$p = \frac{C_2}{C_1},\ r = \left(\frac{1}{G} - 1\right)\frac{C_2}{C_1}a_0,\ q = -\frac{C_2}{LG^2},$$

and still use the symbol t to denote the transformed time τ, then system (3.38) becomes a dimensionless one:

$$\begin{cases} \dot{x}_1 = px_2 + r(x_1 - 2x_1^3), \\ \dot{x}_2 = x_1 - x_2 + x_3, \\ \dot{x}_3 = qx_2. \end{cases} \tag{3.39}$$

When $p = 10$, $r = 10/7$, and $q = -100/7$, system (3.39) has a chaotic attractor, which is shown in Fig. 3.4. Suppose that we want to steer the solution of system (3.39) to a target $g(t) = (g_1(t), g_2(t), g_3(t))^{\mathrm{T}}$. Adding an OPCL control action to (3.39), we get a group of equations about u as follows:

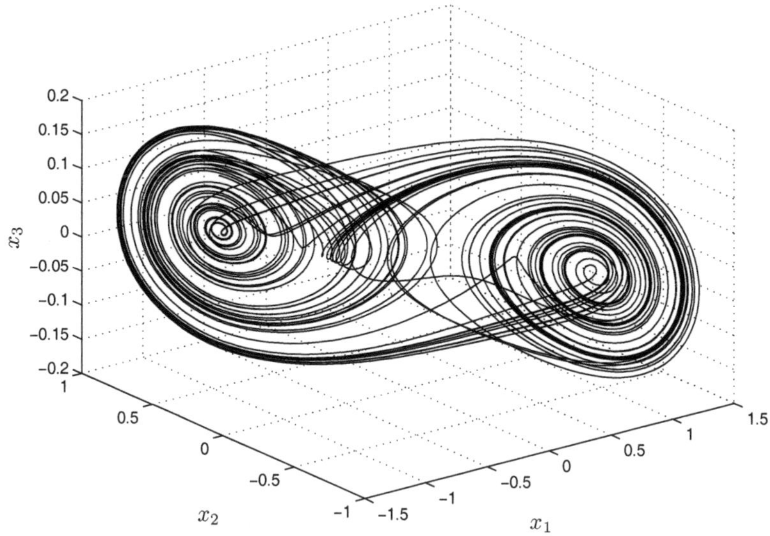

Fig. 3.4 The attractor of Chua's circuit

$$\dot{u}_i = \sum_{j=1}^{3} b_{ij}u_j + \frac{1}{2}\sum_{j=1}^{3}\sum_{k=1}^{3}\frac{\partial^2 F_i(g,t)}{\partial g_j \partial g_k}u_j u_k + \frac{1}{6}\sum_{j=1}^{3}\sum_{k=1}^{3}\sum_{l=1}^{3}\frac{\partial^3 F_i(g,t)}{\partial g_j \partial g_k \partial g_l}u_j u_k u_l$$
$$(i = 1,2,3). \tag{3.40}$$

From (3.39) and (3.40), we get

$$\begin{cases} \dot{u}_1 = b_{11}u_1 - 6rg_1(t)u_1^2 - 2ru_1^3, \\ \dot{u}_2 = b_{22}u_2, \\ \dot{u}_3 = b_{33}u_3, \end{cases} \tag{3.41}$$

where $(b_{ij})_{3\times3}$ is a diagonal matrix with all-negative elements. According to the notation used in [4], the first equation of (3.40) can be written as

$$\dot{u}_1 = -(a + b(t)u_1 + cu_1^2)u_1,$$

where $a = -b_{11}$, $b(t) = 6rg_1(t)$, and $c = 2r$. When $b^2(t) < 4ac$, the $BE(g)$ of Chua's circuit (3.39) is global. When $b^2(t) > 4ac$, the cases of $\dot{g}_1(t) = 0$ and $\dot{g}_1 \neq 0$ should be considered, respectively, in order to determine the $BE(g)$. However, for some target $g(t)$, it becomes more difficult to determine the basin of entrainment because of the uncertainty on the sign of $b^2(t) - 4ac$ [4]. For example, we choose the target as follows:

$$g(t) = (g_1, g_2, g_3)^{\mathrm{T}} = (1 + 0.1\sin t, t\cos t, -10)^{\mathrm{T}}.$$

According to the OPCL design procedure, we have $a = -b_{11}$, $b(t) = 6r(1+0.1\sin t)$, and $c = 2r$. Thus, the sign of $b^2(t) - 4ac$ is undetermined, which results in the difficulty of getting the basin of entrainment.

If the OPNCL action is applied to system (3.39), the control terms are as follows:

$$\begin{aligned} K(g,x,t) &= \dot{g} - F(g,t) + D(g,x,t)(g-x) \\ &= \begin{pmatrix} \dot{g}_1 - F_1(g,t) \\ \dot{g}_2 - F_2(g,t) \\ \dot{g}_3 - F_3(g,t) \end{pmatrix} - \begin{pmatrix} r(1-6g_1^2)(x_1-g_1) + p(x_2-g_2) \\ (x_1-g_1)-(x_2-g_2)+(x_3-g_3) \\ -q(x_2-g_2) \end{pmatrix} \\ &\quad - \begin{pmatrix} -6rg_1u_1^2 - 2ru_1^3 + b_{11}u_1 \\ b_{22}u_2 \\ b_{33}u_3 \end{pmatrix}. \end{aligned}$$

Substituting the above equation into (3.39) and setting $x(t) = g(t) + u(t)$, we get

$$\dot{u}_i = b_{ii}u_i \ \ (i = 1,2,3). \tag{3.42}$$

Choose $b_{ii} = -1$ $(i = 1,2,3)$. It is easy to see that $\|u\|$ will tend to zero quickly without determining the basin of entrainment. The effects of methods of open-loop, linear closed-loop, OPCL, and OPNCL are shown in Figs. 3.5–3.8, in which the controllers, u_i, are those of (3.31), (3.32), (3.35), and (3.42). From these figures we can see that it is impossible for the methods of open-loop and linear feedback

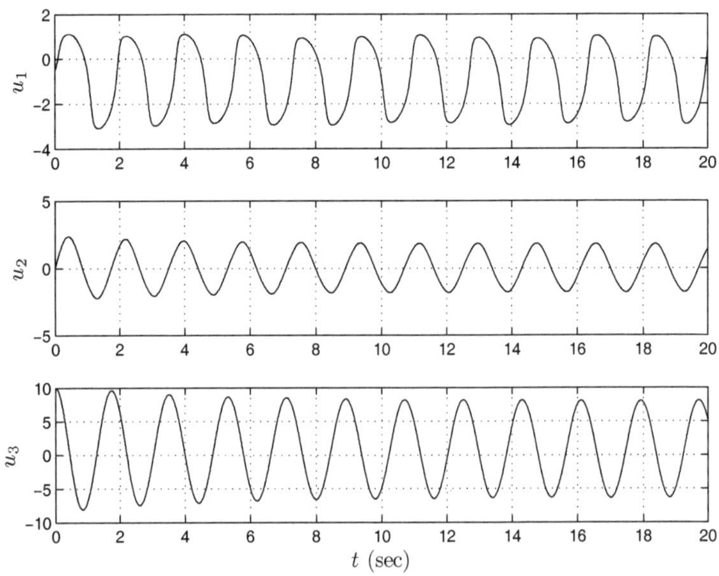

Fig. 3.5 Error curves under the Hübler open-loop action

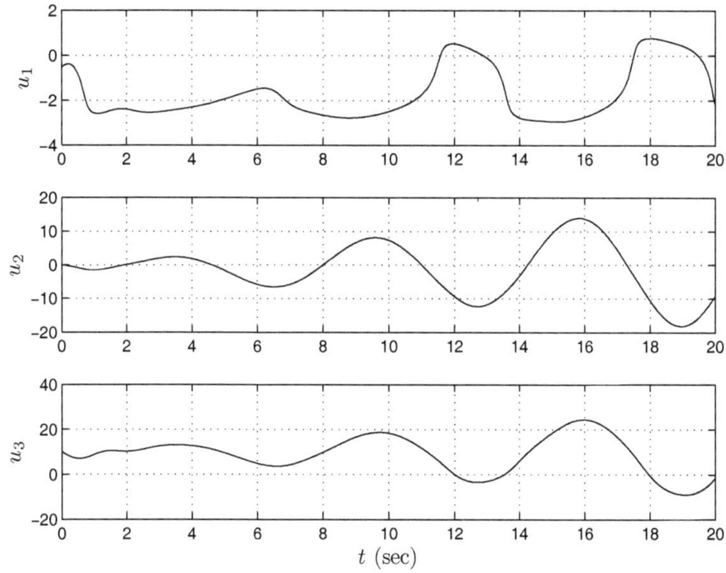

Fig. 3.6 Error curves under the linear feedback action

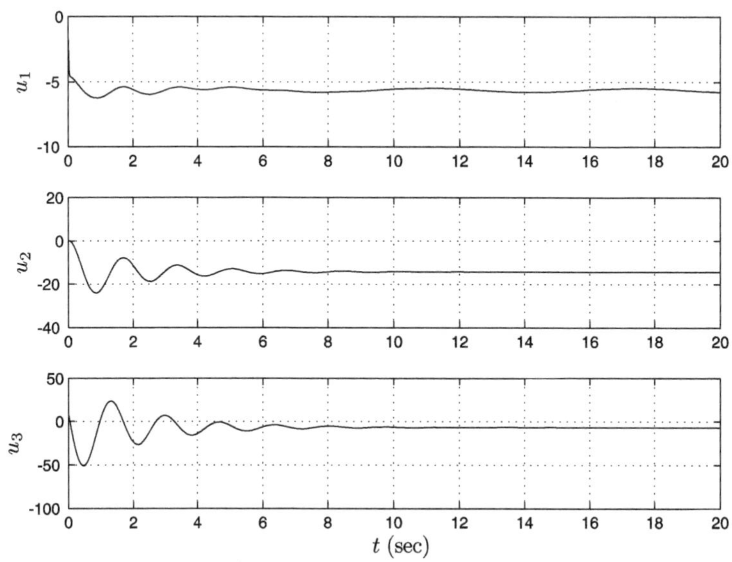

Fig. 3.7 Error curves under the OPCL action

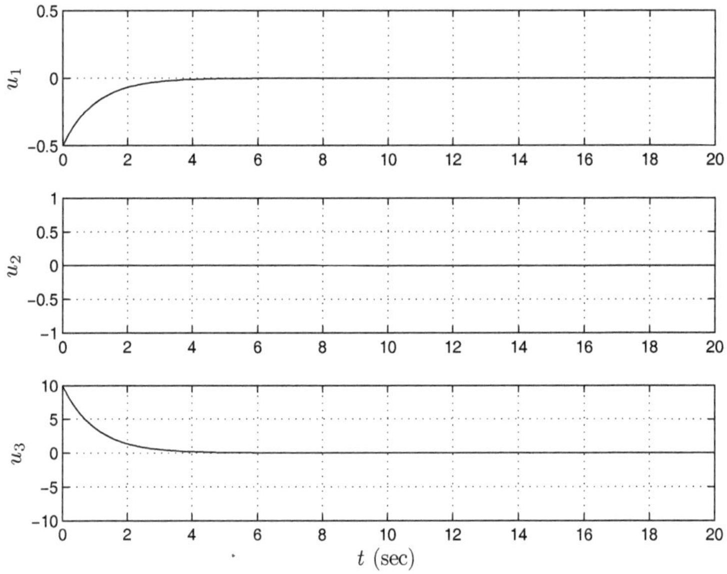

Fig. 3.8 Error curves under the OPNCL action

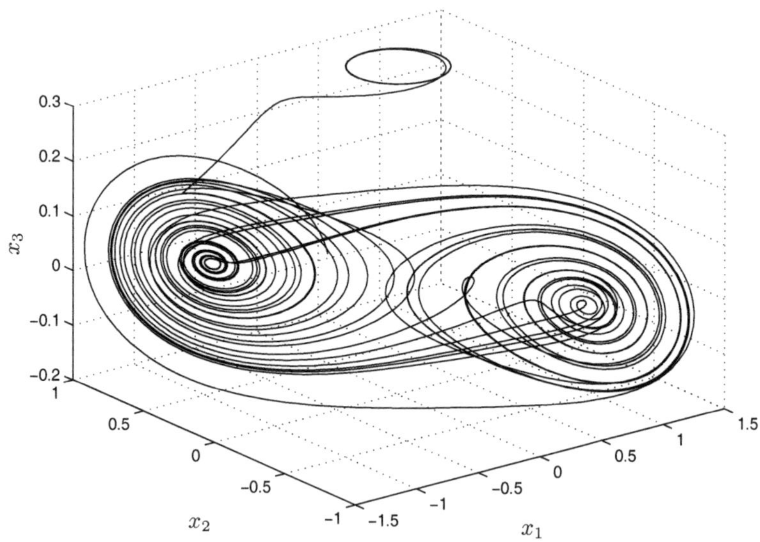

Fig. 3.9 The solution of Chua's circuit system approaches the target orbit quickly under the OP-NCL action

to achieve the goal. The OPCL method can approximately achieve the aim but the error curves are piecewise convergent. However, the OPNCL method can achieve the goal and the basin of entrainment is global. We choose another target $g(t) = (0.5\sin t, 0.3+\mathrm{e}^{-t}, 0.8\cos t)^{\mathrm{T}}$. Under the action of the OPNCL method, the solution of Chua's circuit system is entrained to the target orbit quickly, which is shown in Fig. 3.9.

3.5 Global Control of a Class of Discrete-Time Systems

3.5.1 OPNCL Control for Discrete-Time Systems

In this section, we will study the OPNCL control for a class of discrete-time systems.

For the given mapping

$$x_{k+1} = F(x_k) \quad (x \in \mathbb{R}^p), \tag{3.43}$$

by adding a Hübler open-loop control action on the right-hand side of (3.43), we get

$$x_{k+1} = F(x_k) + [g_{k+1} - F(g_k)]S_k \quad (g_k \in \mathbb{R}^p), \tag{3.44}$$

where the $\{g_k\}$ are some target to be reached and S_k is the switching function ($S_k = 0$, $k < 0$; $0 \leq S_k \leq 1$, $k \geq 0$). This kind of control (when $S_k = 1$) was first proposed by Hübler and was used in the study of the logistic map and nonlinear damped oscillations. Jackson and Hübler studied the global control scale of the logistic map and proposed the concept of the basin of entrainment, denoted by $BE(\{g_k\})$; that is,

$$x_0 \in BE(\{g_k\}),$$

if the solution of (3.44) satisfies the following condition:

$$\lim_{t\to\infty} \|x_k - g_k\| = 0.$$

The determination of $BE(\{g_k\})$ is highly dependent on the given target orbit. In some cases, $BE(\{g_k\})$ even has a fractal structure, which makes it difficult to use in practical applications.

If the target orbit is a point g_0 in the phase space, then this point must be in a convergent region C, which is defined as

$$C = \{x\colon \|\partial F_i/\partial x_j - \delta_{ij}\mu(x)\| = 0,\ \forall |\mu(x)| < 1\},$$

where

$$\delta_{ij} = \begin{cases} 1, & i = j; \\ 0, & i \neq j. \end{cases}$$

If the target orbit is not a point, then it is not easy to determine $BE(\{g_k\})$.

By adding a linear closed-loop control action to the right-hand side of system (3.43), we have

$$x_{k+1} = F(x_k) + D_k(g_k - x_k),$$

where D_k is a suitable selected matrix. For chaotic systems, the target orbit is usually required to be an unstable periodic solution embedded in the chaotic attractor, which satisfies

$$g_{k+1} = F(g_k).$$

Now suppose that there is a discrete-time chaotic system

$$x_{k+1}^{(i)} = F_i\left(x_k^{(1)}, x_k^{(2)}, \ldots, x_k^{(p)}\right) \quad (i = 1, 2, \ldots, p), \tag{3.45}$$

where $F_i(x_k)$ $(x_k \in \mathbb{R}^p, i = 1, 2, \ldots, p)$ is an mth-order polynomial about $x_k^{(1)}, x_k^{(2)}, \ldots, x_k^{(p)}$. If $\{g_k\} = \left\{g_1^{(1)}, \ldots, g_k^{(p)}\right\}$ is the target orbit to be achieved for (3.45), by adding a combination control, $K(g_k, x_k)$, composed of a Hübler open-loop and a nonlinear feedback, to the right-hand side of (3.45), we get

$$x_{k+1}^{(i)} = F_i(x_k) + S_k K_i(g_k, x_k) \quad (i = 1, 2, \ldots, p), \tag{3.46}$$

where

$$K_i(g_k, x_k) = g_{k+1}^{(i)} - F_i(g_k) - \sum_{j=1}^{p} C_{ij}\left(x_k^{(j)} - g_k^{(j)}\right),$$

$$\begin{aligned} C_{ij} = {} & \frac{1}{1!}\frac{\partial F_i(g_k)}{\partial g_k^{(j)}} + \frac{1}{2!}\sum_{l=1}^{p}\frac{\partial^2 F_i(g_k)}{\partial g_k^{(j)}\partial g_k^{(l)}}(x_k^{(l)} - g_k^{(l)}) + \cdots \\ & + \sum_{l,\ldots,n=1}^{p}\frac{1}{m!}\frac{\partial^m F_i(g_k)}{\partial g_k^{(j)}\partial g_k^{(l)}\cdots\partial g_k^{(n)}}(x_k^{(l)} - g_k^{(l)})\cdots(x_k^{(n)} - g_k^{(n)}) - a_{ij}, \end{aligned}$$

and a_{ij} is the element of a suitably selected matrix A. Thus, when $S_k = 1$ the set $BE(\{g_k\}) = \{x_0\colon \lim_{n\to\infty}\|x_n - g_n\| = 0\}$ is not empty, but is global. That is to say, from any initial point the solution of system (3.45) can be entrained to the target orbit $\{g_k\}$ by adding an OPNCL control action.

In fact, when $S_k = 1$, setting $x_k = u_k + g_k$ and noting that $F(x_k)$ is an mth-order polynomial, we know that the expansion of $F_i(g_k + u_k)$ is finite. Substituting $x_k = u_k + g_k$ into (3.46), we get

$$
\begin{aligned}
u_{k+1}^{(i)}+g_{k+1}^{(i)} &= F_i(u_k+g_k)+K_i(g_k,x_k) \\
&= F_i(g_k)+\frac{1}{1!}\sum_{j=1}^{p}\frac{\partial F_i(g_k)}{\partial g_k^{(j)}}u_k^{(j)}+\cdots \\
&\quad +\frac{1}{m!}\sum_{j,\ldots,l=1}^{p}\frac{\partial^m F_i(g_k)}{\partial g_k^{(j)}\cdots\partial g_k^{(l)}}u_k^{(j)}\cdots u_k^{(l)}+K_i(g_k,x_k).
\end{aligned}
$$

Substituting the expression of $K_i(g_k,x_k)$ into the above equation, we get a group of equations with respect to u_k:

$$
u_{k+1}^{(i)}=\sum_{j=1}^{p}a_{ij}u_k^{(j)}. \tag{3.47}
$$

There are many ways to select the matrix A. For simplicity, we can choose $a_{ij}=0\ (i\neq j)$ and $|a_{ii}|<1$. By this choice, the $u_k^{(i)}$ $(i=1,\ldots,p)$ of (3.47) will be convergent to 0, which means that the solution of system (3.45) is entrained to the target orbit $\{g_k\}$ after $S_k=1$.

3.5.2 Simulations

To present the effectiveness of the OPNCL control action we take the logistic map,

$$
x_{k+1}=\lambda x_k(1-x_k), \tag{3.48}
$$

as the first example. The *bifurcation* diagram[2] (it displays how the periodic points of the logistic map vary with λ) and the Lyapunov exponent of (3.48) are shown in Fig. 3.10 and Fig. 3.11. When $\lambda=3.7$ the logistic map is in chaotic state. Set the target orbit as $g_k=0.65+0.15(-1)^k$ and $a_{11}=-0.95$. When $k=2000$, $S_k=1$. The control effect is shown in Fig. 3.12.

The second example is to consider the Hénon map:

$$
\begin{cases}
x_{k+1}=y_k+1-ax_k^2, \\
y_{k+1}=bx_k.
\end{cases} \tag{3.49}
$$

[2] Consider an nth-order continuous-time system

$$
\dot{x}=f(x,\alpha)
$$

with a parameter $\alpha\in\mathbb{R}$. As α changes, the limit sets of the system also change. Typically, a small change in α produces small quantitative changes in a limit set. For instance, perturbing α could change the position of a limit set slightly and, if the limit set is not an equilibrium point, its shape or size could also change. There is also the possibility that a small change in α can cause a limit set to undergo a qualitative change. Such a qualitative change is called a *bifurcation* and the value of α at which a bifurcation occurs is called a *bifurcation value*.

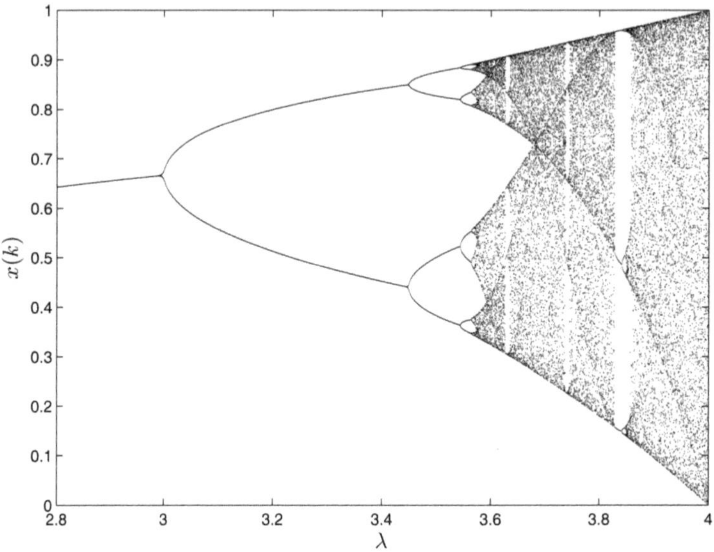

Fig. 3.10 The bifurcation diagram of logistic map

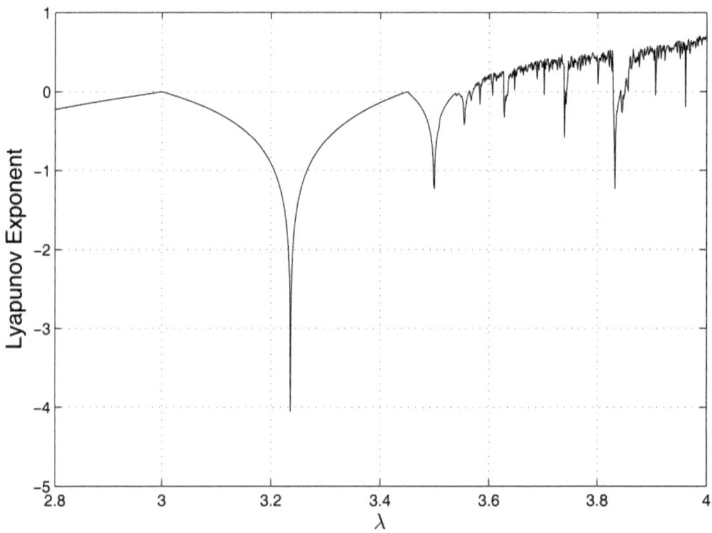

Fig. 3.11 The Lyapunov exponent of logistic map

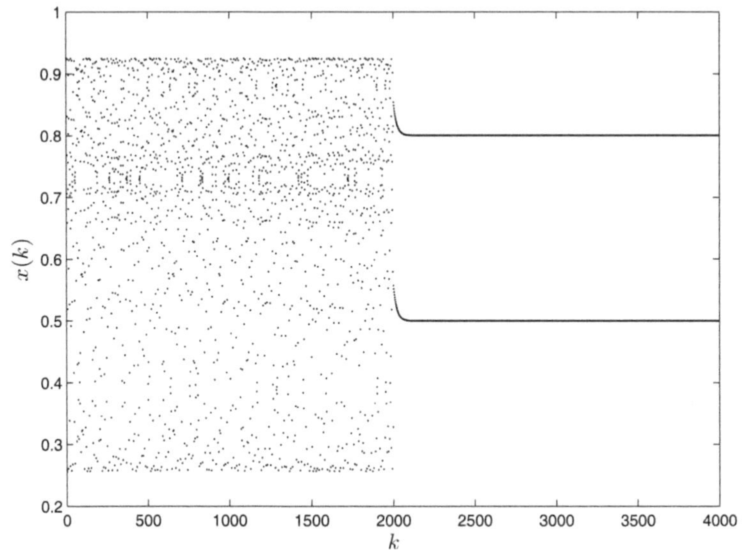

Fig. 3.12 The logistic map under the action of OPNCL control

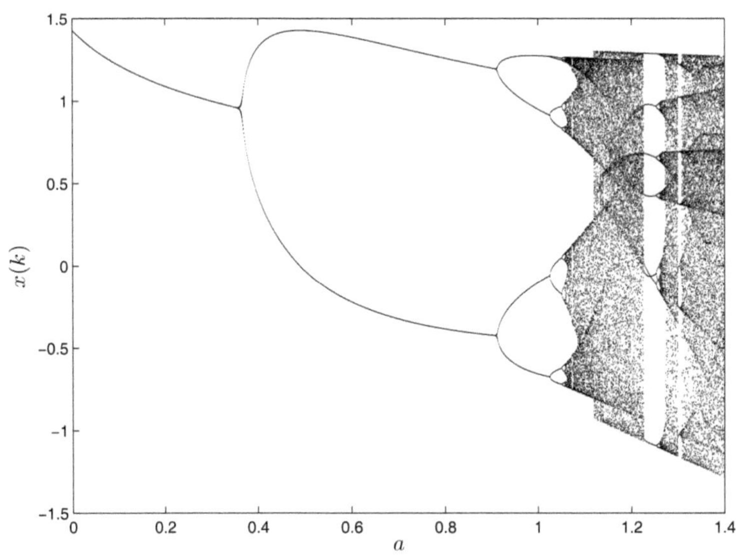

Fig. 3.13 The bifurcation diagram of Hénon map

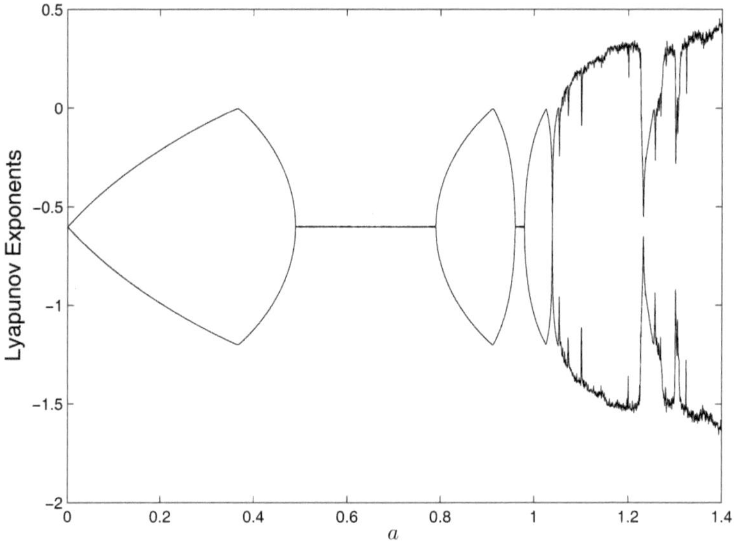

Fig. 3.14 The Lyapunov exponents of Hénon map

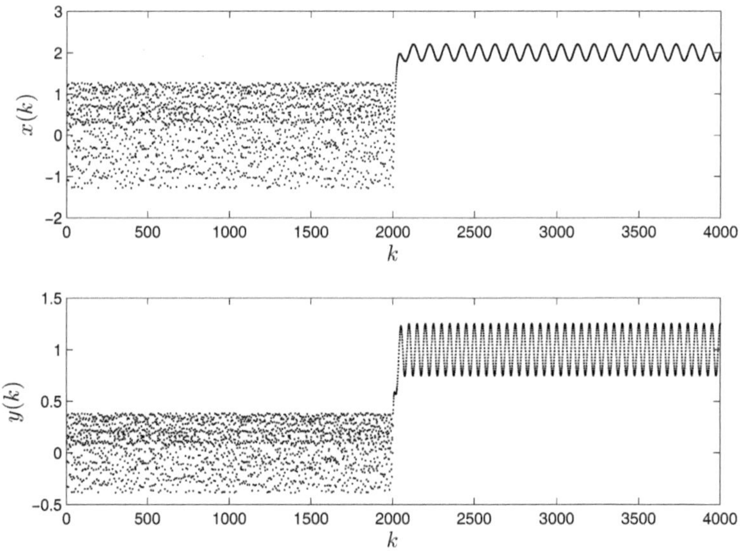

Fig. 3.15 The Hénon map under the action of OPNCL control

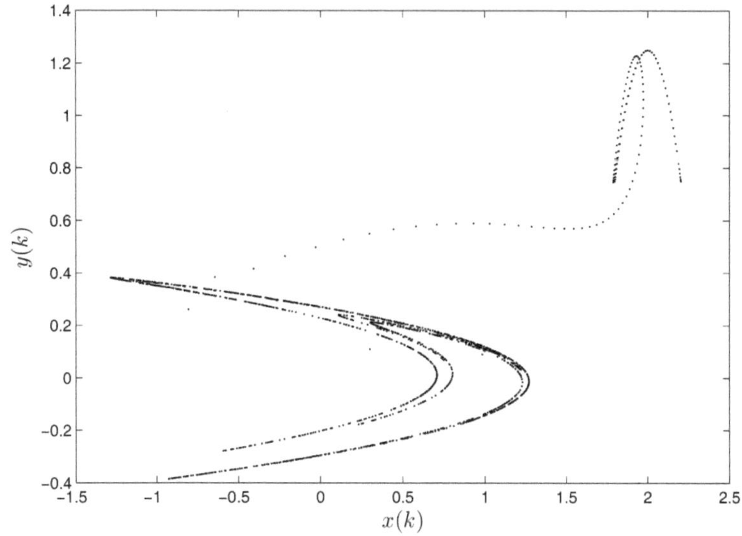

Fig. 3.16 The phase diagram of Hénon map under the action of OPNCL control

When $a = 1.75$ and $b = 0.3$ the Hénon map is in chaotic state. The bifurcation diagram and Lyapunov exponents are shown in Fig. 3.13 and Fig. 3.14. The target orbit is chosen as

$$g_k^{(1)} = 2 + 0.2\sin(0.02k\pi), \;\; g_k^{(2)} = 1 + 0.25\cos(0.04k\pi).$$

When $k = 2000$, $S_k = 1$, the OPNCL controller begins to work. The control effect is shown in Fig. 3.15 and Fig. 3.16, respectively.

Remark 3.2. In theory, the OPNCL method can entrain the solution of the controlled system to any target orbit, even if the target is unbounded. But, in practice, when the target orbit is unbounded it is impossible to design the controller since the output energy of the controller is infinite. □

3.6 Summary

In this chapter, we studied methods for chaos control, entrainment, and migration. At the beginning, we introduced an open-loop entrainment method, and then a modified entrainment method, the OPCL control method, was introduced. Based on the OPCL method, we proposed the OPNCL control method. It can entrain the solution of the controlled system to any bounded target orbit. We applied the OPNCL method to both continuous-time systems and discrete-time systems. Simulation results validated the theoretical analysis.

References

1. Chua LO, Komura M, Matsumoto T (1986) The double scroll family: parts I and II. IEEE Trans Circuits Syst I 33:1073–1118
2. Hu G, Xiao J, Zheng Z (2000) *Chaos Control*. Shanghai Scientific and Technological Education Publishing House, Shanghai
3. Hübler A (1989) Adaptive control of chaotic systems. Helv Phys Acta 62:343–346
4. Jackson EA, Grosu I (1995) An open-plus-closed-loop (OPCL) control of complex dynamic systems. Physica D 85:1–9
5. Wang J, Tian P, Chen C (2000) Global control of continuous polynomial chaotic systems. Control Decis 15:309–313 (in Chinese)

Chapter 4
Feedback Control of Chaotic Systems

Abstract The parameters of a chaotic system play an important role, whose variation will lead to completely different dynamics. Sometimes, we want to design a controller which is optimal in a certain sense. In this chapter, we focus on two kinds of methods of suppressing chaos: the adaptive control method and the inverse optimal control method. We develop two new methods of parametric adaptive control for a class of discrete-time chaotic systems and a class of continuous-time chaotic systems with multiple parameters, respectively. The systems are first assumed to be linear with respect to parameters. Then, systems with nonlinear distributed parameters and uncertain noise are considered. Finally, we apply the inverse optimal control method to stabilize a four-dimensional chaotic system.

4.1 Introduction

In recent years, many conventional and novel control methods have been applied to the suppression of chaotic behaviors, such as adaptive control [14], fuzzy control [8], linear feedback control [13], impulsive control [3], time-delay feedback control [9], optimal control [6], etc.

The parameters of a chaotic system play an important role, whose variation will lead to completely different dynamics. The dynamical mechanism of the system may change qualitatively during a small change of parameters. Such abrupt transitions between dynamical regimes are usually modeled in terms of bifurcations, catastrophes, and crises [2, 4]. In other cases the transitions are not unwanted, but are useful for switching between related phenomena (bistability and hysteresis). Recently, there has been much work done in the area of parametric adaptive control of chaotic systems, most of which usually assumed the parameters to be constants or to be varying very slowly compared to the dynamical time scales of the systems. In real applications the parameters are not always linearly distributed. Sometimes, we want to design a controller which is optimal in a certain sense. However, it is a difficult task to solve the Hamilton–Jacobi–Bellman (HJB) equation when one designs an optimal controller directly along the conventional route. To solve the aforementioned problems, in this chapter, we focus on two kinds of methods of suppressing chaos: the adaptive control method and the inverse optimal control method. We will develop two new methods of parametric adaptive control for a class of discrete-time chaotic systems and a class of continuous-time chaotic systems with multiple parameters, respectively. The systems are first assumed to be linear with respect to parameters. Then, systems with nonlinear distributed parameters and uncertain noise are considered.

Finally, we will apply the inverse optimal control method to stabilize a four-dimensional chaotic system. In the 1990s, Freeman and Kokotovic [1] proposed the inverse optimal control method, in which an optimal controller was constructed through a Lyapunov function. As long as we can find such a Lyapunov function, it can be inferred that the designed controller is optimal with a specified performance index. The merit of this approach is that we do not need to solve the complicated HJB equation.

4.2 Model-Reference Adaptive Control of a Class of Discrete-Time Chaotic Systems

4.2.1 Basic Formulation

Consider a class of discrete-time nonlinear systems as follows:

$$x(k+1) = F(x(k), \mu(k)), \quad k = 1, 2, \ldots, \tag{4.1}$$

where $x(k) \in \mathbb{R}^n$ is the state vector, $\mu(k) = (\mu_1(k), \ldots, \mu_m(k))^{\mathrm{T}}$ is the parameter vector which determines the nature of dynamics and is time dependent under the influence of both disturbances and control, and $F\colon \mathbb{R}^n \times \mathbb{R}^m \to \mathbb{R}^n$ is a nonlinear map. First, suppose that F is linearly parameterized. Many typical discrete-time chaotic systems belong to this case, such as the Hénon map and the logistic map. Then, (4.1) can be rewritten as

$$x(k+1) = A(x(k)) + \sum_{l=1}^{m} B_l(x(k))\mu_l(k)$$

or

$$x_p(k+1) = A_p(x(k)) + \sum_{q=1}^{m} B_{pq}(x(k))\mu_q(k), \quad p = 1, 2, \ldots, n,$$

where $A, B_l\colon \mathbb{R}^n \to \mathbb{R}^n$ $(l = 1, 2, \ldots, m)$, $A_p, B_{pq}\colon \mathbb{R}^n \to \mathbb{R}$ $(p = 1, 2, \ldots, n;\ q = 1, 2, \ldots, m)$.

Consider the following reference model:

$$y(k+1) = F_M(y(k), \mu_g), \tag{4.2}$$

where $F_M: \mathbb{R}^n \times \mathbb{R}^m \to \mathbb{R}^n$, and μ_g is the desired value (target value).

Our goal is to design an appropriate adaptive controller such that the parameter $\mu(k)$ converges to μ_g as soon as possible.

4.2.2 Main Results

In light of the basic idea of adaptive control, we propose a new parametric adaptive algorithm as follows:

$$\mu(k+1) = \mu(k) + \alpha(x(k))G(e(k+1)), \tag{4.3}$$

where $\alpha = \operatorname{diag}\{\alpha_l\}$ $(l = 1,2,\ldots,m)$, $\alpha_l: \mathbb{R}^n \to \mathbb{R}$, $G: \mathbb{R}^n \to \mathbb{R}^m$, and $e(k+1) = x(k+1) - y(k+1)$. G is an arbitrary nonlinear vector function with the error signal $e(k+1)$ as its argument. Noting that there is no coupling between the adjustable system (4.1) and the reference model (4.2), when $x(k)$ or $y(k)$ presents high periodic or even quasiperiodic chaotic behavior, the change of the value of $e(k) = x(k) - y(k)$ will become extremely complicated. In order to effectively control the complicated system, in light of the idea of coupling systems, we redirect the state vector $x(k)$ of (4.1) to the reference model (4.2), i.e.,

$$y(k+1) = F_M(x(k), \mu_g). \tag{4.4}$$

Therefore, we can let the dynamics of the adjustable system redirect to the reference model to predict the dynamics of $y(k+1)$. Let $F_M = F$, and denote $G_l(z) = \sum_{k=1}^{n} z_k$ $(l = 1,2,\ldots,m)$, where $z = (z_1, z_2, \ldots, z_n)^{\mathrm{T}} \in \mathbb{R}^n$. Then, combining (4.1), (4.3), and (4.4) yields

$$\begin{cases} x_p(k+1) = A_p(x(k)) + \sum\limits_{q=1}^{m} B_{pq}(x(k))\mu_q(k), \\ y_p(k+1) = A_p(x(k)) + \sum\limits_{q=1}^{m} B_{pq}(x(k))\mu_{qg}, \\ \mu_l(k+1) = \mu_l(k) + \alpha_l(x(k))\left[\sum\limits_{p=1}^{n}\sum\limits_{q=1}^{m} B_{pq}(x(k))(\mu_q(k) - \mu_{qg})\right], \end{cases} \tag{4.5}$$

where A_p, B_{pq} $(p = 1,2,\ldots,n;\ q = 1,2,\ldots,m)$ are nonlinear functions of $x(k)$ and $l = 1,2,\ldots,m$. For any q, there is at least one of B_{pq} $(p = 1,2,\ldots,n)$ which is not equal to 0. μ_{qg} $(q = 1,2,\ldots,m)$ are target values of $\mu_q(k)$. By choosing $\alpha_l(x(k))$ appropriately, it can be proven that the above parametric adaptive control equations make the parameters converge asymptotically to the desired values for any disturbed parameter initial values. The relevant theorem is presented as follows.

Theorem 4.1 ([11]). For the parametric adaptive control equations (4.5), if we choose

$$\alpha_l(x(k)) = -K(x(k))W_l(x(k)), \tag{4.6}$$

where $W_l(x(k)) = \sum_{p=1}^{n} B_{pl}(x(k))$,

$$K(x(k)) = 2\left(d + \sum_{l=1}^{m} W_l^2(x(k))\right)^{-1}, \tag{4.7}$$

and $d > 0$ is an arbitrary constant, then the parameters of (4.5) will converge asymptotically to the desired values.

Proof. First, we choose a Lyapunov function as

$$V(k) = \sum_{l=1}^{m} \left(\mu_l(k) - \mu_{lg}\right)^2, \tag{4.8}$$

and then we have

$$\begin{aligned} \Delta V(k) &= V(k+1) - V(k) \\ &= \sum_{l=1}^{m} (\mu_l(k+1) - \mu_{lg})^2 - \sum_{l=1}^{m} (\mu_l(k) - \mu_{lg})^2. \end{aligned} \tag{4.9}$$

Substituting (4.5) into (4.9), and denoting $\mu_j(k) - \mu_{jg} = \mu_{k,jg}$ $(1 \le j \le m)$, we derive

$$\Delta V(k) = \sum_{l=1}^{m} \left(\mu_{k,lg} + \alpha_l(x(k)) \left(\sum_{p=1}^{n} \sum_{q=1}^{m} B_{pq}(x(k))\mu_{k,qg}\right)\right)^2 - \sum_{l=1}^{m} \mu_{k,lg}^2. \tag{4.10}$$

Substituting (4.6) into (4.10), we get

$$\Delta V(k) = K(x(k)) \left(K(x(k)) \sum_{l=1}^{m} W_l^2(x(k)) - 2\right) \left(\sum_{q=1}^{m} W_q(x(k))\mu_{k,qg}\right)^2. \tag{4.11}$$

Substituting (4.7) into (4.11) yields

$$\Delta V(k) = -d\left(K(x(k)) \left(\sum_{q=1}^{m} W_q(x(k))(\mu_q(k) - \mu_{qg})\right)\right)^2 \le 0.$$

Since $\Delta V(k)$ is negative semidefinite, $V(k)$ is a nonincreasing function with respect to time t and bounded from below. Therefore, $V(k)$ must converge to a finite value $V(\infty)$. Noting that $V(1)$ is bounded, we have

$$\begin{aligned} \lim_{s\to\infty} \sum_{k=1}^{s} \Delta V(k) &= \lim_{s\to\infty} \sum_{k=1}^{s} (V(k+1) - V(k)) \\ &= V(\infty) - V(1) \\ &= \text{finite value}. \end{aligned}$$

Noting that $\Delta V(k) \leq 0$, we can derive

$$\lim_{k\to\infty} (\mu_q(k) - \mu_{qg}) = 0 \quad (q = 1, 2, \ldots, m).$$

It is clear that system (4.5) is asymptotically stable with respect to $\mu_q(k)$. In other words, the parameters $\mu_q(k)$ will asymptotically converge to the desired values μ_{qg}, $q = 1, 2, \ldots, m$, respectively. □

Next, we will discuss the case that the system parameters are nonlinearly distributed. Moreover, the disturbances of the uncertain noise are taken into account. Consider a class of discrete-time nonlinear systems as follows:

$$x(k+1) = F(x(k), \mu(k)) + \Delta W(k),$$

where $\Delta W(k)$ is the uncertain noise. Consider the same reference model as in (4.4), i.e.,

$$y(k+1) = F_M(x(k), \mu_g).$$

Furthermore, assume that $F = F_M$, and the adaptive control algorithm is designed the same as (4.3), i.e.,

$$\mu(k+1) = \mu(k) + \alpha(x(k))G(e(k+1)),$$

where

$$\begin{aligned} e(k+1) &= x(k+1) - y(k+1) \\ &= F(x(k), \mu(k)) - F(x(k), \mu_g) + \Delta W(k). \end{aligned}$$

In the above, $\alpha(x(k))$ is defined as

$$\alpha_l(x(k)) = -2R_l(x(k)) \left(d + \sum_{l=1}^{m} R_l^2(x(k)) \right)^{-1}, \tag{4.12}$$

where $R_l(x(k)) = \sum_{i=1}^{n} \partial F_i(x(k), \mu(k)) / \partial \mu_{lg}$ and $d > 0$ is an arbitrary positive constant.

Using Taylor expansion of $F(x(k), \mu(k)) - F(x(k), \mu_g)$ around the neighborhood of $\mu(k) = \mu_g$, we can derive

$$\begin{aligned} \mu_l(k+1) = \mu_l(k) + \alpha_l(x(k)) &\left(\sum_{i=1}^{n} \sum_{j=1}^{m} \frac{\partial F_i}{\partial \mu_{jg}} (\mu_j(k) - \mu_{jg}) \right) \\ &+ h_{2l}(x(k), \mu(k) - \mu_g), \end{aligned} \tag{4.13}$$

where

$$h_{2l}(x(k), \mu(k) - \mu_g) = \sum_{i=1}^{n} (\alpha_l(x(k)) h_{1i}(x(k), \mu(k) - \mu_g) + \Delta W_i)$$

and $h_{1i}(x(k),\mu(k)-\mu_g)=o(\mu(k)-\mu_g)$. Now, consider the stability of solutions of the linear part of (4.13), i.e.,

$$\mu_l(k+1)=\mu_l(k)+\alpha_l(x(k))\left(\sum_{i=1}^{n}\sum_{j=1}^{m}\frac{\partial F_i}{\partial \mu_{jg}}(\mu_j(k)-\mu_{jg})\right). \tag{4.14}$$

We choose a Lyapunov function as

$$V(k)=\sum_{l=1}^{m}(\mu_l(k)-\mu_{lg})^2,$$

and then we have

$$\begin{aligned}\Delta V &= V(k+1)-V(k)\\ &= \sum_{l=1}^{m}(\mu_l(k+1)-\mu_{lg})^2-\sum_{l=1}^{m}(\mu_l(k)-\mu_{lg})^2.\end{aligned} \tag{4.15}$$

Substituting (4.14) into (4.15), we get

$$\begin{aligned}\Delta V &= -4d\left\{\left(\sum_{i=1}^{n}\sum_{j=1}^{m}\frac{\partial F_i}{\partial \mu_{jg}}(\mu_j(k)-\mu_{jg})\right)\left(d+\sum_{l=1}^{m}\left(\sum_{i=1}^{n}\frac{\partial F_i}{\partial \mu_{lg}}\right)^2\right)^{-1}\right\}^2\\ &\le 0.\end{aligned} \tag{4.16}$$

From (4.16) we can infer that (4.14) is asymptotically stable. As for the nonlinear equation given in (4.13), as long as the term h_{2l} is small enough, it is stable, too.

4.2.3 Simulations

An illustrative example is used here to demonstrate the effectiveness of the proposed parametric adaptive control method. Consider the Hénon map:

$$\begin{cases} x_1(k+1)=-px_1^2(k)+x_2(k)+1, \\ x_2(k+1)=qx_1(k), \end{cases} \quad k=0,1,\ldots, \tag{4.17}$$

where p and q are real parameters. When $p=1.4$ and $q=0.3$, system (4.17) presents chaotic behaviors. When $p=0.85$ and $q=0.55$, system (4.17) is in the normal state. Suppose that the reference model is given as follows:

$$\begin{cases} y_1(k+1)=-0.85x_1^2(k)+x_2(k)+1, \\ y_2(k+1)=0.55x_1(k). \end{cases}$$

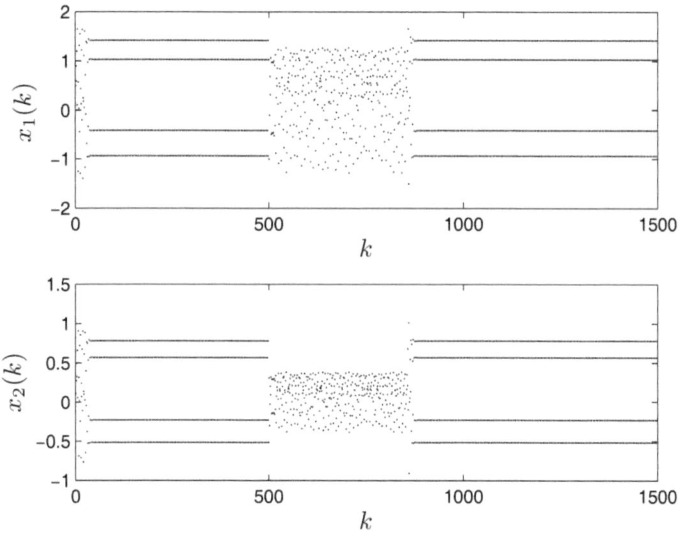

Fig. 4.1 The process of controlling the Hénon map (i)

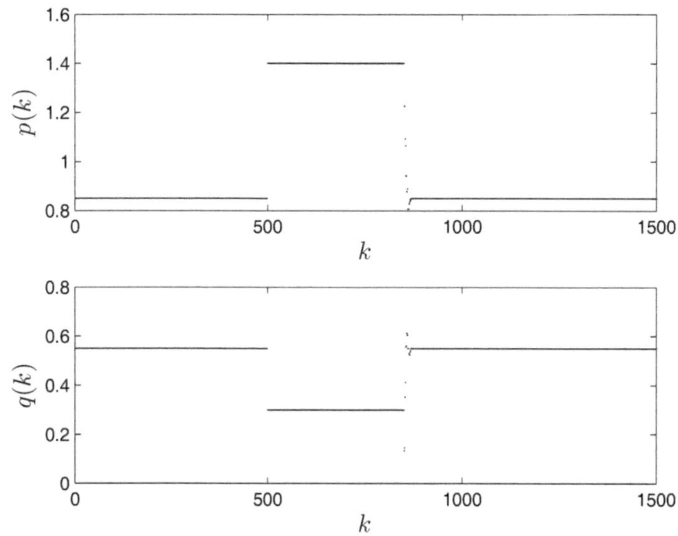

Fig. 4.2 The parameter change process of controlling the Hénon map (i)

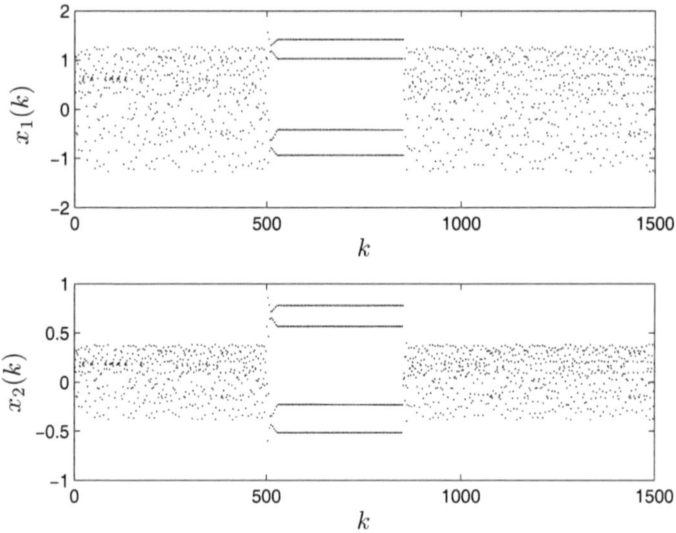

Fig. 4.3 The process of controlling the Hénon map (ii)

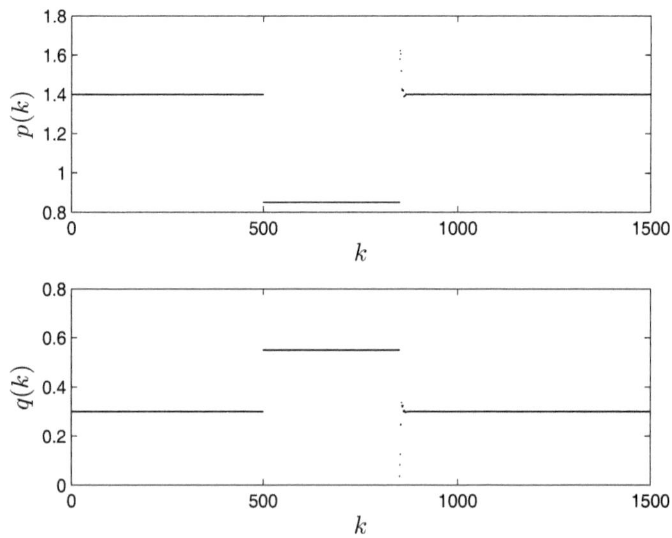

Fig. 4.4 The parameter change process of controlling the Hénon map (ii)

Assume that the parameters p and q of system (4.17) deviate from the desired values ($p_g = 0.85$, $q_g = 0.55$) by a sudden disturbance. The parameters p and q change to $p = 1.4$ and $q = 0.3$ for $k = 501, 502, \ldots, 850$. The disturbed system (4.17) presents chaotic behavior with the changed parameters. The control goal is to make the changed parameters return to the desired values such that the disturbed system (4.17) returns to the normal state. Apply the parametric adaptive control law (4.5), i.e.,

$$p(k+1) = p(k) + 2x_1^2(k)Q\left[x_1(k+1) - y_1(k+1) + x_2(k+1) - y_2(k+1)\right],$$
$$q(k+1) = q(k) - 2x_1(k)Q\left[x_1(k+1) - y_1(k+1) + x_2(k+1) - y_2(k+1)\right],$$

where $Q = (d + x_1^4(k) + x_1^2(k))^{-1}$, $d = 2$, and $k = 851, \ldots, 1500$. The simulation results are given in Figs. 4.1 and 4.2. From the two figures, we can see that the disturbed parameters converge to the desired values rapidly. This control method can be applied to many practical discrete-time systems.

On the other hand, assume that the normal state of system (4.17) is chaotic. The system presents nonchaotic behavior with external disturbances. Now the control goal is to make the disturbed system return to the chaotic state. Still consider the Hénon map, but the disturbed parameters are given as $p = 0.85$, $q = 0.55$, and $k = 501, \ldots, 850$. Suppose that the reference model is given as follows:

$$\begin{cases} y_1(k+1) = -1.4x_1^2(k) + x_2(k) + 1, \\ y_2(k+1) = 0.3x_1(k). \end{cases}$$

Apply the parametric adaptive control law (4.5), where $d = 2$ and $k = 851, \ldots, 1500$. The simulation results are given in Figs. 4.3 and 4.4. From the two figures, we can see that the disturbed parameters converge to the desired values rapidly.

4.3 Model-Reference Adaptive Control of a Class of Continuous-Time Chaotic Systems

4.3.1 Basic Formulation

Consider a class of continuous-time nonlinear dynamical systems:

$$\dot{x} = F(x, t, \mu), \tag{4.18}$$

where $x \in \mathbb{R}^n$ is the state vector, $\mu \in \mathbb{R}^m$ is the parameter vector which determines the nature of the dynamics, and $F\colon \mathbb{R}^n \times \mathbb{R} \times \mathbb{R}^m \to \mathbb{R}^n$. First, suppose that F is linear with respect to parameters, and then (4.18) can be described by the following equation:

$$\dot{x} = \phi(x, t) + \sum_{k=1}^{m} \psi_k(x, t)\mu_k, \tag{4.19}$$

where $\phi, \psi_k \colon \mathbb{R}^n \times \mathbb{R} \to \mathbb{R}^n$ $(k = 1,2,\ldots,m)$, and $\mu = (\mu_1, \mu_2, \ldots, \mu_m)^{\mathrm{T}}$ is the parameter vector. The parameter vector determines the asymptotic behavior of a typical trajectory. In spite of any sources of noise or accidental fluctuations, the purpose of the parametric adaptive control is to maintain the parameter vector μ at a predetermined desired value. The disturbed parameters are updated by the control algorithm. Under the influence of the fluctuations as well as the control, the parameter μ becomes time-dependent quantity $\mu(t)$. The action of the control draws them toward a desired value μ_g. One of the demands on the control method is that the restoration should be as brief as possible. A further requirement is the ability of the control method to control the system for any initial condition $x(0)$ in the state space with respect to any desired parameter value μ_g, i.e., the parametric adaptive control is global.

If the instantaneous values of the parameters are known, then their deviations from the desired values could readily be calculated. It would then be trivial to control the system by updating the parameters with their desired values. In reality, researchers do not generally know the changed value of the parameters at any instant. Without any information about parameter changes, the problem of detecting deviations from the desired values becomes much more difficult. The parameters can only be identified by the properties of the global behavior of the system. To detect deviations in the parameter setting, one can use the vector x, the deviation between the vector x of and the reference vector y, to estimate the values of system parameters.

4.3.2 Main Results

First, consider a reference model $\dot{y} = F_M(y, t, \mu)$, which has the same dynamics as the system (4.19) at the desired value $\mu = \mu_g$, i.e.,

$$\dot{y} = F_M(y, t, \mu_g). \tag{4.20}$$

The present adaptive control law is given as follows:

$$\dot{\mu} = \beta(x, t) G(\dot{e}), \tag{4.21}$$

where $\beta(x,t) = \operatorname{diag}\{\beta_l(x,t)\}$ $(l = 1,2,\ldots,m)$ is a diagonal matrix, $e = x - y$ is the error vector, and $G \colon \mathbb{R}^n \to \mathbb{R}^m$ is a continuous vector function with the error signal as its argument. The difficulty lies in how to choose an appropriate G.

The systems (4.19)–(4.21) constitute a set of control equations. If the parameter μ of system (4.19) is equal to μ_g, and system behavior tends to an equilibrium or low-periodic behavior, the reference model (4.20) can be considered as a simple pattern of behavior. Thus, the control strategy containing (4.19), (4.20), and (4.21) is a successful set of parametric adaptive control equations [7, 10]. However, when $\mu = \mu_g$, if system (4.19) presents high periodic, quasiperiodic, or even chaotic behavior, then the error signal e or $\dot{e}$ becomes an extremely complicated varying value. In order to effectively control more complicated systems, we will redirect the output $x(t)$ of system (4.19) to system (4.20). We then derive the coupling equation between the system and the model:

$$y = F_M(x,t,\mu_g).$$

Therefore, the proposed method of control consists of a combination of equations (4.19), (4.20), and (4.21), or

$$\begin{cases} \dot{x}_k = F_k(x,t,\mu), \\ \dot{y}_k = F_{Mk}(x,t,\mu_g), \\ \dot{\mu}_l = \beta_l(x,t)G_l(\dot{e}), \end{cases} \tag{4.22}$$

where $\dot{e} = \dot{x} - \dot{y}$, $k = 1,2,\ldots,n$, and $l = 1,2,\ldots,m$. We assume that the model dynamics is the same as that of the system's, i.e., $F_{Mk} = F_k$ $(k = 1,2,\ldots,n)$. Then, the control equations (4.22) are globally stable with respect to μ_l if we choose G_l appropriately. The relevant theorem is given as follows.

Theorem 4.2 ([11]). The parameters of the parametric adaptive control equations (4.22) can converge asymptotically to the desired values for any disturbed parameter initial values if we choose

$$\beta_l(x,t) = -\sum_{k=1}^{n} \psi_{lk}(x,t)$$

and

$$G_l(\dot{e}) = \sum_{k=1}^{n} \dot{e}_k, \tag{4.23}$$

where $l = 1,2,\ldots,m$.

Proof. Suppose that $F(x,t,\mu) = \phi(x,t) + \sum_{l=1}^{m} \psi_l(x,t)\mu_l$, in which $\phi, \psi_l : \mathbb{R}^n \times \mathbb{R} \to \mathbb{R}^n$, $\phi(x,t) = (\phi_1(x,t), \phi_2(x,t), \ldots, \phi_n(x,t))^{\mathrm{T}}$, $\psi_l(x,t) = (\psi_{l1}(x,t), \psi_{l2}(x,t), \ldots, \psi_{ln}(x,t))^{\mathrm{T}}$, and $\psi_{lk}(x,t)$ $(l = 1,2,\ldots,m;\ k = 1,2,\cdots,n)$ are uniformly continuous real functions. Then, the adaptive control algorithm (4.22) can be transformed to the following:

$$\begin{cases} \dot{x}_k = \phi_k(x,t) + \sum\limits_{l=1}^{m} \psi_{lk}(x,t)\mu_l, \\ \dot{y}_k = \phi_k(x,t) + \sum\limits_{l=1}^{m} \psi_{lk}(x,t)\mu_{lg}, \\ \dot{\mu}_l = -\left(\sum\limits_{k=1}^{n} \psi_{lk}(x,t)\right)\left(\sum\limits_{p=1}^{m}\sum\limits_{q=1}^{n} \psi_{pq}(x,t)\left(\mu_p - \mu_{pg}\right)\right). \end{cases} \tag{4.24}$$

Now, we choose a Lyapunov function as

$$V(t) = \frac{1}{2}\sum_{l=1}^{m}(\mu_l - \mu_{lg})^2,$$

and then we have

$$\begin{aligned}\dot{V}(t) &= \sum_{l=1}^{m}(\mu_l-\mu_{lg})\dot{\mu}_l\\ &= \sum_{l=1}^{m}(\mu_l-\mu_{lg})\left\{-\left(\sum_{k=1}^{n}\psi_{lk}(x,t)\right)\left(\sum_{p=1}^{m}\sum_{q=1}^{n}\psi_{pq}(x,t)(\mu_p-\mu_{pg})\right)\right\}\\ &= \sum_{l=1}^{m}(\mu_l-\mu_{lg})\left\{-\left(\sum_{k=1}^{n}\psi_{lk}(x,t)\right)\sum_{q=1}^{n}\psi_{lq}(x,t)(\mu_l-\mu_{lg})\right.\\ &\quad\left.-\left(\sum_{k=1}^{n}\psi_{lk}(x,t)\right)\sum_{p\neq l}^{m}\sum_{q=1}^{n}\psi_{pq}(x,t)(\mu_p-\mu_{pg})\right\}\\ &= -\sum_{l=1}^{m}\left\{\left(\sum_{k=1}^{n}\psi_{lk}(x,t)\right)^2(\mu_l-\mu_{lg})^2\right.\\ &\quad\left.+\left(\sum_{k=1}^{n}\psi_{lk}(x,t)\right)(\mu_l-\mu_{lg})\sum_{p\neq l}^{m}\sum_{q=1}^{n}\psi_{pq}(x,t)(\mu_p-\mu_{pg})\right\}\\ &= -\left\{\sum_{l=1}^{m}\sum_{k=1}^{n}\psi_{lk}(x,t)(\mu_l-\mu_{lg})\right\}^2 \le 0.\end{aligned}$$

It is shown that $\dot{V}(t)$ is a negative semidefinite function. Therefore, $V(t)$ is a non-increasing function of time which is also bounded from below and hence converges to a finite value $V(\infty)$. Since the initial value $\mu(0)$ is bounded, $V(0)$ is bounded. In such a case we have

$$\lim_{t\to\infty}\int_0^t \dot{V}(t)\mathrm{d}t = V(\infty)-V(0),$$

which is a finite value. Since $\psi_{lk}(x,t)$ $(l=1,2,\ldots,m;\, k=1,2,\ldots,n)$ are uniformly continuous, $\dot{V}(t)$ is also uniformly continuous. Then, we find that

$$\lim_{t\to\infty}\dot{V}(t)=0$$

or

$$\lim_{x\to\infty}\sum_{l=1}^{m}\sum_{k=1}^{n}\psi_{lk}(x,t)(\mu_l-\mu_{lg})=0.$$

Note that $\sum_{k=1}^{n}\psi_{lk}(x,t)$ $(l=1,2,\ldots,m)$ do not approach constants as $t\to\infty$. Then, we obtain

$$\lim_{t\to\infty}(\mu_l-\mu_{lg})=0 \quad (l=1,2,\ldots,m).$$

Therefore, we have proven that the parametric adaptive control equations (4.24) are efficient. In other words, the parameter μ_l is asymptotically stable, which achieves the stability match between the adjustable system and the reference model. □

Next, we will discuss the case in which the system is nonlinearly parameterized. Here, we cannot adopt the same adaptive control method as (4.24). Consider the following equations:

$$\begin{cases} \dot{x}_k = F_k(x,t,\mu) + \Delta W_k(t), \\ \dot{y}_k = F_{Mk}(x,t,\mu_g), \\ \dot{\mu}_l = \beta_l(x,t) G_l(\dot{e}), \end{cases} \tag{4.25}$$

where $l = 1,2,\ldots,m$ and $\Delta W_k(t)$ $(k = 1,2,\ldots,n)$ are disturbed terms of external uncertain noise. Suppose that there is no modeling error, i.e., $F_k = F_{Mk}$ $(k = 1,2,\ldots,n)$, and G_l is defined by (4.23). Then, the third equation of (4.25) can be rewritten as

$$\dot{\mu}_l = \beta_l(x,t)\left(\sum_{k=1}^{n} (F_k(x,t,\mu) - F_k(x,t,\mu_g)) + \Delta W_k(t)\right). \tag{4.26}$$

Expanding $(F_k(x,t,\mu) - F_k(x,t,\mu_g))$ around a neighborhood of $\mu = \mu_g$, one obtains

$$F_k(x,t,\mu) - F_k(x,t,\mu_g) = \sum_{l=1}^{m} \frac{\partial F_k(x,t,\mu)}{\partial \mu_{lg}}(\mu_l - \mu_{lg}) + h_{1k}(x,t,\mu - \mu_g), \tag{4.27}$$

where $\partial F_k(x,t,\mu)/\partial \mu_{lg} = (\partial F_k(x,t,\mu)/\partial \mu)|_{\mu=\mu_g}$ and $h_{1k}(x,t,\mu - \mu_g)$ is the higher-order infinitesimal term of $\mu - \mu_g$. Substituting (4.27) into (4.26), we get

$$\begin{aligned} \dot{\mu}_l &= \beta_l(x,t)\sum_{k=1}^{n}\left\{\sum_{l=1}^{m}\frac{\partial F_k(x,t,\mu)}{\partial \mu_{lg}}(\mu_l - \mu_{lg}) + h_{1k}(x,t,\mu - \mu_g) + \Delta W_k(t)\right\} \\ &= \beta_l(x,t)\sum_{l=1}^{m}\sum_{k=1}^{n}\frac{\partial F_k(x,t,\mu)}{\partial \mu_{lg}}(\mu_l - \mu_{lg}) + h_2(x,t,\mu - \mu_g), \end{aligned} \tag{4.28}$$

where $h_2(x,t,\mu - \mu_g) = \sum_{k=1}^{n} \beta_l(x,t)(h_{1k}(x,t,\mu - \mu_g) + \Delta W_k(t))$. Let

$$\beta_l(x,t) = -\sum_{k=1}^{n}\frac{\partial F_k(x,t,\mu)}{\partial \mu_{lg}}. \tag{4.29}$$

Substituting (4.29) into (4.28), we get

$$\begin{aligned} \dot{\mu}_l = &-\left(\sum_{k=1}^{n}\frac{\partial F_k(x,t,\mu)}{\partial \mu_{lg}}\right)\left(\sum_{p=1}^{m}\sum_{q=1}^{n}\frac{\partial F_q(x,t,\mu)}{\partial \mu_{pg}}(\mu_p - \mu_{pg})\right) \\ &+ h_2(x,t,\mu - \mu_g). \end{aligned} \tag{4.30}$$

Since the term $h_2(x,t,\mu - \mu_g)$ includes the information of the unknown disturbance $\Delta W_k(t)$, the parametric adaptive law (4.30) is not feasible. Therefore, let us consider the following parametric adaptive law instead of (4.30):

$$\dot{\mu}_l = -\left(\sum_{k=1}^{n}\frac{\partial F_k(x,t,\mu)}{\partial \mu_{lg}}\right)\left(\sum_{p=1}^{m}\sum_{q=1}^{n}\frac{\partial F_q(x,t,\mu)}{\partial \mu_{pg}}(\mu_p - \mu_{pg})\right). \tag{4.31}$$

Now we analyze the stability of the solution of (4.31). We still choose a Lyapunov function as

$$V(t) = \frac{1}{2}\sum_{l=1}^{m}(\mu_l - \mu_{lg})^2,$$

and then we get

$$\begin{aligned}
\dot{V}(t) &= \sum_{l=1}^{m}(\mu_l - \mu_{lg})\,\dot{\mu}_l \\
&= \sum_{l=1}^{m}(\mu_l - \mu_{lg})\left\{-\left(\sum_{k=1}^{n}\frac{\partial F_k(x,t,\mu)}{\partial \mu_{lg}}\right)\left(\sum_{p=1}^{m}\sum_{q=1}^{n}\frac{\partial F_q(x,t,\mu)}{\partial \mu_{pg}}(\mu_p - \mu_{pg})\right)\right\} \\
&= \sum_{l=1}^{m}(\mu_l - \mu_{lg})\left\{-\left(\sum_{k=1}^{n}\frac{\partial F_k(x,t,\mu)}{\partial \mu_{lg}}\right)\sum_{q=1}^{n}\frac{\partial F_q(x,t,\mu)}{\partial \mu_{lg}}(\mu_l - \mu_{lg})\right. \\
&\quad \left. -\left(\sum_{k=1}^{n}\frac{\partial F_k(x,t,\mu)}{\partial \mu_{lg}}(\mu_l - \mu_{lg})\right)\left(\sum_{p\neq l}^{m}\sum_{q=1}^{n}\frac{\partial F_q(x,t,\mu)}{\partial \mu_{lg}}(\mu_p - \mu_{pg})\right)\right\} \\
&= -\sum_{l=1}^{m}\left\{\left(\sum_{k=1}^{n}\frac{\partial F_k(x,t,\mu)}{\partial \mu_{lg}}\right)^2(\mu_l - \mu_{lg})^2\right. \\
&\quad \left. +\left(\sum_{k=1}^{n}\frac{\partial F_k(x,t,\mu)}{\partial \mu_{lg}}(\mu_l - \mu_{lg})\right)\sum_{p\neq l}^{m}\sum_{q=1}^{n}\frac{\partial F_q(x,t,\mu)}{\partial \mu_{lg}}(\mu_p - \mu_{pg})\right\} \\
&= -\left\{\sum_{l=1}^{m}\sum_{k=1}^{n}\frac{\partial F_k(x,t,\mu)}{\partial \mu_{lg}}(\mu_l - \mu_{lg})\right\}^2 \\
&\leq 0.
\end{aligned}$$

Similar to the discussion of $V(t)$ earlier, one obtains that $\lim_{t\to\infty}\dot{V}(t) = 0$ and

$$\lim_{t\to\infty}(\mu_l - \mu_{lg}) = 0 \quad (l = 1,2,\ldots,m).$$

We have proven that (4.31) is asymptotically stable. For the nonlinear equation given in (4.30), if the term $h_2(x,t,\mu - \mu_g)$ is sufficiently small, then it is also stable.

4.3.3 Simulations

To visualize the effectiveness of the proposed parametric adaptive control method, two illustrative examples are presented below. First, consider the Duffing equation:

$$\begin{cases} \dot{x}_1 = x_2, \\ \dot{x}_2 = -p_2x_1 - x_1^3 - p_1x_2 + q\cos(1.8t), \end{cases} \tag{4.32}$$

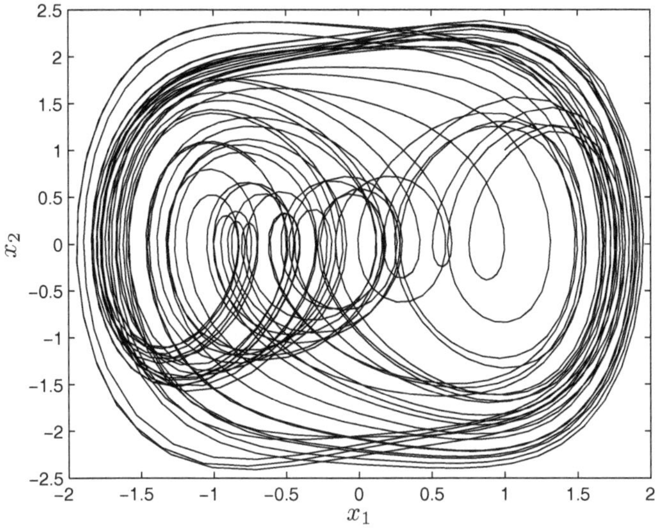

Fig. 4.5 The chaotic attractor of the Duffing equation described by (4.32)

where p_1, p_2, and q are system parameters. When $p_1 = 0.4$, $p_2 = -1.1$, $q = 1.8$, and the initial state is given as $x(0) = (1,1)^{\mathrm{T}}$, system (4.32) presents chaotic behavior, which is illustrated in Fig. 4.5.

Suppose that the reference model is

$$\begin{cases} \dot{x}_1 = x_2, \\ \dot{x}_2 = 0.8x_1 - x_1^3 - 0.7x_2 + 0.6\cos(1.8t), \end{cases}$$

i.e., the desired values of the system parameters are $p_{1g} = 0.7$, $p_{2g} = -0.8$, and $q_g = 0.6$. The reference model corresponds to a normal state. Suppose that the parameters are disturbed to $p_1 = 0.4$, $p_2 = -1.1$, and $q = 1.8$, which make system (4.32) chaotic. Our goal is to make the disturbed system return to the normal state. Applying the parametric adaptive control law (4.24), we get

$$\begin{cases} \dot{p}_1 = -(-x_2)\left[-x_2(p_1 - 0.7) - x_1(p_2 + 0.8) + (q - 0.6)\cos(1.8t)\right], \\ \dot{p}_2 = -(-x_1)\left[-x_2(p_1 - 0.7) - x_1(p_2 + 0.8) + (q - 0.6)\cos(1.8t)\right], \\ \dot{q} = -\cos(1.8t)\left[-x_2(p_1 - 0.7) - x_1(p_2 + 0.8) + (q - 0.6)\cos(1.8t)\right]. \end{cases}$$

The simulation results are shown in Figs. 4.6–4.10 which show that the disturbed parameters converge quickly to their desired values after the transient. Figs. 4.6 and 4.7 show the moving trajectory of the undisturbed normal state, disturbed chaotic state, and controlled desired state of system (4.32). From Figs. 4.8–4.10 we can see that parametric adaptive control law (4.24) is very powerful in stabilizing the system. The control method can be applied to electric power systems in a chaotic

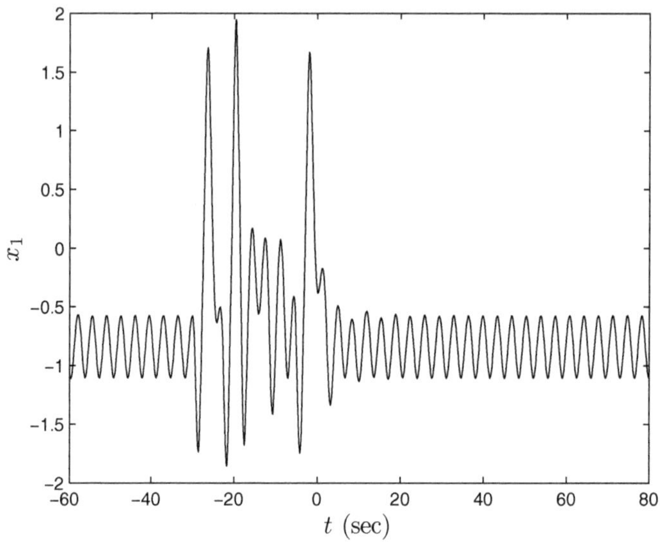

Fig. 4.6 The changed state response curve of $x_1(t)$, which is disturbed to be chaotic at $t=-30$, and approaches the desired state when the controller is applied at $t=0$

state after the external disturbance, which automatically adjusts to a normal state and guarantees the security of the electric power network.

Now, we assume that an external disturbance $\Delta W(t)$ acts on the Duffing equation when $t=0$. The parametric adaptive law remains unchanged. Choosing $\Delta W(t)$ as a random background noise and $\max_t\{|\Delta W(t)|\}=0.1$, the adaptive control law is remarkably robust (see Figs. 4.11–4.15).

Next, consider the Rössler system:

$$\begin{cases} \dot{x}_1 = -x_2 - x_3, \\ \dot{x}_2 = x_1 + \mu_1 x_2, \\ \dot{x}_3 = 0.2 + \mu_2 x_3(x_1 - 5.7), \end{cases} \tag{4.33}$$

where μ_1 and μ_2 are system parameters. When the desired parameter values are $\mu_{1g}=0.2$ and $\mu_{2g}=1$, system (4.33) corresponds to a chaotic state (see Fig. 4.16). Suppose that the reference model of system (4.33) is chosen as

$$\begin{cases} \dot{x}_1 = -x_2 - x_3, \\ \dot{x}_2 = x_1 + 0.2x_2, \\ \dot{x}_3 = 0.2 + x_3(x_1 - 5.7). \end{cases}$$

Suppose that the parameters μ_1 and μ_2 of system (4.33) are disturbed from the target values ($\mu_{1g}=0.2$, $\mu_{2g}=1$), and set to $\mu_1=-0.8$, $\mu_2=2.8$, which corresponds to a nonchaotic state. The control automatically changes the parameters until the system

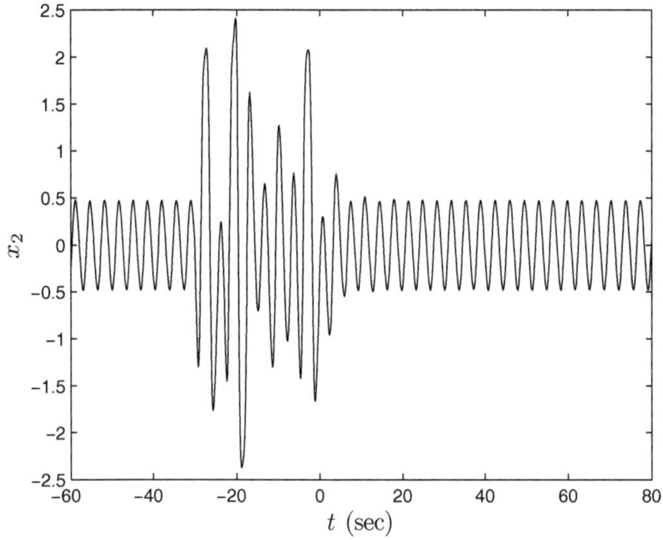

Fig. 4.7 The changed state response curve of $x_2(t)$, which is disturbed to be chaotic at $t = -30$, and approaches the desired state when the controller is applied at $t = 0$

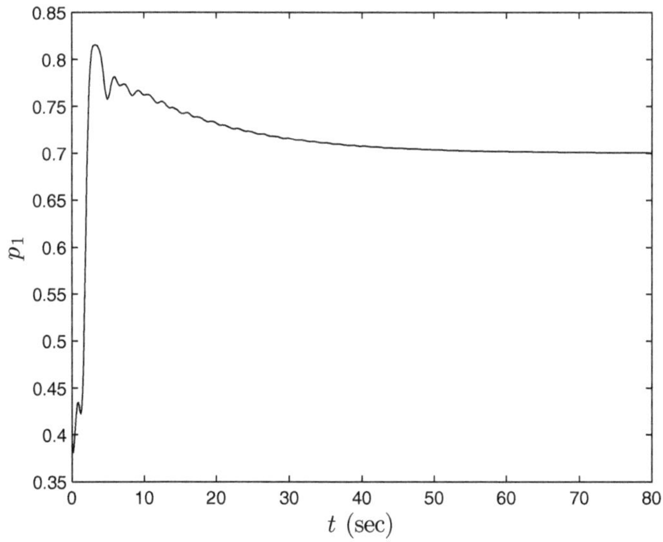

Fig. 4.8 Parameter p_1 (disturbed initial value $p_1(0) = 0.4$) approaches the desired value $p_{1g} = 0.7$

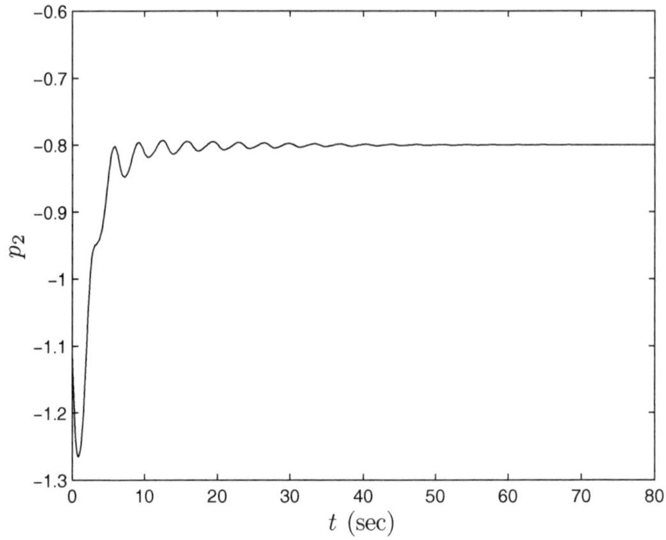

Fig. 4.9 Parameter p_2 (disturbed initial value $p_2(0) = -1.1$) approaches the desired value $p_{2g} = -0.8$

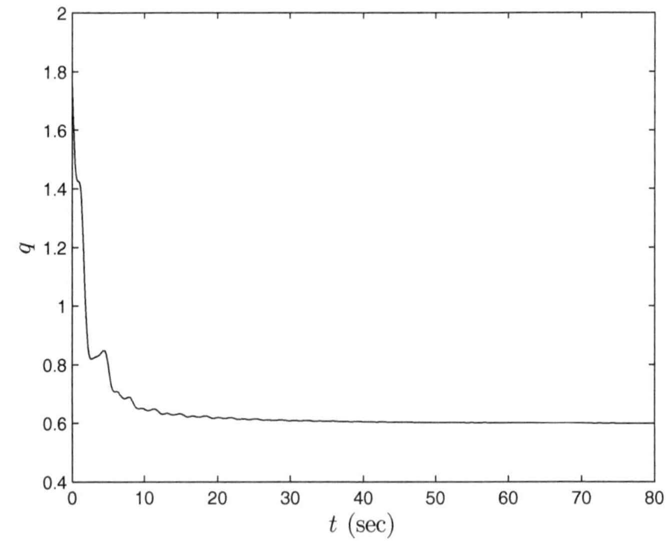

Fig. 4.10 Parameter q (disturbed initial value $q(0) = 1.8$) approaches the desired value $q_g = 0.6$

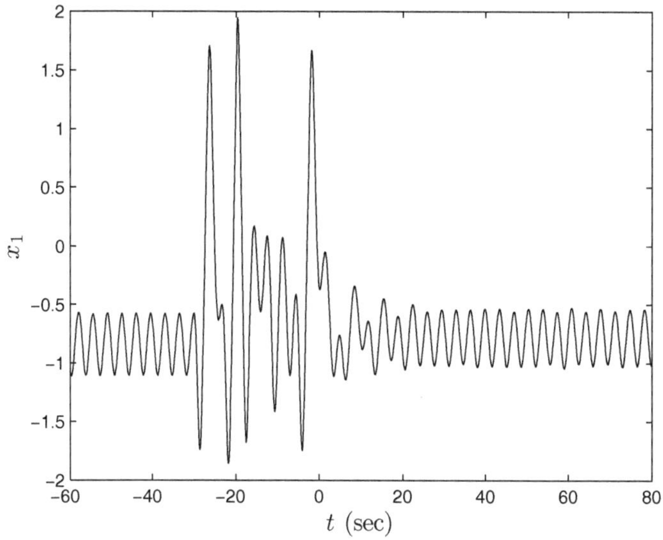

Fig. 4.11 External disturbance on the Duffing equation. The changed state response curve of $x_1(t)$, which is disturbed to be chaotic at $t = -30$, and approaches the desired state when the controller is applied at $t = 0$

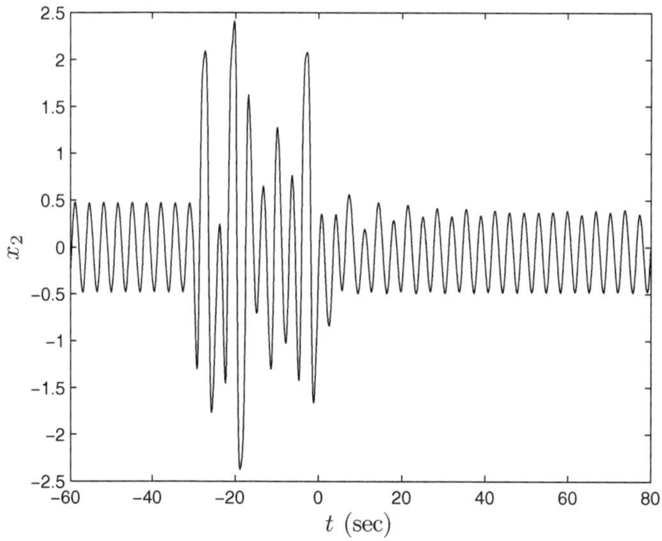

Fig. 4.12 External disturbance on the Duffing equation. The changed state response curve of $x_2(t)$, which is disturbed to be chaotic at $t = -30$, and approaches the desired state when the controller is applied at $t = 0$

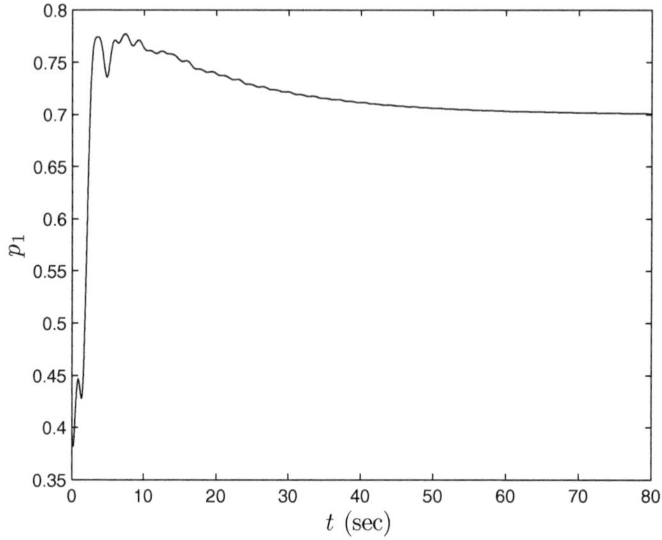

Fig. 4.13 External disturbance on the Duffing equation. Parameter p_1 (disturbed initial value $p_1(0) = 0.4$) approaches the desired value $p_{1g} = 0.7$

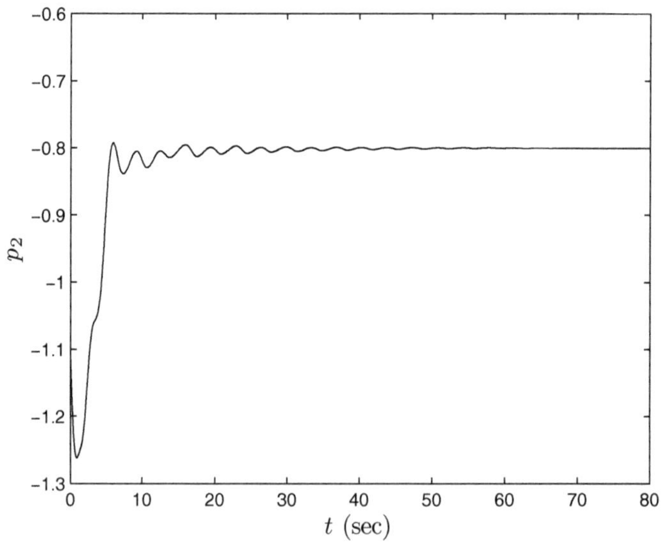

Fig. 4.14 External disturbance on the Duffing equation. Parameter p_2 (disturbed initial value $p_2(0) = -1.1$) approaches the desired value $p_{2g} = -0.8$

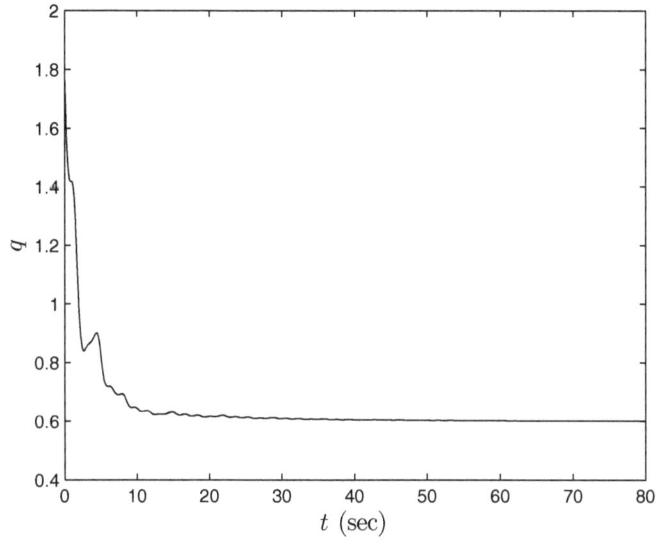

Fig. 4.15 External disturbance on the Duffing equation. Parameter q (disturbed initial value $q(0) = 1.8$) approaches the desired value $q_g = 0.6$

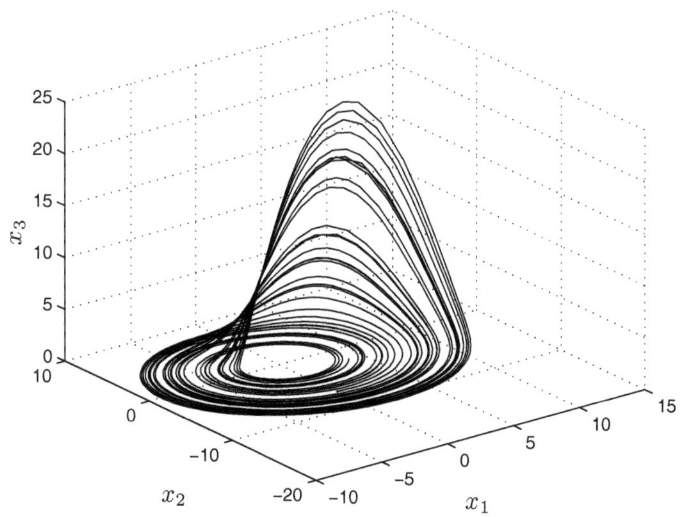

Fig. 4.16 The chaotic attractor of the Rössler system

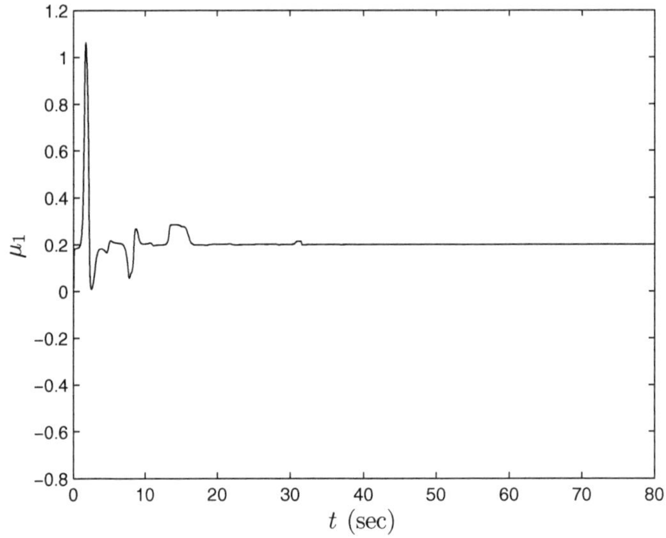

Fig. 4.17 Parameter μ_1 (disturbed initial value $\mu_1(0) = -0.8$) approaches the desired value $\mu_{1g} = 0.2$

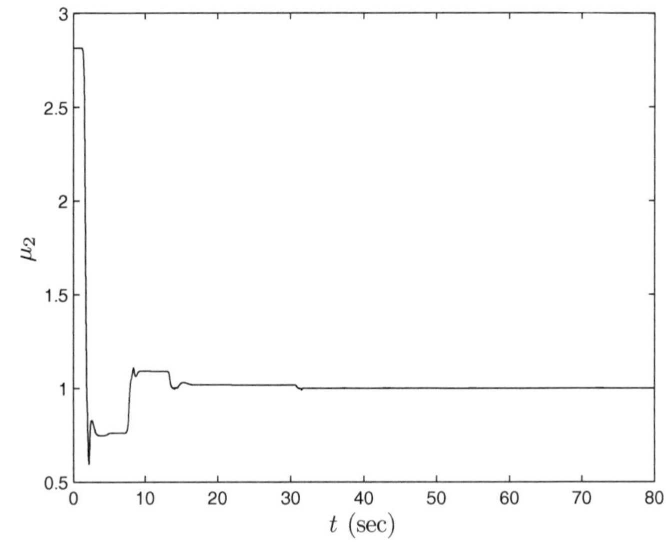

Fig. 4.18 Parameter μ_2 (disturbed initial value $\mu_2(0) = 2.8$) approaches the desired value $\mu_{2g} = 1$

returns to the initial target state. Using the parametric adaptive law (4.24), we get

$$\begin{cases} \dot{\mu}_1 = -x_2[x_2(\mu_1 - 0.2) + x_3(x_1 - 5.7)(\mu_2 - 1)], \\ \dot{\mu}_2 = -x_3(x_1 - 5.7)[x_2(\mu_1 - 0.2) + x_3(x_1 - 5.7)(\mu_2 - 1)]. \end{cases} \tag{4.34}$$

The simulation results are shown in Figs. 4.17 and 4.18. The figures show that the disturbed parameters converge quickly to the desired parameter values after the transient, which indicates that the parametric adaptive control law (4.24) is very powerful in stabilizing the system. This example makes it clear that the control method can be applied to control a nonchaotic state to a chaotic state. For example, a chaotic diode resonator should be kept in a chaotic state under external disturbances.

4.4 Control of a Class of Chaotic Systems Based on Inverse Optimal Control

4.4.1 Problem Description

Consider a four-dimensional chaotic system [5] as follows:

$$\begin{cases} \dot{x}_1 = a(x_2 - x_1) + x_2x_3x_4, \\ \dot{x}_2 = b(x_1 + x_2) - x_1x_3x_4, \\ \dot{x}_3 = -cx_3 + x_1x_2x_4, \\ \dot{x}_4 = -dx_4 + x_1x_2x_3, \end{cases} \tag{4.35}$$

where $x = (x_1, x_2, x_3, x_4)^{\mathrm{T}} \in \mathbb{R}^4$ is the state vector a, b, c, and d are known parameters of the system (4.35), which satisfy the following conditions: $a > 0$, $b > 0$, $c > 0$, $d > 0$, and $b - (a + c + d) < 0$.

By choosing various parameters, system (4.35) has 65 unstable equilibria: S_0: $(0, 0, 0, 0)^{\mathrm{T}}$, S_k: $(x_1^i, x_2^i, x_3^j, x_4^j)^{\mathrm{T}}$, $k = 1, \ldots, 64$, $i = 1, \ldots, 8$, $j = 1, \ldots, 8$, where

$$\begin{aligned}
x_1^1 &= \frac{1}{\sqrt{2}a}\sqrt{a(a+b+p)q}, \quad x_1^2 = -\frac{1}{\sqrt{2}a}\sqrt{a(a+b+p)q}, \\
x_1^3 &= \frac{1}{2a}\sqrt{-2a(a+b-p)q}, \quad x_1^4 = -\frac{1}{2a}\sqrt{-2a(a+b-p)q}, \\
x_1^5 &= \frac{1}{\sqrt{2}a}\sqrt{a(a+b-p)q}, \quad x_1^6 = -\frac{1}{\sqrt{2}a}\sqrt{a(a+b-p)q}, \\
x_1^7 &= \frac{1}{2a}\sqrt{-2a(a+b+p)q}, \quad x_1^8 = -\frac{1}{2a}\sqrt{-2a(a+b+p)q}, \\
x_2^i &= \pm\frac{q}{x_1^i}, \quad x_3^j = \pm\frac{1}{q}\sqrt{ab\left(x_1^j\right)^2 - adq}, \quad x_4^j = \pm\frac{c}{q}x_3^j, \\
q &= \sqrt{cd}, \quad p = \sqrt{a^2 + 6ab + b^2}.
\end{aligned}$$

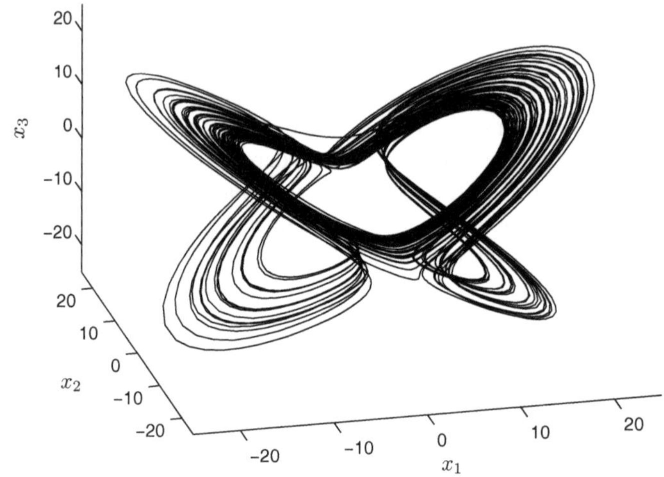

Fig. 4.19 The projection of the chaotic attractor of system (4.35) in the space of x_1–x_2–x_3 with $a = 30$, $b = 10$, $c = 37$, and $d = 10$

The chaotic behavior of the system (4.35) is shown in Fig. 4.19, when the parameters of the system (4.35) are chosen as $a = 30$, $b = 10$, $c = 37$, and $d = 10$ and the initial condition of the system (4.35) is chosen as $x(0) = (-10, 10, 10, 20)^{\mathrm{T}}$.

4.4.2 Inverse Optimal Controller Design

Suppose that the controlled chaotic system is

$$\begin{cases} \dot{x}_1 = a(x_2 - x_1) + x_2x_3x_4, \\ \dot{x}_2 = b(x_1 + x_2) - x_1x_3x_4 + \dfrac{1}{3}u, \\ \dot{x}_3 = -cx_3 + x_1x_2x_4, \\ \dot{x}_4 = -dx_4 + x_1x_2x_3, \end{cases} \tag{4.36}$$

where u is the control input. We have the following theorem.

Theorem 4.3 ([12]). If the linear state feedback controller is designed as

$$u = -\left[6b + \frac{(a+3b)^2}{2a} + 1\right] x_2, \tag{4.37}$$

then the zero equilibrium of the controlled four-dimensional chaotic system (4.36) is globally asymptotically stable.

Proof. Construct a Lyapunov function as

$$V = \frac{1}{2}\left(x_1^2 + 3x_2^2 + x_3^2 + x_4^2\right).$$

Then, the time derivative of V along the trajectory of system (4.36) is given as follows:

$$\begin{aligned}
\dot{V} &= x_1\dot{x}_1 + 3x_2\dot{x}_2 + x_3\dot{x}_3 + x_4\dot{x}_4 \\
&= a(x_2 - x_1)x_1 + x_1x_2x_3x_4 + 3b(x_1 + x_2)x_2 - 3x_1x_2x_3x_4 \\
&\quad - cx_3^2 + x_1x_2x_3x_4 - dx_4^2 + x_1x_2x_3x_4 + x_2u \\
&= -a\left(x_1 - \frac{(a+3b)}{2a}x_2\right)^2 - cx_3^2 - dx_4^2 \\
&\quad + \left(3b + \frac{(a+3b)^2}{4a}\right)x_2^2 + x_2u \\
&= L_fV + (L_gV)u,
\end{aligned} \tag{4.38}$$

where

$$\begin{cases} L_fV = -a\left(x_1 - \dfrac{(a+3b)}{2a}x_2\right)^2 - cx_3^2 - dx_4^2 + \left(3b + \dfrac{(a+3b)^2}{4a}\right)x_2^2, \\ L_gV = x_2. \end{cases} \tag{4.39}$$

From (4.38), we can easily obtain $L_fV \leq 0$ if $L_gV = 0$.

Next, we design a simple linear state feedback controller as follows:

$$u = -\beta R^{-1}(x)(L_gV) = -\left(3b + \frac{3(a+3b)^2}{4a} + k_0\right)x_2, \tag{4.40}$$

where $\beta > 0$ is a constant, $R^{-1}(x) = \dfrac{1}{\beta}\left(3b + \dfrac{3(a+3b)^2}{4a} + k_0\right)$, and $k_0 > 0$ is a constant to be determined later.

Substituting (4.40) into (4.38), we obtain

$$\dot{V} = -a\left(x_1 - \frac{(a+3b)}{2a}x_2\right)^2 - cx_3^2 - dx_4^2 - \left(k_0 + \frac{(a+3b)^2}{2a}\right)x_2^2,$$

which implies that $\dot{V} < 0$ for all $x \neq 0$, i.e., the controller (4.40) can globally asymptotically stabilize the closed-loop system (4.36) .

Based on the idea of inverse optimal control, in order to determine the value of k_0, we define the following performance index:

$$J(u) = \lim_{t\to\infty}\left\{2\beta V(x(t)) + \int_0^t \left(l(x(\tau)) + u^{\mathrm{T}}(x(\tau))R(x(\tau))u(x(\tau))\right)\mathrm{d}\tau\right\}, \tag{4.41}$$

where

$$l(x) = -2\beta L_f V + \beta^2 R^{-1}(x)(L_g V)^2 > 0\,. \tag{4.42}$$

Substituting (4.39) into (4.42), we can derive

$$\begin{aligned} l(x) = {} & 2a\beta\left(x_1 - \frac{(a+3b)}{2a}x_2\right)^2 + 2c\beta x_3^2 + 2d\beta x_4^2 \\ & - 2\beta\left(3b + \frac{(a+3b)^2}{4a}\right)x_2^2 + \beta\left(3b + \frac{3(a+3b)^2}{4a} + k_0\right)x_2^2\,. \end{aligned} \tag{4.43}$$

Let $l(x) > 0$ and

$$k_0 = 3b - \frac{(a+3b)^2}{4a} + 1\,. \tag{4.44}$$

From (4.43), we obtain

$$l(x) = 2a\beta\left(x_1 - \frac{(a+3b)}{2a}x_2\right)^2 + 2c\beta x_3^2 + 2d\beta x_4^2 + \beta x_2^2 > 0.$$

Substituting (4.40) into (4.38), we get

$$\dot V = L_f V - \beta R^{-1}(x)(L_g V)^2\,. \tag{4.45}$$

Multiplying -2β on both sides of (4.45), and noting (4.42), we have

$$-2\beta\dot V = l(x) + \beta^2 R^{-1}(x)(L_g V)^2\,. \tag{4.46}$$

From (4.40) we have

$$u^{\mathrm{T}}R(x)u = \beta^2 R^{-1}(x)(L_g V)^2\,. \tag{4.47}$$

Substituting (4.47) into (4.46), we derive

$$l(x) + u^{\mathrm{T}}R(x)u = -2\beta\dot V\,. \tag{4.48}$$

Substituting (4.48) into (4.41), we have

$$J(u) = \lim_{t\to\infty}\left\{2\beta V(x(t)) + \int_0^t (-2\beta\dot V)\,\mathrm{d}\tau\right\} = 2\beta V(x(0))\,.$$

Therefore, the controller (4.40) is optimal with performance index (4.41). Substituting (4.44) into (4.40), we get the optimal controller (4.37). □

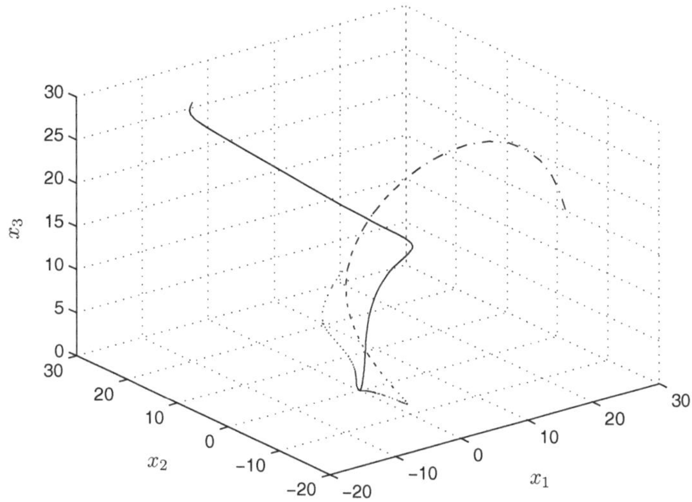

Fig. 4.20 Globally asymptotically stable system with $a = 30$, $b = 10$, $c = 37$, and $d = 10$

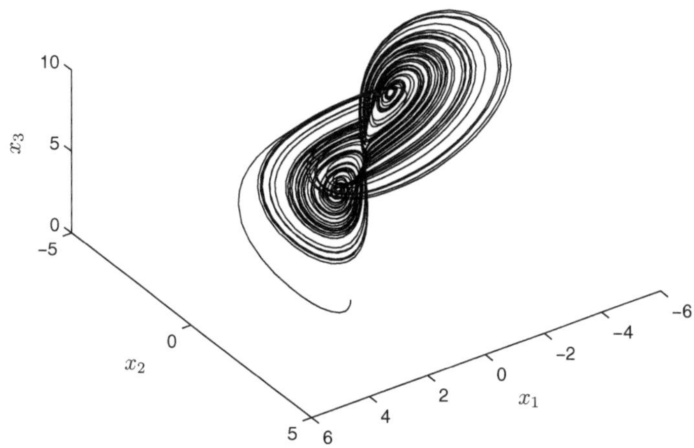

Fig. 4.21 The projection of the chaotic attractor of system (4.35) in the space of x_1–x_2–x_3 with $a = 35$, $b = 10$, $c = 1$, and $d = 10$

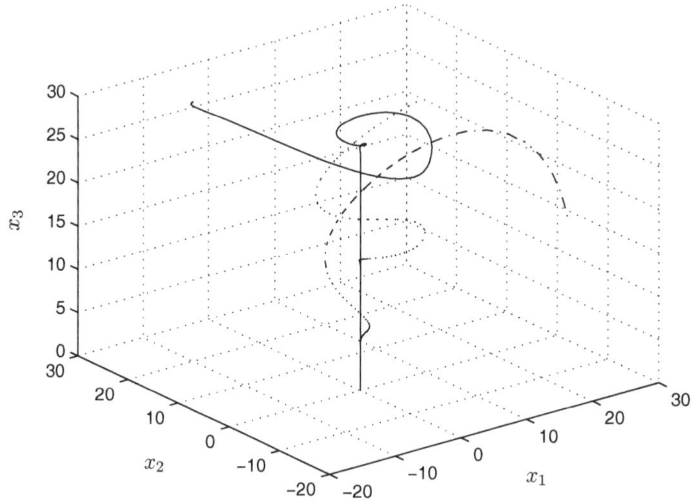

Fig. 4.22 Globally asymptotically stable system with $a = 35$, $b = 10$, $c = 1$, and $d = 10$

4.4.3 Simulations

The following illustrative example is used to demonstrate the effectiveness of the above method.

According to Theorem 4.3, we get $u = -121x_2$. As shown in Fig. 4.20, the chaotic trajectory of the system can be controlled rapidly to its zero equilibrium under different initial conditions. If the parameters of system (4.35) are chosen as $a = 35$, $b = 10$, $c = 1$, and $d = 10$, and the initial condition of the system (4.35) is selected as $x(0) = (3,3,3,3)^{\mathrm{T}}$, the chaotic behavior of the system (4.35) is shown in Fig. 4.21. According to Theorem 4.3, we get $u = -1699x_2/14$. As shown in Fig. 4.22, the chaotic trajectory of the system can be controlled rapidly to its zero equilibrium under different initial conditions, too. The above numerical simulation results show the effectiveness of the controller.

4.5 Summary

In this chapter, we investigated the problem of feedback control of discrete-time or continuous-time chaotic systems via different control methods. In Sect. 4.2, we developed a parametric adaptive controller to control discrete-time chaotic systems with multiple parameters. In Sect. 4.3, another parametric adaptive control method was developed to control continuous-time chaotic systems. The considered systems are both linearly parameterized and nonlinearly parameterized, respectively. In Sect. 4.4, we studied how to stabilize a chaotic system by inverse optimal control theory in detail. This method can avoid the difficulty caused by solving the HJB equation, and hence is very efficient in practice when the controller is demanded to be optimal in a certain sense.

References

1. Freeman RA, Kokotovic PV (1996) Inverse optimality in robust stabilization. J Control Optim 34:1365–1391
2. Guckenheimer J, Holmes P (1983) *Nonlinear Oscillations, Dynamical Systems and Bifurcations of vector Fields*. Spinger, New York
3. Khadra A, Liu XZ, Shen XM (2005) Impulsive control and synchronization of spatiotemporal chaos. Chaos Solitons Fractals 26:615–636
4. Poston T, Steward I (1978) *Catastrophe Theory and its Application*. Pitman, London
5. Qi GY, Du SZ, Chen GR, Chen ZQ, Yuan ZZ (2005) On a four-dimensional chaotic system. Chaos Solitons Fractals 23:1671–1682
6. Rafikov M, Balthazar JM (2004) On an optimal control design for Rössler system. Phys Lett A 333:241–245
7. Sinha S, Ramaswamy R, Rao JS (1990) Adaptive control in nonlinear dynamics. Physica D 43:118–128
8. Tanaka K, Ikeda T, Wang HO (1998) A unified approach to controlling chaos via an LMI-based fuzzy control system design. IEEE Trans Circuits Syst I 45:1021–1040
9. Tian YP (2005) Delayed feedback control of chaos in a switched arrival system. Phys Lett A 339:446–454
10. Vassiliadis D (1994) Parametric adaptive control and parameter identification of low-dimensional chaotic systems. Physica D 71:319–341
11. Wang J, Wang X (1998) Parametric adaptive control in nonlinear dynamical systems. Int J Bifurc Chaos 8:2215–2223
12. Xie YH (2007) Researches on control and synchronization of several kinds of nonlinear chaotic systems with time delay. Doctoral dissertation, Northeastern University, Shenyang
13. Yassen MT (2005) Controlling chaos and synchronization for new chaotic system using linear feedback control. Chaos Solitons Fractals 26:913–920
14. Yau HT, Chen CL (2007) Chaos control of Lorenz systems using adaptive controller with input saturation. Chaos Solitons Fractals 34:1567–1574

Chapter 5
Synchronizing Chaotic Systems Based on Feedback Control

Abstract The goal of this chapter is to present some methods that can be used to synchronize a rather wide class of chaotic systems. Considering that a chaotic system in nature is a nonlinear system, the theories and methods developed for controlling nonlinear systems could be utilized for synchronizing chaotic systems. The content of this chapter is about the synchronization of chaotic systems which can be described by ordinary differential equations and ordinary difference equations. The method is mainly based on the technology of feedback linearization. In this chapter we first develop synchronization methods for continuous-time chaotic systems, and after that we study the problem of synchronizing discrete-time chaotic systems. We also study the situation of synchronizing two different chaotic systems with adaptive and nonadaptive controllers.

5.1 Introduction

Although many methods have been proposed [1, 3] for synchronizing two or more chaotic systems since the landmark work by Pecora and Carroll [2], a systematic framework under which the synchronizing controller can be designed is still not constructed. It is usually the case that one method is only applicable to some specified chaotic systems. Our goal of this chapter is to present some methods that work for a rather wide class of chaotic systems. Considering that a chaotic system in nature is a nonlinear system, the theories and methods developed for controlling nonlinear systems could be utilized for the synchronization of chaotic systems. The content of this chapter is about the synchronization of chaotic systems which can be described by ordinary differential equations and ordinary difference equations. The method is based on the technology of feedback linearization. In this chapter we first develop synchronization methods for continuous-time chaotic systems, and after that we study the problem of synchronizing discrete-time chaotic systems.

5.2 Synchronization of Continuous-Time Chaotic Systems with a Single Input

5.2.1 Basic Formulation

Consider the following single-input single-output (SISO) nonlinear system:

$$\begin{cases} \dot{x} = f(x) + g(x)u, \\ y = h(x), \end{cases} \tag{5.1}$$

where $x \in \mathbb{R}^n$ represents the state variables, $f, g\colon \mathbb{R}^n \mapsto \mathbb{R}^n$ are vector fields, $u \in \mathbb{R}$ is the input, $y \in \mathbb{R}$ is the output, and $h\colon \mathbb{R}^n \mapsto \mathbb{R}$ is the output function. Denote $B^\rho(x_0) = \{x \in \mathbb{R}^n\colon \|x - x_0\| < \rho,\ \rho > 0\}$, the neighborhood of x_0 in $\mathbb{R}^n$. We need the following definition.

Definition 5.1. System (5.1) is said to have relative degree r if

(i) $L_g L_f^k h(x) = 0$, for $0 \le k < r-1$ and for all $x \in B^\rho(x_0)$;

(ii) $L_g L_f^{r-1} h(x) \neq 0$.

Here, $L_f h(x) := \dfrac{\partial h(x)}{\partial x} f(x)$, $L_f^k h(x) := L_f L_f^{k-1} h(x) = \dfrac{\partial L_f^{k-1} h(x)}{\partial x} f(x)$, $L_g L_f^k h(x) := \dfrac{\partial L_f^k h(x)}{\partial x} g(x)$, and $L_f^0 h(x) := h(x)$. □

Theorem 5.1 ([9]). If system (5.1) has relative degree r at x_0, then $r \le n$. Set

$$\begin{cases} \phi_1(x) = h(x), \\ \phi_2(x) = L_f h(x), \\ \vdots \\ \phi_r(x) = L_f^{r-1} h(x). \end{cases} \tag{5.2}$$

If $r < n$, it is always possible to find $n - r$ additional functions $\phi_{r+1}(x), \ldots, \phi_n(x)$ such that the mapping

$$\Phi(x) = \begin{pmatrix} \phi_1(x) \\ \vdots \\ \phi_n(x) \end{pmatrix}$$

has a Jacobian matrix which is nonsingular at x_0 and therefore qualifies as a local coordinate transformation in a neighborhood of x_0. The value of these additional functions at x_0 can be fixed arbitrarily. Moreover, it is always possible to choose $\phi_{r+1}(x), \ldots, \phi_n(x)$ in such a way that

$$L_g \phi_i(x) = 0 \ \text{ for all } r+1 \le i \le n \text{ and all } x \text{ around } x_0.$$ □

Theorem 5.2 ([9]). If system (5.1) has relative degree r at x_0, then by transformation $z = \Phi(x)$ system (5.1) can be transformed to

$$\begin{cases} \dot{z}_1 = z_2, \\ \dot{z}_2 = z_3, \\ \quad \vdots \\ \dot{z}_{r-1} = z_r, \\ \dot{z}_r = b(z) + a(z)u, \\ \dot{z}_{r+1} = q_{r+1}(z), \\ \quad \vdots \\ \dot{z}_n = q_n(z), \\ y = z_1, \end{cases} \tag{5.3}$$

where $a(z) = L_g L_f^{r-1} h(x)$, $b(z) = L_f^r h(x)$, and $q_i(z) = L_f \phi_i(\Phi^{-1}(z))$ for all $r+1 \leq i \leq n$. □

System (5.3) can be written in a compact form:

$$\begin{cases} \dot{z}^1 = \Lambda z^1 + E(b(z) + a(z)u), \\ \dot{z}^2 = Q(z^1, z^2), \\ y = z_1, \end{cases} \tag{5.4}$$

where $z^1 = (z_1, z_2, \ldots, z_r)^{\mathrm{T}}$, $z^2 = (z_{r+1}, z_{r+2}, \ldots, z_n)^{\mathrm{T}}$,

$$\Lambda = \begin{pmatrix} 0 & 1 & 0 & \cdots & 0 \\ 0 & 0 & 1 & \cdots & 0 \\ \vdots & \vdots & \vdots & \ddots & \vdots \\ 0 & 0 & 0 & \cdots & 1 \\ 0 & 0 & 0 & \cdots & 0 \end{pmatrix}$$

is an $r \times r$ matrix, $E = (0, 0, \ldots, 0, 1)^{\mathrm{T}}_{r \times 1}$, and $Q(z^1, z^2) = (q_{r+1}(z), \ldots, q_n(z))^{\mathrm{T}}$. When $z^1 = 0$, the second equation of (5.4) becomes

$$\dot{z}^2 = Q(0, z^2). \tag{5.5}$$

System (5.5) is called the zero dynamics which reflects the internal stability of system (5.4).

Remark 5.1. If $r = n$, then system (5.3) becomes

$$\begin{cases} \dot{z}_1 = z_2, \\ \dot{z}_2 = z_3, \\ \quad \vdots \\ \dot{z}_{n-1} = z_n, \\ \dot{z}_n = b(z) + a(z)u, \\ y = z_1. \end{cases} \tag{5.6}$$

If $a(z) \neq 0$ for any $\Phi^{-1}(z) \in B^{\rho}(x_0)$, then by setting

$$u = \frac{1}{a(z)}(-b(z) + v),$$

the nonlinear control system (5.6) becomes a linear control system given by

$$\begin{cases} \dot{z}_1 = z_2, \\ \dot{z}_2 = z_3, \\ \quad \vdots \\ \dot{z}_{n-1} = z_n, \\ \dot{z}_n = v, \\ y = z_1. \end{cases}$$

□

5.2.2 Main Results

Suppose that we have the following chaotic system:

$$\begin{cases} \dot{x} = f(x), \\ y = h(x). \end{cases} \tag{5.7}$$

We construct the following controlled response system:

$$\begin{cases} \dot{\hat{x}} = f(\hat{x}) + g(\hat{x})u, \\ \hat{y} = h(\hat{x}), \end{cases} \tag{5.8}$$

such that system (5.8) has relative degree r $(r \leq n)$. Set $\hat{z} = \Phi(\hat{x})$, where Φ is defined as (5.2). Thus, system (5.8) can be transformed to

$$\begin{cases} \dot{\hat{z}}^1 = \Lambda \hat{z}^1 + E(b(\hat{z}) + a(\hat{z})u), \\ \dot{\hat{z}}^2 = Q(\hat{z}^1, \hat{z}^2), \\ \hat{y} = \hat{z}_1. \end{cases}$$

Similarly, by $z = \Phi(x)$ system (5.7) is transformed to

$$\begin{cases} \dot{z}^1 = \Lambda z^1 + Eb(z), \\ \dot{z}^2 = Q(z^1, z^2), \\ y = z_1. \end{cases}$$

Defining $e = (e^1, e^2)^{\mathrm{T}} = (\hat{z}^1 - z^1, \hat{z}^2 - z^2)^{\mathrm{T}}$, we can obtain the following error system:

$$\begin{cases} \dot{e}^1 = \Lambda e^1 + E(b(\hat{z}^1, \hat{z}^2) + a(\hat{z}^1, \hat{z}^2)u - b(z^1, z^2)), \\ \dot{e}^2 = Q(\hat{z}^1, \hat{z}^2) - Q(z^1, z^2). \end{cases} \tag{5.9}$$

Theorem 5.3 ([15]). Denote $C = (c_0, c_1, \ldots, c_{r-1})$, where $c_0, \ldots, c_{r-1}$ are coefficients of the Hurwitz polynomial $P(s) = s^r + c_{r-1}s^{r-1} + \cdots + c_1 s + c_0$. If $a(\hat{z}) \neq 0$ for all $z \in B^{\rho}(\Phi(x_0))$ and the zero dynamics of system (5.8) is asymptotically stable, then with the controller

$$u = -\frac{1}{a(\hat{z}^1, \hat{z}^2)}(b(\hat{z}^1, \hat{z}^2) - b(z^1, z^2) + Ce) \tag{5.10}$$

the origin of system (5.9) is asymptotically stable.

Proof. Since Φ is nonsingular and the zero dynamics of system (5.9) is asymptotically stable, z and $\hat{z}$ are both bounded. Therefore, the output signal of the controller is also bounded. Substituting (5.10) into the first equation of (5.9), we have

$$\dot{e}^1 = \tilde{\Lambda} e^1, \tag{5.11}$$

where $\tilde{\Lambda} = \begin{pmatrix} 0 & 1 & 0 & \cdots & 0 \\ 0 & 0 & 1 & \cdots & 0 \\ \vdots & \vdots & \vdots & \ddots & \vdots \\ 0 & 0 & \cdots & 0 & 1 \\ -c_0 & -c_1 & -c_2 & \cdots & -c_{r-1} \end{pmatrix}$. By the selection of C we know that all eigenvalues of $\tilde{\Lambda}$ have real parts which are less than 0, and therefore the origin of (5.11) is asymptotically stable. The proof is completed. □

Remark 5.2. The above theorem guarantees that outputs of system (5.7) and system (5.8) can be synchronized asymptotically, i.e., $\lim_{t\to\infty} |y - \hat{y}| = 0$. □

5.2.3 Simulations

We take both the Rössler system and the Lorenz system as examples to show how the method developed above is used.

5.2.3.1 Synchronization of the Rössler System

The state equations of the Rössler system are as follows:

$$\begin{cases} \dot{x} = -y - z, \\ \dot{y} = x + ay, \\ \dot{z} = b + z(x - c). \end{cases} \tag{5.12}$$

When $a = 0.2$, $b = 0.2$, and $c = 5.7$, this system has a chaotic attractor, which is shown in Fig. 5.1. Taking system (5.12) as the drive system, we construct the following response system:

$$\begin{cases} \dot{\hat{x}} = -\hat{y} - \hat{z}, \\ \dot{\hat{y}} = \hat{x} + a\hat{y}, \\ \dot{\hat{z}} = b + \hat{z}(\hat{x} - c) + u, \end{cases} \tag{5.13}$$

where u is the controller and $g(\hat{x}) = (0,0,1)^{\mathrm{T}}$. Choosing the second state of system (5.12) as the output, we can easily verify that the system (5.13) has relative degree 3. The transformation can be chosen as $(\hat{y},\ \hat{x} + a\hat{y},\ a\hat{x} + (a^2 - 1)\hat{y} - \hat{z})^{\mathrm{T}}$. Select $C =$

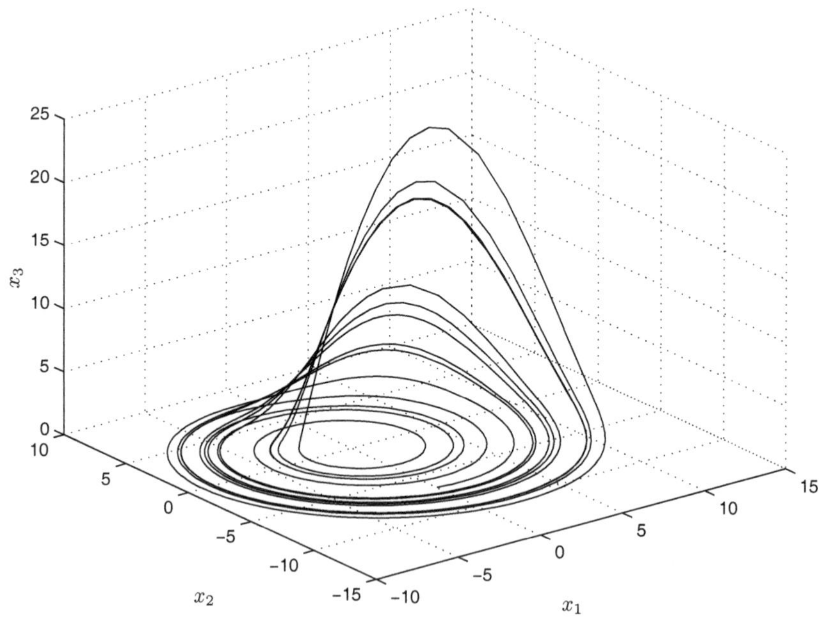

Fig. 5.1 The attractor of Rössler system

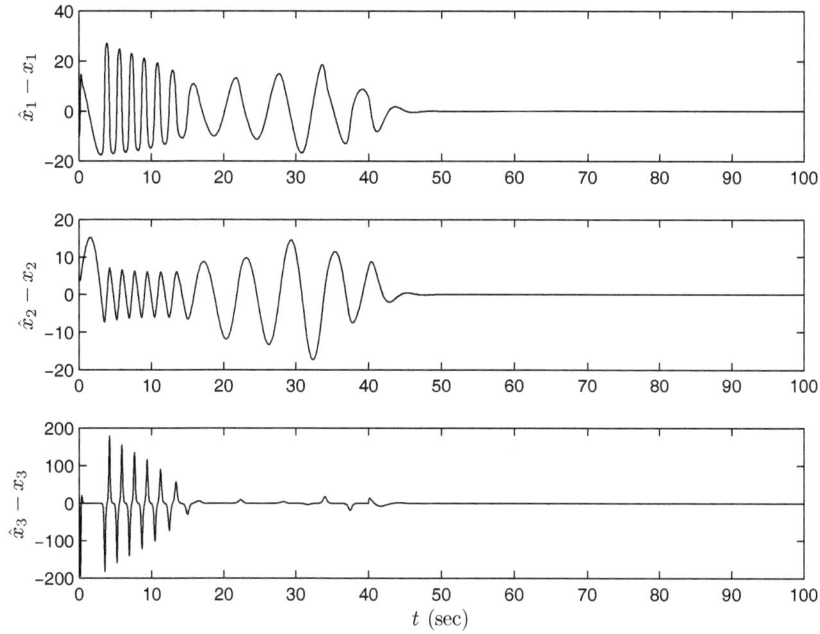

Fig. 5.2 The error curves of system (5.12) and system (5.13)

$(27,27,9)$. Initial values of system (5.12) and system (5.13) are set as $(30,10,10)^{\mathrm{T}}$ and $(20,12,10)^{\mathrm{T}}$, respectively. When $t = 40$ sec the controller is turned on. Fig. 5.2 presents the error curves of the synchronization between systems (5.12) and (5.13). Since the coordinate transformation is linear, all three state variables of the system (5.13) will converge to those of the system (5.12).

5.2.3.2 Synchronization of the Lorenz System

The Lorenz system is described by the following differential equations:

$$\begin{cases} \dot{x} = \sigma(y-x), \\ \dot{y} = \rho x - y - xz, \\ \dot{z} = xy - \beta z. \end{cases} \tag{5.14}$$

When $\sigma = 10$, $\beta = 8/3$, and $\rho = 28$, the Lorenz system has a chaotic attractor, which is the well-known butterfly attractor and is shown in Fig. 5.3.

We take system (5.14) as the drive system and construct the controlled response system as follows:

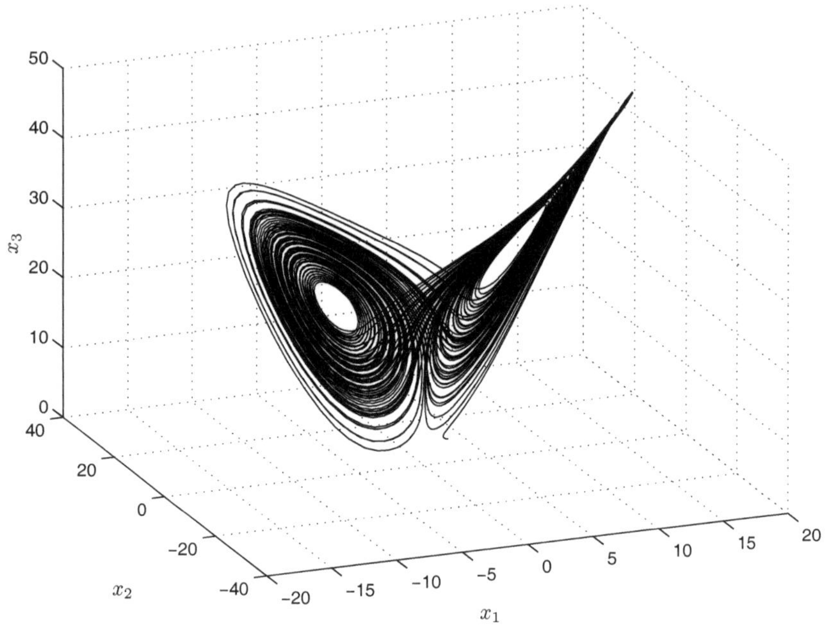

Fig. 5.3 The attractor of Lorenz system

$$\begin{cases} \dot{\hat{x}} = \sigma(\hat{y} - \hat{x}), \\ \dot{\hat{y}} = \rho\hat{x} - \hat{y} - \hat{x}\hat{z} + u, \\ \dot{\hat{z}} = \hat{x}\hat{y} - \beta\hat{z}. \end{cases} \tag{5.15}$$

Here, the output is chosen as $h(\hat{x}) = \hat{x}$ and $g = [0,1,0]^{\mathrm{T}}$. By simple calculation we have $L_g h(\hat{x}) = 0$, $L_f h(\hat{x}) = \sigma(\hat{y} - \hat{x})$, and $L_g L_f h(\hat{x}) = \sigma \neq 0$. Therefore, the relative degree of system (5.15) is 2. Choosing the coordinate transformation as

$$\underline{\hat{z}} = (\hat{z}_1, \hat{z}_2, \hat{z}_3)^{\mathrm{T}} = (h(\hat{x}), L_f h(\hat{x}), \hat{z})^{\mathrm{T}}, \tag{5.16}$$

we can transform system (5.15) into the following form:

$$\begin{cases} \dot{\hat{z}}_1 = \hat{z}_2, \\ \dot{\hat{z}}_2 = L_f^2 h(\underline{\hat{z}}) + L_g L_f h(\underline{\hat{z}}) u, \\ \dot{\hat{z}}_3 = \dfrac{1}{\sigma}(\sigma\hat{z}_1^2 + \hat{z}_1\hat{z}_2) - \beta\hat{z}_3. \end{cases}$$

Similarly, system (5.14) can be changed to

$$\begin{cases} \dot{z}_1 = z_2, \\ \dot{z}_2 = L_f^2 h(\underline{z}), \\ \dot{z}_3 = \dfrac{1}{\sigma}(\sigma z_1^2 + z_1 z_2) - \beta z_3. \end{cases}$$

Here, $L_f^2 h(\underline{z}) = (\sigma^2 + \sigma\rho)x - (\sigma^2 + \sigma)y - \sigma xz$, $L_g L_f h(\underline{z}) = -\sigma$, and $\underline{z} = (z_1, z_2, z_3)^T$. Set $e = \hat{\underline{z}} - \underline{z}$. The error system can be obtained as

$$\begin{cases} \dot{e}_1 = e_2, \\ \dot{e}_2 = L_f^2 h(\hat{\underline{z}}) - L_f^2 h(\underline{z}) + L_g L_f h(\hat{\underline{z}})u, \\ \dot{e}_3 = e_1(\hat{z}_1 + z_1) + \dfrac{1}{\sigma}(\hat{z}_2 e_1 + z_1 e_2) - \beta e_3. \end{cases} \tag{5.17}$$

The zero dynamics equation of system (5.17) is

$$\dot{e}_3 = -\rho e_3,$$

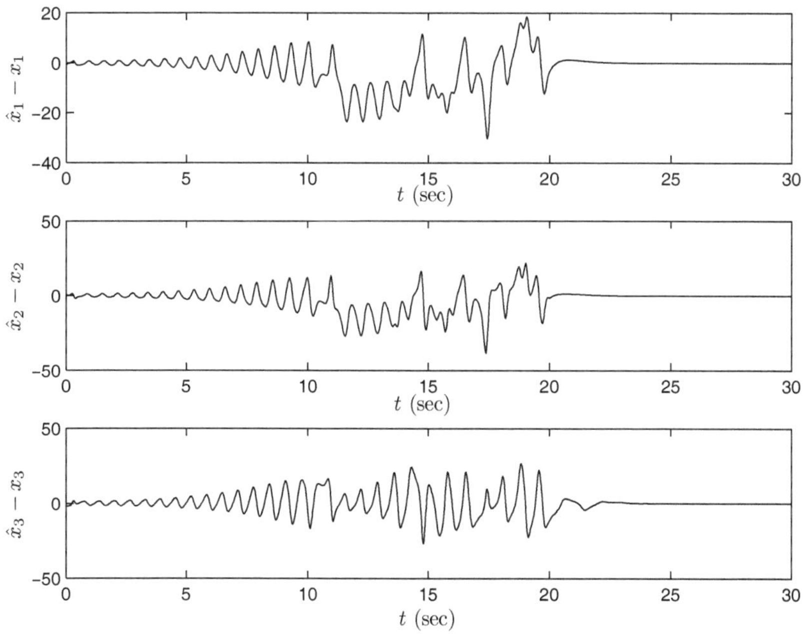

Fig. 5.4 The error curves of system (5.14) and system (5.15)

which is asymptotically stable. Select $C = (4,4)$. Under the action of controller $u = -\frac{1}{L_g L_f h(\hat{\underline{z}})}(L_f^2 h(\hat{\underline{z}}) - L_f^2 h(\underline{z}) + Ce)$ the origin of system (5.17) is asymptotically stable. Furthermore, for any $\hat{x} \in \mathbb{R}^3$ it is always true that $L_g L_f h(\hat{x}) \neq 0$, and the origin $e = 0$ is also globally and asymptotically stable. Since the controller is added on the second equation of (5.15), if $\hat{x}$ tends to x it implies that $\hat{y}$ tends to y. Recall that $e_3 = \hat{z} - z$. The three state variables of system (5.15) will be synchronized to those of system (5.14). Fig. 5.4 validates the above analysis.

5.3 Synchronization of Multi-Signals in Continuous-Time Chaotic Systems

The method proposed in the last section can only synchronize a single signal between the drive system and the response system. In this section we study how to synchronize multi-signals simultaneously, which is especially useful in secret communications when channel resources are not enough.

5.3.1 Basic Formulation

To synchronize multi-output signals we first generalize the notion of relative degree so that the multi-input multi-output (MIMO) nonlinear control theory can be used to design the controller.

Definition 5.2. Suppose that f_1 and f_2 are smooth vector fields of $\mathbb{R}^n$. The Lie bracket of f_1 and f_2 is defined as

$$[f_1, f_2](x) = \frac{\partial f_2}{\partial x} f_1(x) - \frac{\partial f_1}{\partial x} f_2(x),$$

where x is in the domain of f_1 and f_2. □

Definition 5.3. Given k smooth vector fields $f_1, \ldots, f_k$ defined in $U \in \mathbb{R}^n$, the notion

$$\Delta = \text{span}\{f_1, \ldots, f_k\}$$

is called the distribution spanned by $f_1, \ldots, f_k$. For any $x \in U$,

$$\Delta(x) = \text{span}\{f_1(x), \ldots, f_k(x)\}. \quad □$$

Remark 5.3. For each $x \in U$, $\Delta(x)$ is actually a vector space in $\mathbb{R}^n$. □

Definition 5.4. A distribution Δ is involutive if the Lie bracket $[f_1, f_2]$ of any pair of vector fields f_1 and f_2 belonging to Δ is a vector field which belongs to Δ, i.e.,

$$f_1 \in \Delta, f_2 \in \Delta \Rightarrow [f_1, f_2] \in \Delta. \qquad \square$$

Consider the following MIMO system:

$$\begin{cases} \dot{x} = f(x) + g(x)u, \\ y = h(x), \end{cases} \tag{5.18}$$

where $f\colon \mathbb{R}^n \to \mathbb{R}^n$, $f \in C^\infty$, $u = (u_1, \ldots, u_m)^{\mathrm{T}}$, $y = (y_1, \ldots, y_m)^{\mathrm{T}}$, $h = (h_1, \ldots, h_m)^{\mathrm{T}}$, $g(x) = (g_1(x), \ldots, g_m(x))$, $g_i\colon \mathbb{R}^n \to \mathbb{R}^n$, $h_i\colon \mathbb{R}^n \to \mathbb{R}$, $h_i \in C^\infty$, and $u_i \in \mathbb{R}$, $i = 1, \ldots, m$, $1 \le m \le n$.

Definition 5.5. MIMO system (5.18) has a vector relative degree $\{r_1, \ldots, r_m\}$ at x_0 if

(i) $L_{g_j} L_f^k h_i(x) = 0$, $1 \le \forall i, j \le m$, $\forall k \le r_i - 1$, $\forall x \in B^\rho(x_0)$;
(ii) the $m \times m$ matrix

$$\begin{pmatrix} L_{g_1} L_f^{r_1-1} h_1(x) & \cdots & L_{g_m} L_f^{r_1-1} h_1(x) \\ L_{g_1} L_f^{r_2-1} h_2(x) & \cdots & L_{g_m} L_f^{r_2-1} h_2(x) \\ \vdots & \vdots & \vdots \\ L_{g_1} L_f^{r_m-1} h_m(x) & \cdots & L_{g_m} L_f^{r_m-1} h_m(x) \end{pmatrix}$$

is nonsingular at $x = x_0$. $\square$

Theorem 5.4 ([9]). Suppose that system (5.18) has a vector relative degree $\{r_1, \ldots, r_m\}$ at x_0. Then,

$$r_1 + \cdots + r_m \le n.$$

For $1 \le i \le m$, set

$$\begin{cases} \phi_1^i(x) = h_i(x), \\ \phi_2^i(x) = L_f h_i(x), \\ \vdots \\ \phi_{r_i}^i(x) = L_f^{r_i-1} h_i(x). \end{cases}$$

If $r = r_1 + \cdots + r_m$ is strictly less than n, it is always possible to find $n - r$ additional functions $\phi_{r+1}(x), \ldots, \phi_n(x)$ such that the mapping

$$\Phi(x) = [(\phi_1^1(x), \ldots, \phi_{r_1}^1, \ldots, \phi_1^m(x), \ldots, \phi_{r_m}^m(x), \phi_{r+1}(x), \ldots, \phi_n(x))^{\mathrm{T}} \tag{5.19}$$

has a Jacobian matrix which is nonsingular at x_0 and therefore qualifies as a local coordinate transformation in a neighborhood of x_0. Moreover, if the distribution

$$G = \mathrm{span}\{g_1, \ldots, g_m\}$$

is involutive near x_0, it is always possible to choose $\phi_{r+1}(x), \ldots, \phi_n(x)$ in such a way that

$$L_{g_j}\phi_i(x) = 0$$

for all $r+1 \le i \le n$, all $1 \le j \le m$, and all x around x_0. □

Remark 5.4. By transformation Φ system (5.18) is changed to

$$\begin{cases} \dot{\xi} = \Lambda\xi + E(A(\xi,\eta)u + B(\xi,\eta)), \\ \dot{\eta} = Q(\xi,\eta), \\ y = \xi, \end{cases} \tag{5.20}$$

where $\xi \in \mathbb{R}^r$, $\xi = (\xi^1, \ldots, \xi^m)^{\mathrm{T}}$, $\xi^i = (\xi_1^i, \ldots, \xi_{r_i}^i)^{\mathrm{T}} = (\phi_1^i, \ldots, \phi_{r_i}^i)^{\mathrm{T}}$, $\eta \in \mathbb{R}^{n-r}$, and $\eta = (\phi_{r+1}, \ldots, \phi_n)^{\mathrm{T}}$. Λ and E are block diagonal matrices, i.e.,

$$\Lambda = \mathrm{diag}\{\Lambda_i\} \;\text{ and }\; E = \mathrm{diag}\{E_i\},$$

where

$$\Lambda_i = \begin{pmatrix} 0 & 1 & 0 \cdots & 0 \\ 0 & 0 & 1 \cdots & 0 \\ \vdots & \vdots & \vdots \ddots & \vdots \\ 0 & 0 & 0 \cdots & 1 \\ 0 & 0 & 0 \cdots & 0 \end{pmatrix}_{r_i \times r_i}$$

and

$$E_i = (0, \ldots, 0, 1)_{r_i \times 1}^{\mathrm{T}}$$

for all $1 \le i \le m$,

$$A(\xi,\eta) = \begin{pmatrix} L_{g_1}L_f^{r_1-1}h_1(\xi,\eta) & \cdots & L_{g_m}L_f^{r_1-1}h_1(\xi,\eta) \\ L_{g_1}L_f^{r_2-1}h_2(\xi,\eta) & \cdots & L_{g_m}L_f^{r_2-1}h_2(\xi,\eta) \\ \vdots & \cdots & \vdots \\ L_{g_1}L_f^{r_m-1}h_m(\xi,\eta) & \cdots & L_{g_m}L_f^{r_m-1}h_m(\xi,\eta) \end{pmatrix}_{m \times m,}$$

and

$$B(\xi,\eta) = (L_f^{r_1}h_1(\xi,\eta), L_f^{r_2}h_2(\xi,\eta), \ldots, L_f^{r_m}h_m(\xi,\eta))^{\mathrm{T}}.$$

When $\xi = 0$, the second equation of (5.20) becomes

$$\dot{\eta} = Q(0,\eta)$$

and it is called the zero dynamics. □

Theorem 5.5. Suppose that system (5.18) has relative degree r and asymptotically stable zero dynamics. Then, with the controller

$$u = -\frac{1}{A^{-1}(\xi,\eta)}(B(\xi,\eta) + C\xi)$$

the origin of system (5.20) is asymptotically stable. Here, $C = \mathrm{diag}\{C_i\}$ and

$$C_i = (c_0^i, c_1^i, \ldots, c_{r_i-1}^i)$$

are the coefficients of the Hurwitz polynomial

$$P_i(s) = s^{r_i-1} + c_{r_i-1}^i s^{r_i-1} + c_{r_i-2}^i s^{r_i-2} + \cdots + c_0^i. \qquad \square$$

5.3.2 Main Results

Suppose that the drive chaotic system has the following form:

$$\begin{cases} \dot{x} = f(x), \\ y = h(x), \end{cases} \tag{5.21}$$

where f and h are smooth vector fields satisfying $f\colon \mathbb{R}^n \to \mathbb{R}^n$, $f(0) = 0$, and $h\colon \mathbb{R}^n \to \mathbb{R}^m$. We construct the controlled response system as

$$\begin{cases} \dot{\hat{x}} = f(\hat{x}) + g(\hat{x})u, \\ \hat{y} = h(\hat{x}). \end{cases} \tag{5.22}$$

If the system (5.22) has relative degree r, then by nonsingular coordinate transformation (5.19), system (5.22) is changed to

$$\begin{cases} \dot{\hat{\xi}} = \Lambda\hat{\xi} + E(A(\hat{\xi}, \hat{\eta})u + B(\hat{\xi}, \hat{\eta})), \\ \dot{\hat{\eta}} = Q(\hat{\xi}, \hat{\eta}), \\ \hat{y} = \hat{\xi}. \end{cases} \tag{5.23}$$

Similarly, system (5.21) is changed to

$$\begin{cases} \dot{\xi} = \Lambda\xi + EB(\xi, \eta), \\ \dot{\eta} = Q(\xi, \eta), \\ y = \xi, \end{cases}$$

where the definitions of Λ, E, $A(\cdot,\cdot)$, $B(\cdot,\cdot)$, and C are the same as those in Remark 5.4. Denote $e = \hat{\xi} - \xi$ and $\varepsilon = \hat{\eta} - \eta$. The error system can then be obtained as

$$\begin{cases} \dot{e} = \Lambda e + E(A(\hat{\xi}, \hat{\eta})u + B(\hat{\xi}, \hat{\eta}) - B(\xi, \eta)), \\ \dot{\varepsilon} = Q(\hat{\xi}, \hat{\eta}) - Q(\xi, \eta) := F(e, \varepsilon, \xi, \eta). \end{cases} \tag{5.24}$$

Theorem 5.6 ([15]). If MIMO system (5.22) has relative degree r in $U \subset \mathbb{R}^n$ as well as asymptotically stable zero dynamics, then the following controller:

$$u = -\frac{1}{A^{-1}(\hat{x})}(\overline{B}(\hat{x}) - \overline{B}(x) + Ce) \tag{5.25}$$

stabilizes the first equation of system (5.24) at $e = 0$, where $e = (e_1, e_2, \ldots, e_m)^{\mathrm{T}}$,

$$e_i = (h_i(\hat{x}) - h_i(x), L_f h_i(\hat{x}) - L_f h_i(x), \ldots, L_f^{r_i-1} h_i(\hat{x}) - L_f^{r_i-1} h_i(x))^{\mathrm{T}},$$

$\overline{B}(\hat{x}) = B(\Phi^{-1}(\hat{\xi}, \hat{\eta}))$, and $\overline{B}(x) = B(\Phi^{-1}(\xi, \eta))$

Proof. Since the zero dynamics of (5.22) is stable, $\hat{x}$ is bounded. Because (5.21) is a chaotic system, x is also bounded. By the smoothness of f and h we know that for each $0 \le k \le r$ and $1 \le i \le m$, $L_f^k h_i(x)$ is also bounded. Therefore, the controller (5.25) is realizable. Substituting the controller into the first equation of (5.24), we get

$$\dot{e} = \underline{\Lambda} e, \tag{5.26}$$

where $\underline{\Lambda} = \mathrm{diag}(\Lambda_i)$ and

$$\Lambda_i = \begin{pmatrix} 0 & 1 & 0 & \cdots & 0 \\ 0 & 0 & 1 & \cdots & 0 \\ \vdots & \vdots & \vdots & \ddots & \vdots \\ 0 & 0 & 0 & \cdots & 1 \\ -c_0^i & -c_1^i & -c_2^i & \cdots & -c_{r_i-1}^i \end{pmatrix}_{r_i \times r_i}.$$

Therefore, we have

$$\det(\lambda I - \underline{\Lambda}) = \prod_{i=1}^{m} \det(\lambda I_i - \Lambda_i).$$

From the selection of C we know that the real parts of all eigenvalues of $\underline{\Lambda}$ are less than 0. Thus, the proof is completed. □

5.3.3 *Simulations*

We take the Rössler hyperchaotic system [14] as an example to validate the effectiveness of the proposed method. The system equation is as follows:

$$\begin{cases} \dot{x}_1 = -(x_2 + x_3), \\ \dot{x}_2 = x_1 + 0.25x_2 + x_4, \\ \dot{x}_3 = 3.0 + x_1 x_3, \\ \dot{x}_4 = 0.05x_4 - 0.5x_3. \end{cases} \tag{5.27}$$

The projection into $\mathbb{R}^3$ of the chaotic attractor is presented in Fig. 5.5. This system has two positive Lyapunov exponents, i.e., $\lambda_1 = 0.16$ and $\lambda_2 = 0.03$, which means that the trajectory of the system has two divergent directions. This fact im-

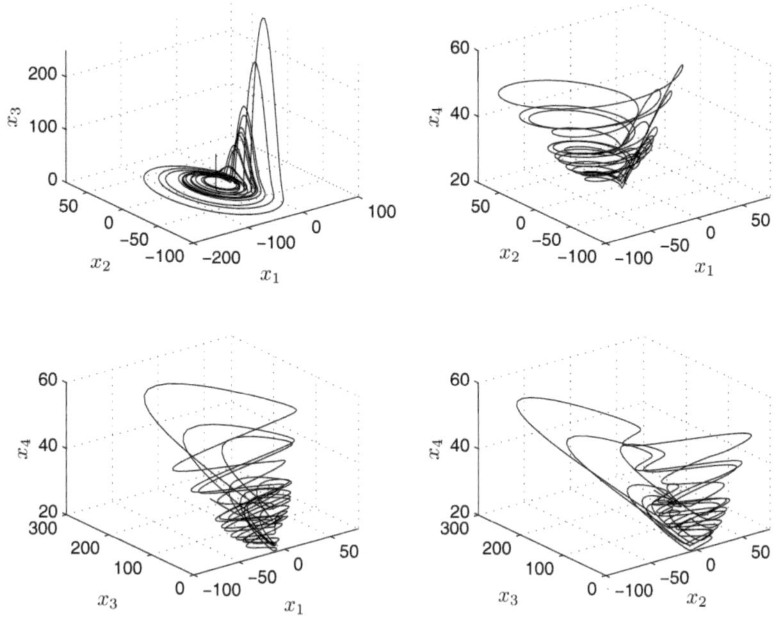

Fig. 5.5 The chaotic attractor of system (5.27)

plies that one cannot even stabilize the hyperchaotic system to the origin with only one scalar controller. In the following, we will design a vector controller to synchronize two Rössler hyperchaotic systems.

Taking system (5.27) as the drive system, we construct the following controlled response system:

$$\begin{cases} \dot{\hat{x}}_1 = -(\hat{x}_2 + \hat{x}_3), \\ \dot{\hat{x}}_2 = \hat{x}_1 + 0.25\hat{x}_2 + \hat{x}_4 + u_1, \\ \dot{\hat{x}}_3 = 3.0 + \hat{x}_1\hat{x}_3 + u_2, \\ \dot{\hat{x}}_4 = 0.05\hat{x}_4 - 0.5\hat{x}_3. \end{cases} \tag{5.28}$$

This means that $g_1 = (0,1,0,0)^{\mathrm{T}}$ and $g_2 = (0,0,1,0)^{\mathrm{T}}$. Select outputs as $h_1(\hat{x}) = \hat{x}_1$ and $h_2(\hat{x}) = \hat{x}_4$. By simple calculation, we have

$$L_{g_1}h_1(\hat{x}) = L_{g_1}h_2(\hat{x}) = L_{g_2}h_1(\hat{x}) = L_{g_2}h_2(\hat{x}) = 0$$

and

$$A(\hat{x}) = \begin{pmatrix} L_{g_1}L_f h_1(\hat{x}) & L_{g_2}L_f h_1(\hat{x}) \\ L_{g_1}L_f h_2(\hat{x}) & L_{g_2}L_f h_2(\hat{x}) \end{pmatrix} = \begin{pmatrix} -1 & -1 \\ 0 & -0.5 \end{pmatrix}, \tag{5.29}$$

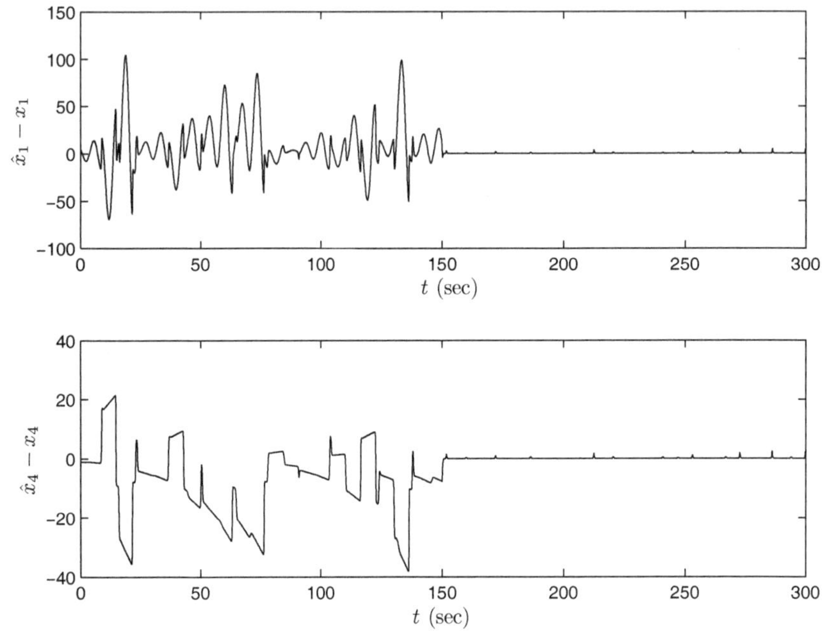

Fig. 5.6 The error curves of the synchronized Rössler hyperchaotic systems

which is nonsingular. Therefore, system (5.28) has relative degree $(2,2)$. Choose a coordinate transformation as

$$\Phi(\hat{x}) = (\hat{z}_1, \hat{z}_2, \hat{z}_3, \hat{z}_4)^{\mathrm{T}} = (h_1(\hat{x}), L_f h_1(\hat{x}), h_2(\hat{x}), L_f h_2(\hat{x}))^{\mathrm{T}}.$$

System (5.28) is changed to

$$\dot{\hat{z}} = \Lambda \hat{z} + E(B(\hat{z}) + A(\hat{z})u),$$

where

$$\Lambda = \begin{pmatrix} 0 & 1 & 0 & 0 \\ 0 & 0 & 1 & 0 \\ 0 & 0 & 0 & 1 \\ 0 & 0 & 0 & 0 \end{pmatrix}, \quad E = \begin{pmatrix} 0 & 0 \\ 1 & 0 \\ 0 & 0 \\ 0 & 1 \end{pmatrix}, \quad B(\hat{z}) = \begin{pmatrix} L_f^2 h_1(\hat{z}) \\ L_f^2 h_2(\hat{z}) \end{pmatrix}, \quad \text{and } u = \begin{pmatrix} u_1 \\ u_2 \end{pmatrix}.$$

$A(\hat{z})$ is equal to (5.29). Applying the same transformation to system (5.27), we get

$$\dot{z} = \Lambda z + EB(z).$$

Defining $e = \hat{z} - z$, we get the error system as

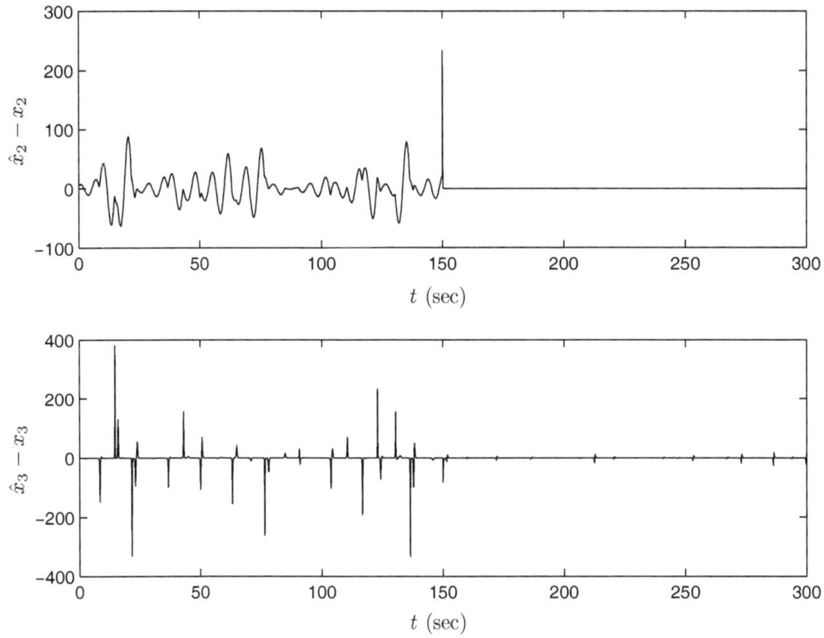

Fig. 5.7 The error curves of the synchronized Rössler hyperchaotic systems

$$\dot{e} = \Lambda e + E(B(\hat{z}) - B(z) + A(\hat{z})u).$$

Choose

$$C = \mathrm{diag}\{C_1, C_2\} = \begin{pmatrix} 625 & 50 & 0 & 0 \\ 0 & 0 & 625 & 50 \end{pmatrix}.$$

Initial values are set to

$$(x_1, x_2, x_3, x_4)^{\mathrm{T}} = (-50, -15, 70, 35)^{\mathrm{T}},$$

$$(\hat{x}_1, \hat{x}_2, \hat{x}_3, \hat{x}_4)^{\mathrm{T}} = (-45, -10, 68, 34)^{\mathrm{T}}.$$

Simulation results are shown in Figs. 5.6 and 5.7. When $t = 150$ sec, the controller is turned on. From the two figures we find that while the two controlled states go to synchronization with the respective states of system (5.27), so do the other two states. This fact can be explained as follows: since the controllers are added on both the second and the third equations of system (5.28), from the forth equation of the system we can see that if $\hat{x}_4$ tends to x_4 it must imply that $\hat{x}_3$ tends to x_3. With this in mind and from the first equation of system (5.28) we see that if $\hat{x}_1$ tends to x_1 then $\hat{x}_2$ must tend to x_2.

5.4 Synchronization of Different Continuous-Time Chaotic Systems

In the last section we assumed that the synchronized chaotic systems have the identical structure and the same parameters, but in practice the structures of the drive system and the response system may be different and effects of noise (such as parameters' shift and measurement errors, etc.) should not be ignored. So, in this section, we consider the problem of synchronizing chaotic systems with different structures and parameter perturbations.

5.4.1 Synchronization of Different Chaotic Systems with Parameter Perturbations

Suppose that the drive system has the following form:

$$\dot{x} = f(x) + F(x)\alpha, \tag{5.30}$$

where $x \in U_1 \subset \mathbb{R}^n$ is the state, $\alpha \in \mathbb{R}^m$ represents the known parameters, $f(x) \in \mathbb{R}^n$, and $F(x) \in \mathbb{R}^{n\times m}$. The controlled response system is as follows:

$$\dot{\tilde{x}} = g(\tilde{x}) + G(\tilde{x})\beta + u, \tag{5.31}$$

where $\tilde{x} \in U_2 \subset \mathbb{R}^n$ is the state, $\beta \in \mathbb{R}^q$ represents the known parameters, $g(\tilde{x}) \in \mathbb{R}^n$, $G(\tilde{x}) \in \mathbb{R}^{n\times q}$, and $u \in \mathbb{R}^n$ is the controller. Denote $e = \tilde{x} - x$. From (5.30) and (5.31) we get the error system

$$\dot{e} = g(\tilde{x}) + G(\tilde{x})\beta - f(x) - F(x)\alpha + u. \tag{5.32}$$

If $\lim_{t\to\infty} \|e\| = \lim_{t\to\infty} \|\tilde{x}(t,\tilde{x}_0) - x(t,x_0)\| = 0$ we say that system (5.30) and system (5.31) are synchronized. In the rest of this section we suppose that each element of $F(x)$ and $G(\tilde{x})$ is in L_∞. This is a reasonable assumption since the systems considered here are chaotic.

Theorem 5.7 ([17]). With the controller

$$u = f(x) + F(x)\tilde{\alpha} - g(\tilde{x}) - G(\tilde{x})\tilde{\beta} - ke \tag{5.33}$$

and parameter update laws

$$\begin{cases} \dot{\tilde{\alpha}} = -\delta F^{\mathrm{T}}(x)e, \\ \dot{\tilde{\beta}} = \gamma G^{\mathrm{T}}(\tilde{x})e, \end{cases} \tag{5.34}$$

system (5.30) and system (5.31) can be synchronized globally and asymptotically. Here, $k > 0$, $\delta > 0$, and $\gamma > 0$ are constants and $\tilde{\alpha}$ and $\tilde{\beta}$ are estimates of α and β.

Proof. From (5.32) and (5.33), we have

$$\dot{e} = F(x)(\tilde{\alpha} - \alpha) - G(\tilde{x})(\tilde{\beta} - \beta) - ke. \tag{5.35}$$

Let $\hat{\alpha} = \tilde{\alpha} - \alpha$ and $\hat{\beta} = \tilde{\beta} - \beta$. A Lyapunov function is chosen as

$$V(e, \hat{\alpha}, \hat{\beta}) = \frac{1}{2}\left[e^{\mathrm{T}} e + \frac{1}{\delta}\hat{\alpha}^{\mathrm{T}}\hat{\alpha} + \frac{1}{\gamma}\hat{\beta}^{\mathrm{T}}\hat{\beta}\right],$$

and its derivative along system (5.35) is

$$\begin{aligned}\dot{V}(e, \hat{\alpha}, \hat{\beta}) &= e^{\mathrm{T}}\dot{e} + \frac{1}{\delta}\hat{\alpha}^{\mathrm{T}}\dot{\hat{\alpha}} + \frac{1}{\gamma}\hat{\beta}^{\mathrm{T}}\dot{\hat{\beta}} \\ &= [F(x)\hat{\alpha} - G(\tilde{x})\hat{\beta} - ke]^{\mathrm{T}} e - \hat{\alpha}^{\mathrm{T}} F^{\mathrm{T}}(x) e + \hat{\beta}^{\mathrm{T}} G^{\mathrm{T}}(\tilde{x}) e \\ &= -ke^{\mathrm{T}} e \\ &\leq 0.\end{aligned}$$

Since $e, \hat{\alpha}, \hat{\beta} \in L_\infty$, from $\dot{V} \leq -ke^{\mathrm{T}} e$ we have

$$\int_0^t \|e\|^2 \mathrm{d}t = \frac{1}{k}(V(0) - V(t)) \leq \frac{V(0)}{k},$$

which means that $e \in L_2$. Because we have assumed that the elements of $F(x)$ and $G(\tilde{x})$ belong to L_∞, from (5.35) we have $\dot{e} \in L_\infty$. Therefore, according to the Barbalat lemma [10] we know that $\lim_{t\to\infty} e = 0$; that is, $\lim_{t\to\infty} \|e\| = \lim_{t\to\infty} \|\tilde{x}(t, \tilde{x}_0) - x(t, x_0)\| = 0$. The proof is completed. □

Remark 5.5. Many chaotic systems have the form of (5.30), such as Lorenz, Chen, Lü, Rössler, Chua, Van der Pol, and Duffing, etc. □

5.4.2 Simulations

In this section we will show the effectiveness of the presented method by simulation.

Suppose that the drive system is the Lorenz system

$$\begin{cases} \dot{x}_1 = \alpha_1(x_2 - x_1), \\ \dot{x}_2 = \alpha_2 x_1 - x_1 x_3 - x_2, \\ \dot{x}_3 = x_1 x_2 - \alpha x_3. \end{cases}$$

Take the controlled Lü system as the response system

$$\begin{cases} \dot{\tilde{x}}_1 = \beta_1(\tilde{x}_2 - \tilde{x}_1) + u_1, \\ \dot{\tilde{x}}_2 = -\tilde{x}_1\tilde{x}_3 + \beta_2\tilde{x}_2 + u_2, \\ \dot{\tilde{x}}_3 = \tilde{x}_1\tilde{x}_2 - \beta_3\tilde{x}_3 + u_3. \end{cases}$$

In this case $x = (x_1, x_2, x_3)^{\mathrm{T}}$, $\alpha = (\alpha_1, \alpha_2, \alpha_3)^{\mathrm{T}}$, $\tilde{x} = (\tilde{x}_1, \tilde{x}_2, \tilde{x}_3)^{\mathrm{T}}$, $\beta = (\beta_1, \beta_2, \beta_3)^{\mathrm{T}}$, $u = (u_1, u_2, u_3)^{\mathrm{T}}$,

$$f(x) = \begin{pmatrix} 0 \\ -x_1x_3 - x_2 \\ x_1x_2 \end{pmatrix}, \quad F(x) = \begin{pmatrix} x_2 - x_1 & 0 & 0 \\ 0 & x_1 & 0 \\ 0 & 0 & -x_3 \end{pmatrix},$$

$$g(\tilde{x}) = \begin{pmatrix} 0 \\ -\tilde{x}_1\tilde{x}_3 \\ \tilde{x}_1\tilde{x}_2 \end{pmatrix}, \quad \text{and} \quad G(\tilde{x}) = \begin{pmatrix} \tilde{x}_2 - \tilde{x}_1 & 0 & 0 \\ 0 & \tilde{x}_2 & 0 \\ 0 & 0 & -\tilde{x}_3 \end{pmatrix}.$$

The controller and parameter update laws are

$$\begin{cases} u = \begin{pmatrix} \tilde{\alpha}_1(x_2 - x_1) - \tilde{\beta}_1(\tilde{x}_2 - \tilde{x}_1) - ke_1 \\ \tilde{\alpha}_2x_1 - x_1x_3 - x_2 + \tilde{x}_1\tilde{x}_3 - \tilde{\beta}_2\tilde{x}_2 - ke_2 \\ x_1x_2 - \tilde{\alpha}_3x_3 - \tilde{x}_1\tilde{x}_2 + \tilde{\beta}_3\tilde{x}_3 - ke_3 \end{pmatrix}, \\ \dot{\tilde{\alpha}} = (\delta(x_1 - x_2)e_1, -\delta x_1e_2, \delta x_3e_3)^{\mathrm{T}}, \\ \dot{\tilde{\beta}} = (\gamma(\tilde{x}_2 - \tilde{x}_1)e_1, \gamma\tilde{x}_2e_2, -\gamma\tilde{x}_3e_3)^{\mathrm{T}}. \end{cases}$$

In the simulation, 'unknown' parameters are chosen as

$$\alpha = (10, 28, 8/3)^{\mathrm{T}} \quad \text{and} \quad \beta = (36, 20, 3)^{\mathrm{T}}.$$

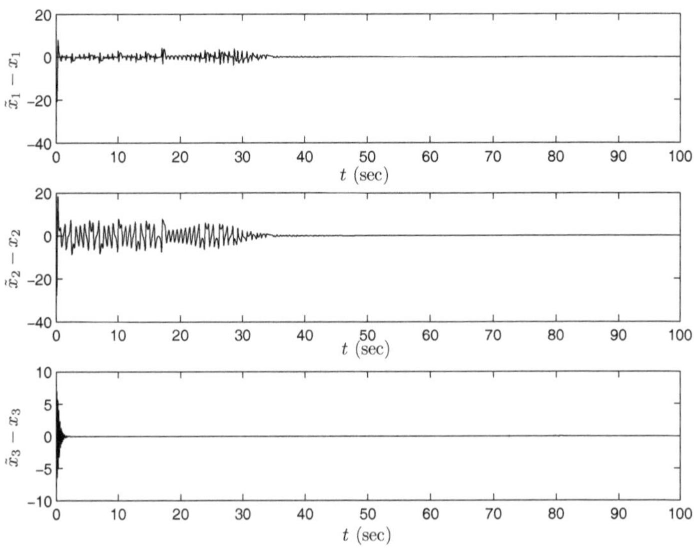

Fig. 5.8 The error curves of the synchronized Lorenz system and Lü system when $k = 1$ and $\delta = \gamma = 1$

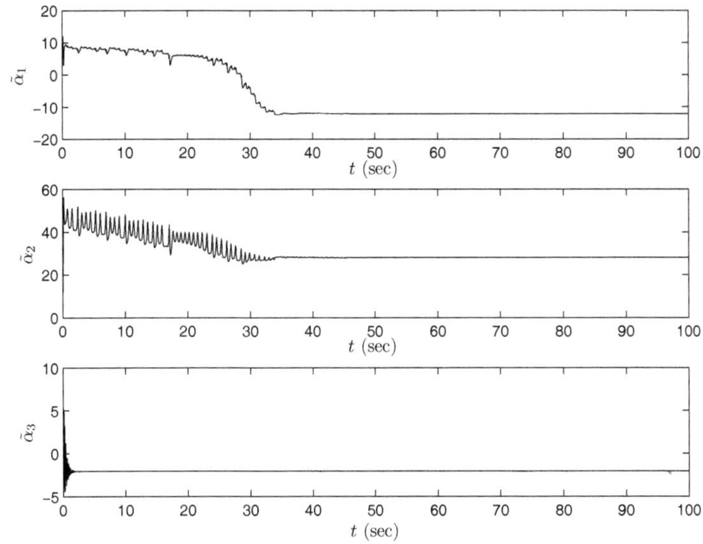

Fig. 5.9 The $\tilde{\alpha}$ curves when $k = 1$ and $\delta = \gamma = 1$

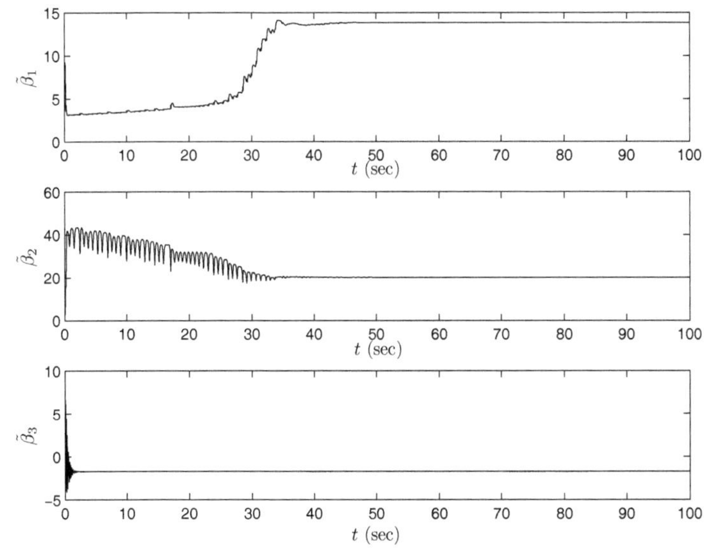

Fig. 5.10 The $\tilde{\beta}$ curves when $k = 1$ and $\delta = \gamma = 1$

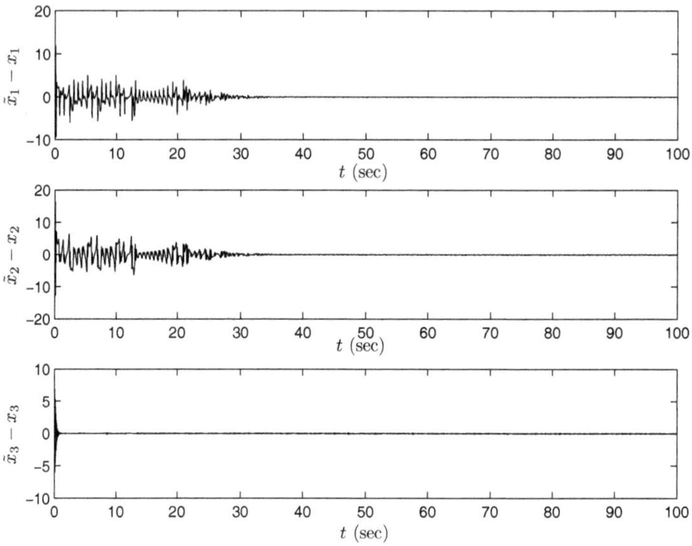

Fig. 5.11 The error curves of the synchronized Lorenz system and Lü system when $k = 1$ and $\delta = \gamma = 4$

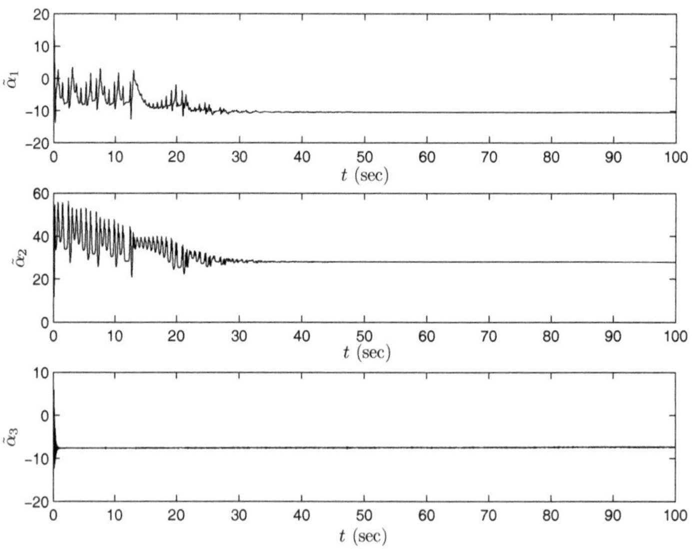

Fig. 5.12 The $\tilde{\alpha}$ curves when $k = 1$ and $\delta = \gamma = 4$

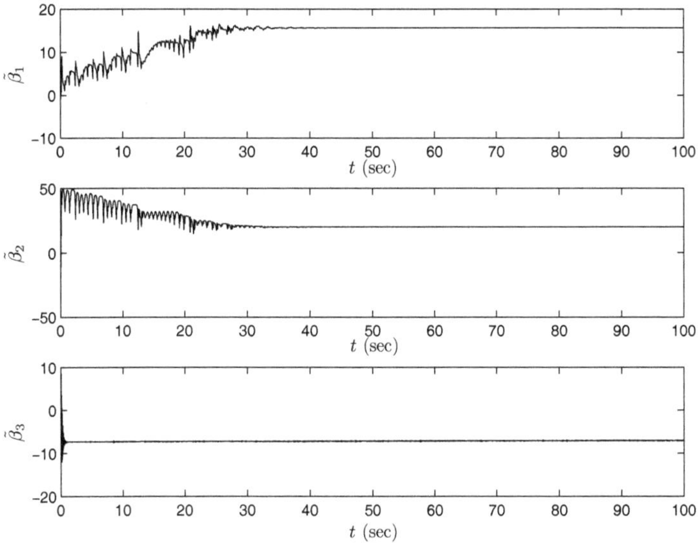

Fig. 5.13 The $\tilde{\beta}$ curves when $k = 1$ and $\delta = \gamma = 4$

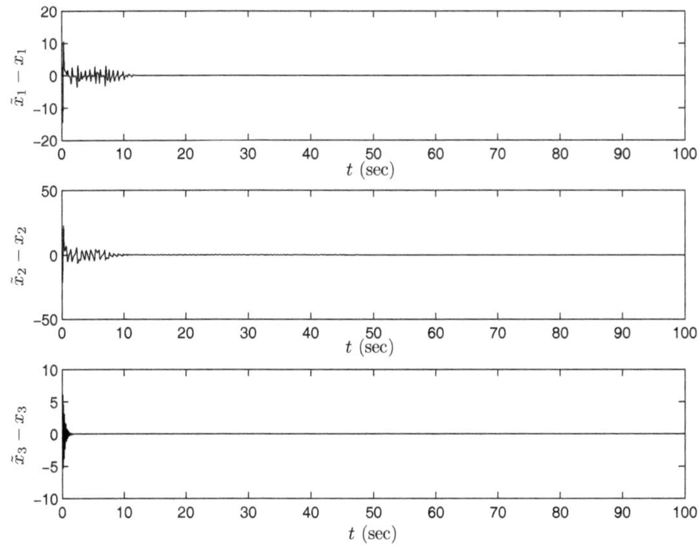

Fig. 5.14 The error curves of the synchronized Lorenz system and Lü system when $k = 3$ and $\delta = \gamma = 1$

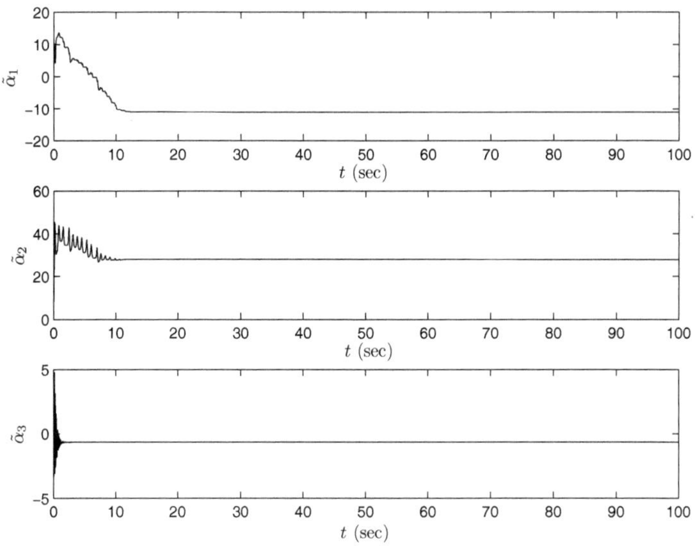

Fig. 5.15 The $\tilde{\alpha}$ curves when $k = 3$ and $\delta = \gamma = 1$

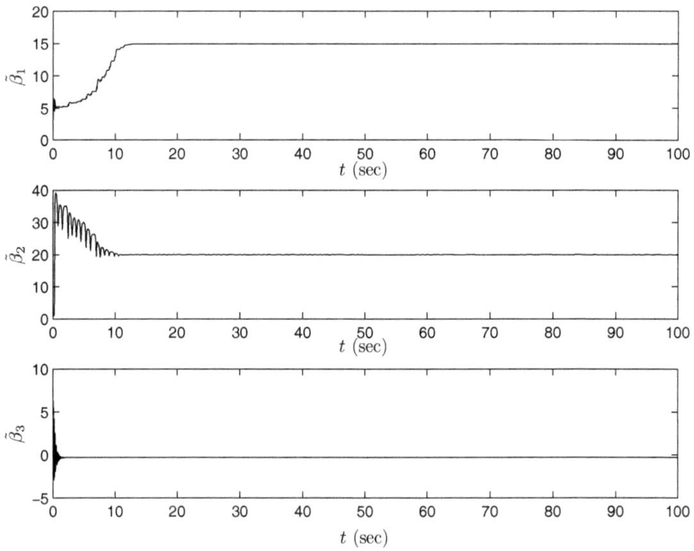

Fig. 5.16 The $\tilde{\beta}$ curves when $k = 3$ and $\delta = \gamma = 1$

The initial values of the systems' states are

$$x(0) = (10, 10, 10)^{\mathrm{T}} \text{ and } \tilde{x}(0) = (2, 2, 2)^{\mathrm{T}}.$$

Parameters' initial values are $\tilde{\alpha}(0) = \tilde{\beta}(0) = (3, 3, 3)^{\mathrm{T}}, k = 1$, and $\delta = \gamma = 1$. Simulation results are shown in Figs. 5.8–5.16. From Figs. 5.11–5.13, it is easy to see that increasing δ and γ does not shorten the transient time. On the contrary, the amplitude of the oscillation becomes bigger. However, from Figs. 5.14–5.16, we find that increasing k can shorten the transient time and decrease the amplitude of oscillation.

5.5 Synchronization of Discrete-Time Chaotic Systems

With the popularization of personal computers more and more people have realized the importance of data security. Compared to continuous-time chaotic systems, discrete-time chaotic systems are more suitable for digital computers in applications of secure communications. Two reasons can be used to explain why discrete-time chaotic systems have more advantages than continuous-time chaotic systems. First, chaos can be found in very simple discrete-time systems even if the dimension of the dynamical equation is only one. But, in a continuous-time autonomous system, the dimension of the phase space must be greater than two in order to see chaos. Second, a discrete-time chaotic system can be directly utilized using a computer program while a continuous-time chaotic system needs to be discretized before it is used. In the process of discretization one must pay careful attention to the chosen method since false discretization may cause deviation dynamics between the discretized system and the original continuous-time system.

Although there are already some works on the control and synchronization of discrete-time chaotic systems [4, 5, 6, 12], most of them assume that parameters of the drive and the response systems are identical. In this section we first propose a control method to synchronize two chaotic systems with the same parameters, after that an adaptive algorithm is designed to tune the controller's parameters when parameter disturbance exists. To show the effectiveness of the proposed methods, a computer simulation is provided in the last part of the section.

5.5.1 Basic Formulation

Consider the following SISO discrete-time nonlinear control systems:

$$\begin{cases} x(k+1) = f(x(k), u(k)), \\ y(k) = h(x(k)), \end{cases} \tag{5.36}$$

where $x \in \mathbb{R}^n$, $u \in \mathbb{R}$, $y \in \mathbb{R}$, $f\colon U \to U$ and $h\colon U \to L$ are smooth mappings satisfying $f(0,0) = 0$ and $h(0) = 0$, and U and L are subsets of $\mathbb{R}^n$ and $\mathbb{R}$, respectively.

Denote $h \circ f = h(f(x,u))$ and $f^s = \underbrace{f \circ f \circ \cdots \circ f}_{s \text{ times}}$, i.e.,

$$f^s = f(f(\cdots f(x,u_s)\cdots,u_2),u_1).$$

Definition 5.6. If there exists a minimal positive integer r such that for any $0 \le i \le r-1$ the following conditions are satisfied:

(i) $\dfrac{\partial(h \circ f^i)}{\partial u} = 0,$

(ii) $\dfrac{\partial(h \circ f^r)}{\partial u} \neq 0,$

then we say that system (5.36) has relative degree r. □

Remark 5.6. System (5.36) having relative degree r means that inputs at $k = 0$ do not affect the system's output until the instant $k = r$. □

Theorem 5.8 ([11]). If system (5.36) has relative degree $r = n$, then there is a coordinate transformation $z = \Phi(x)$ such that, in the new coordinates, system (5.36) has the following form:

$$\begin{cases} z(k+1) = Az(k) + Bv(k), \\ y(k) = Cz(k), \end{cases}$$

where

$$\begin{aligned} z &= (z_1,\ldots,z_n)^{\mathrm{T}} \\ &= (h(x), h \circ f(x,u), \ldots, h \circ f^{n-1}(x,u))^{\mathrm{T}}, \end{aligned} \tag{5.37}$$

$$v = h \circ f^n(x,u), \tag{5.38}$$

$$A = \begin{pmatrix} 0 & 1 & 0 & \cdots & 0 & 0 \\ 0 & 0 & 1 & \cdots & 0 & 0 \\ \vdots & \vdots & \vdots & \ddots & \vdots & \vdots \\ 0 & 0 & 0 & \cdots & 0 & 1 \\ 0 & 0 & 0 & \cdots & 0 & 0 \end{pmatrix}_{n\times n}, \tag{5.39}$$

$$\begin{aligned} B &= (0,\ldots,0,1)^{\mathrm{T}}_{n\times 1}, \\ C &= (1,0,\ldots,0)_{1\times n}. \end{aligned} \tag{5.40}$$

□

5.5.2 Main Results

5.5.2.1 Synchronization of Discrete-Time Chaotic Systems with the Same Parameters

Before stating the main results we need the following lemma.

Lemma 5.1. If system (5.36) has relative degree n and is of the following form:

$$\begin{cases} x(k+1) = f(x(k),u(k)) = F(x(k)) + gu(k), \\ y(k) = h(x(k)) = d^{\mathrm{T}}x(k), \end{cases} \tag{5.41}$$

then it can be transformed to

$$\begin{cases} z(k+1) = Az(k) + B(a(z(k)) + bu(k)), \\ y(k) = Cz(k), \end{cases} \tag{5.42}$$

where g, d are constant vectors, A, B, C are defined as those in (5.39) and (5.40), $a(\cdot)\colon \mathbb{R}^n \to \mathbb{R}$ is a continuous function, and b is a nonzero constant.

Proof. Define mappings $\psi = F(x)$ and $\phi_u = x + gu$. Then, $f(x,u) = \phi_u \circ \psi(x)$. Since system (5.41) has relative degree n, we have

$$\begin{aligned} \frac{\partial}{\partial u}(h \circ f^i) &= \frac{\partial}{\partial u}(h \circ (\phi_u \circ \psi)^i) = \frac{\partial h}{\partial (\phi_u \circ \psi)^i} \frac{\partial (\phi_u \circ \psi)^i}{\partial u} = 0,\ 0 \le i \le n-1, \\ \frac{\partial}{\partial u}(h \circ f^n) &= \frac{\partial}{\partial u}(h \circ (\phi_u \circ \psi)^n) = \frac{\partial h}{\partial (\phi_u \circ \psi)^n} \frac{\partial (\phi_u \circ \psi)^n}{\partial u} \neq 0, \end{aligned}$$

which means that $f^n = F^n + gu$. Noticing that $h(x) = d^{\mathrm{T}}x$, we have $h \circ f^n = h \circ F^n + h(gu) = d^{\mathrm{T}}F + d^{\mathrm{T}}gu$. Let $a(x) = d^{\mathrm{T}}F(x)$ and $b = d^{\mathrm{T}}g$. We have $h \circ f^n(x) = a(x) + bu$. Thus, the proof is completed. □

Suppose that the drive system is as follows:

$$\begin{cases} x(k+1) = f(x(k),0), \\ y(k) = h(x(k)), \end{cases} \tag{5.43}$$

where $x \in \mathbb{R}^n$, $y \in \mathbb{R}$ is the output, and $f\colon \mathbb{R}^n \times \mathbb{R} \to \mathbb{R}^n$ and $h\colon \mathbb{R}^n \to \mathbb{R}$ are smooth mappings satisfying $f(0,0) = 0$ and $h(0) = 0$. Construct the following controlled response system:

$$\begin{cases} \hat{x}(k+1) = f(\hat{x}(k),u), \\ \hat{y}(k) = h(\hat{x}(k)), \end{cases} \tag{5.44}$$

where u is a controller. If

$$\lim_{k\to\infty} \|\hat{y}(k) - y(k)\| = 0,$$

then system (5.43) and system (5.44) are said to be in synchronization. Before designing the controller, we assume that the following two conditions hold for system (5.44):

(i) system (5.44) has relative degree n in the neighborhood of zero $B^{\rho}(0)$;
(ii) $h \circ f^n(x,u) = a(x) + bu$.

For some mechanical systems [12], the first condition can be satisfied. At least for affine control systems the second condition can also be satisfied.

Considering the above assumptions and Lemma 5.1, we know that through the nonlinear coordinate transformation

$$\hat{z} = \Phi(\hat{x}) = (h(\hat{x}), h \circ f(\hat{x},u), \ldots, h \circ f^{n-1}(\hat{x},u))^{\mathrm{T}}$$

system (5.44) can be transformed into

$$\begin{cases} \hat{z}(k+1) = A\hat{z}(k) + B(a(\hat{z}(k)) + bu(k)), \\ \hat{y}(k) = C\hat{z}(k). \end{cases}$$

With the same transformation, system (5.43) becomes

$$\begin{cases} z(k+1) = Az(k) + Ba(z(k)), \\ y(k) = Cz(k). \end{cases}$$

Set $e = \hat{z} - z$. Then, we get the error system as

$$e(k+1) = Ae(k) + B(a(\hat{z}) - a(z) + bu). \tag{5.45}$$

Theorem 5.9 ([16]). With the controller

$$u = -b^{-1}(a(\hat{z}) - a(z) + Qe), \tag{5.46}$$

the origin of system (5.45) is asymptotically stable, where $Q = (q_0, q_1, \ldots, q_n)^{\mathrm{T}}$ are coefficients of a polynomial $P(s) = s^n + q_{n-1}s^{n-1} + \cdots + q_1 s + q_0$ such that the absolute values of all roots of $P(s)$ are less than 1.

Proof. Substituting controller (5.46) into system (5.45), we get

$$e(k+1) = \Lambda e(k),$$

where

$$\Lambda = \begin{pmatrix} 0 & 1 & 0 & \cdots & 0 & 0 \\ 0 & 0 & 1 & \cdots & 0 & 0 \\ \vdots & \vdots & \vdots & \ddots & \vdots & \vdots \\ 0 & 0 & 0 & \cdots & 0 & 1 \\ -q_0 & -q_1 & -q_2 & \cdots & -q_{n-2} & -q_{n-1} \end{pmatrix}.$$

By the selection of Q we know that all eigenvalues of Λ have absolute values less than 1. The proof is completed. □

5.5.2.2 Synchronization of Discrete-Time Chaotic Systems with Parameter Perturbation

In this section we will study how to synchronize two discrete-time chaotic systems whose parameters are not identical. Before stating the main results we need two lemmas.

For a general discrete-time system

$$y(k) = \phi^{\mathrm{T}}(k-1)\beta_0, \tag{5.47}$$

where $y(k) \in \mathbb{R}$ and $\phi(k-1)$ is a linear or nonlinear function of output $Y = (y(k-1),\ldots,y(k-p))^{\mathrm{T}}$ and input $U = (u(k-1),\ldots,u(k-m))^{\mathrm{T}}$. $\hat{y}(k)$, the estimate of $y(k)$, can be obtained by

$$\hat{y}(k) = \phi^{\mathrm{T}}(k-1)\beta(k), \tag{5.48}$$

where $\beta(k)$ is the estimate of β_0.

Lemma 5.2 ([7]). Calculate $\beta(k)$ according to the following algorithm:

$$\begin{cases} \beta(k) = \beta(k-1) + \dfrac{P(k-2)\phi(k-1)}{\phi^{\mathrm{T}}(k-1)P(k-2)\phi(k-1)}[y(k) - \phi^{\mathrm{T}}(k-1)\beta(k-1)], \\ \qquad\qquad \text{if } \phi^{\mathrm{T}}(k-1)P(k-2)\phi(k-1) \neq 0, \\ \beta(k) = \beta(k-1), \ \text{if } \phi^{\mathrm{T}}(k-1)P(k-2)\phi(k-1) = 0, \\ P(k-1) = P(k-2) - \dfrac{P(k-2)\phi(k-1)\phi^{\mathrm{T}}(k-1)P(k-2)}{\phi^{\mathrm{T}}(k-1)P(k-2)\phi(k-1)}, \\ \qquad\qquad \text{if } \phi^{\mathrm{T}}(k-1)P(k-2)\phi(k-1) \neq 0, \\ P(k-1) = P(k-2), \ \text{if } \phi^{\mathrm{T}}(k-1)P(k-2)\phi(k-1) = 0, \\ P(0) = I. \end{cases}$$

If the rank of $(\phi(1),\ldots,\phi(s))$ equals the dimension of β_0, then $\beta(k)$ will converge to β_0 within s steps. □

Setting $G \subset \mathbb{R}$, we state the discrete version of the LaSalle invariance principle.

Lemma 5.3 ([13]). For a discrete-time system $x(k+1) = f(x(k))$, where $f\colon \mathbb{R}^n \to \mathbb{R}^n$ is a smooth mapping, if the following two conditions are satisfied:

(i) there is a function $V \in C[G \subset \mathbb{R}^n, \mathbb{R}]$ such that along the solutions of the system, we have

$$\Delta V(x(k)) = V(x(k+1)) - V(x(k)) \leq 0;$$

(ii) there exists a bounded solution $x(k)$ such that $x(k) \subset G$ for all $k \geq 0$, then there is a constant c such that $x(k) \to M \cap V^{-1}(c)$ when $k \to \infty$, where M denotes the maximal invariance set in $E = \{x\colon \Delta V(x) = 0, x \in \overline{G}\}$, $V^{-1}(c) = \{x\colon V(x) = c, x \in \mathbb{R}^m\}$, and $\overline{G}$ is the closure of G. □

In the following we will design an adaptive controller to synchronize two discrete-time chaotic systems when parameter perturbation exists. Suppose that the drive system has the following form:

$$\begin{cases} x(k+1) = f^{\mathrm{T}}(x(k))\theta_0, \\ y(k) = Dx(k), \end{cases} \tag{5.49}$$

where $x \in \mathbb{R}^n$, $f\colon \mathbb{R}^n \to \mathbb{R}^{q \times n}$ is a smooth mapping, $D \in \mathbb{R}^{1 \times n}$, and $\theta_0 \in \mathbb{R}^q$ is the unknown parameter matrix. We construct the following controlled response system:

$$\begin{cases} \hat{x}(k+1) = f^{\mathrm{T}}(\hat{x}(k))\theta(k) + gu, \\ \hat{y}(k) = D\hat{x}(k), \end{cases} \tag{5.50}$$

where $g \in \mathbb{R}^n$, $u \in \mathbb{R}$, and $\theta(k)$ is the estimate of θ_0. Suppose that system (5.50) has relative degree n in $U \subset \mathbb{R}^n$. Then, the input–output form of system (5.43) is as follows:

$$y(k) = Df^{\mathrm{T}}(x(k-1))\theta_0. \tag{5.51}$$

We construct an auxiliary system

$$\tilde{y}(k) = Df^{\mathrm{T}}(x(k-1))\theta(k-1).$$

Since system (5.50) has relative degree n in U, we can find a local transformation $\hat{z} = \Phi(x)$ such that, in the new coordinates, system (5.50) becomes

$$\begin{cases} \hat{z}(k+1) = A\hat{z}(k) + B(a(\hat{z}(k), \theta(k)) + bu(k)), \\ \hat{y}(k) = C\hat{z}(k). \end{cases}$$

With the same transformation, system (5.49) becomes

$$\begin{cases} z(k+1) = Az(k) + Ba(z(k), \theta_0), \\ y(k) = Cz(k), \end{cases}$$

where A, B, and C have the same form as in (5.39) and (5.40). Setting $e(k) = \hat{z}(k) - z(k)$, we get the error system as

$$e(k+1) = Ae(k) + B(a(\hat{z}(k), \theta(k)) - a(z(k), \theta_0) + bu). \tag{5.52}$$

Theorem 5.10 ([16]). With the following controller:

$$u = -b^{-1}(a(\hat{z}(k), \theta(k)) - a(z(k), \theta(k)) + ce) \tag{5.53}$$

and parameter update laws

$$\begin{cases} \theta(k) = \theta(k-1) + \dfrac{P(k-2)Df^{\mathrm{T}}(k-1)[y(k) - Df^{\mathrm{T}}(k-1)\theta(k-1)]}{Df^{\mathrm{T}}(k-1)P(k-2)f(k-1)D^{\mathrm{T}}}, \\ \qquad\qquad \text{if } Df^{\mathrm{T}}(k-1)P(k-2)f(k-1)D^{\mathrm{T}} \neq 0, \\ \theta(k) = \theta(k-1), \text{ if } Df^{\mathrm{T}}(k-1)P(k-2)f(k-1)D^{\mathrm{T}} = 0, \\ P(k-1) = P(k-2) - \dfrac{P(k-2)f(k-1)D^{\mathrm{T}}Df^{\mathrm{T}}(k-1)P(k-2)}{Df^{\mathrm{T}}(k-1)P(k-2)f(k-1)D^{\mathrm{T}}}, \\ \qquad\qquad \text{if } Df^{\mathrm{T}}(k-1)P(k-2)f(k-1)D^{\mathrm{T}} \neq 0, \\ P(k-1) = P(k-2), \text{ if } Df^{\mathrm{T}}(k-1)P(k-2)f(k-1)D^{\mathrm{T}} = 0, \\ P(0) = I, \end{cases}$$

the origin of system (5.52) is asymptotically stable in $U \in \mathbb{R}^n$, where $Q = (q_0, \ldots, q_{n-1})^{\mathrm{T}}$ are coefficients of a polynomial $P(s) = s^n + q_{n-1}s^{n-1} + \cdots + q_1 s + q_0$ whose roots have absolute values less than 1.

Proof. Substituting (5.53) into (5.52), we have

$$\begin{aligned} e(k+1) &= Ae(k) + B(a(\hat{z}(k), \theta(k)) - a(z(k), \theta_0) \\ &\quad - a(\hat{z}(k), \theta(k)) + a(z(k), \theta(k)) - Qe) \\ &= \Lambda e(k) + B(a(z(k), \theta(k)) - a(z(k), \theta_0)), \end{aligned} \tag{5.54}$$

where

$$\Lambda = \begin{pmatrix} 0 & 1 & 0 & \cdots & 0 & 0 \\ 0 & 0 & 1 & \cdots & 0 & 0 \\ \vdots & \vdots & \vdots & \ddots & \vdots & \vdots \\ 0 & 0 & 0 & \cdots & 0 & 1 \\ -q_0 & -q_1 & -q_2 & \cdots & q_{n-2} & -q_{n-1} \end{pmatrix}_{n \times n.}$$

For system (5.54) we choose the following Lyapunov function:

$$V(e(k)) = \sum_{i=1}^{n} |e_i(k)|,$$

where $e_i(k)$ is the ith element of e. Thus, we have

$$\Delta V = V(e(k+1)) - V(e(k)) = -|e_1(k)| + |a(z(k), \theta(k)) - a(z(k), \theta_0)|.$$

Since system (5.49) is a chaotic system, there must exist a large $N > 0$ such that the condition of Lemma 5.2 is satisfied. Therefore, $\theta(k)$ will equal θ_0 at most in N steps. Thus, we have

$$\Delta V|_{k>N} = -|e_1(k)| \leq 0.$$

When $k > N$ the error system becomes $e(k+1) = \Lambda e(k)$. If $\Delta V|_{k>N} = 0$, it means that $e_1(k) = 0$ $(k > N)$. From the error system equations we have $e_i(k) = 0, 2 \leq i \leq n$. By Lemma 5.3 we know that system (5.52) is asymptotically stable in U. The proof is completed. □

5.5.3 Simulations

In this section we take the Hénon system [8] as an example to illustrate the proposed methods. The difference equations of the Hénon system are as follows:

$$\begin{cases} x_1(k+1) = 1 - p_0 x_1^2(k) + x_2(k), \\ x_2(k+1) = q_0 x_1(k). \end{cases} \tag{5.55}$$

When $p = 1.4$ and $q = 0.3$, the Hénon system is in chaotic state and its attractor is shown in Fig. 5.17. The system has two fixed points:

$$X_{f_1} = \left(\frac{q_0 - 1 + \sqrt{(q_0-1)^2 + 4p_0}}{2p_0}, \ q_0 \frac{q_0 - 1 + \sqrt{(q_0-1)^2 + 4p_0}}{2p_0} \right)^{\mathrm{T}},$$

$$X_{f_2} = \left(\frac{q_0 - 1 - \sqrt{(q_0-1)^2 + 4p_0}}{2p_0}, \ q_0 \frac{q_0 - 1 + \sqrt{(q_0-1)^2 + 4p_0}}{2p_0} \right)^{\mathrm{T}}.$$

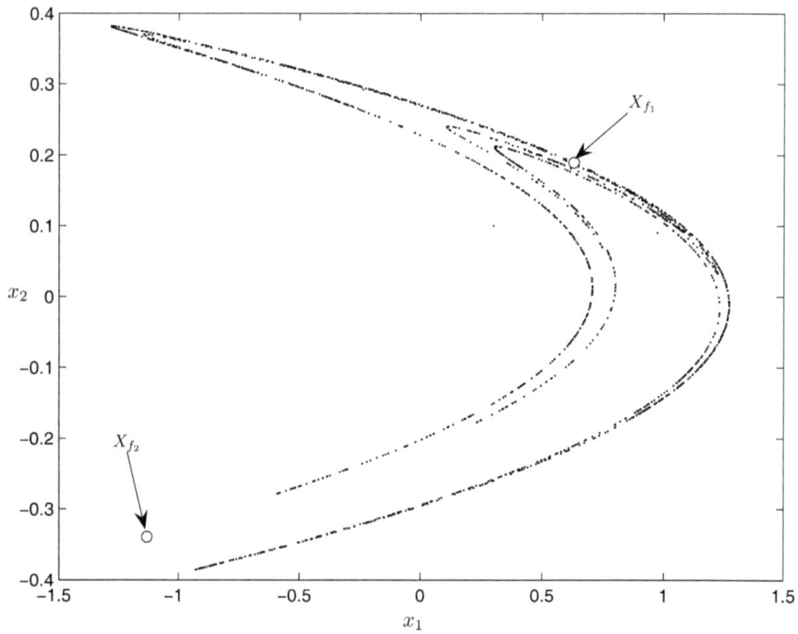

Fig. 5.17 The chaotic attractor of Hénon system (5.55)

From Fig. 5.17 we can see that X_{f_1} is in the attractor. To apply the proposed method we should first displace the origin of system (5.55) to X_{f_1}. For simplicity we will use the same symbols to denote the transformed system. Using the new coordinates, the Hénon system is as follows:

$$\begin{cases} x_1(k+1) = -p_0 x_1^2(k) - 2X_{f_1} p_0 x_1(k) + x_2(k), \\ x_2(k+1) = q_0 x_1(k). \end{cases}$$

Let x_1 be the output. Then, the drive system has the following form:

$$\begin{cases} x_1(k+1) = -p_0 x_1^2(k) - 2X_{f_1} p_0 x_1(k) + x_2(k), \\ x_2(k+1) = q_0 x_1(k), \\ y(k) = x_1(k). \end{cases} \tag{5.56}$$

Similarly, for the controlled response system

$$\begin{cases} \hat{x}_1(k+1) = 1 - p(k)\hat{x}_1^2(k) + \hat{x}_2(k), \\ \hat{x}_2(k+1) = q(k)\hat{x}_1(k) + u, \\ \hat{y}(k) = \hat{x}_1(k) \end{cases}$$

we can transform it into the following form:

$$\begin{cases} \hat{x}_1(k+1) = -p(k)\hat{x}_1^2(k) - 2X_{f_1} p(k)\hat{x}_1(k) + \hat{x}_2(k), \\ \hat{x}_2(k+1) = q(k)\hat{x}_1(k) + u, \\ \hat{y}(k) = \hat{x}_1(k). \end{cases} \tag{5.57}$$

Notice that system (5.55) can be transformed into the input–output form as

$$y_1(k+1) = 1 - p_0 x_1^2(k) + q_0 x_1(k-1) = \phi^{\mathrm{T}}(k)\theta_0.$$

Thus, we can construct an auxiliary system

$$\tilde{y}(k+1) = 1 - p(k)x_1^2(k) + q(k)x_1(k-1) = \phi^{\mathrm{T}}(k)\theta(k).$$

So, parameter update laws (5.34) can be used. It is easy to validate that system (5.57) has relative degree 2. By the following coordinate transformation:

$$\Phi(\hat{x}) = (\hat{z}_1, \hat{z}_2) = (h(\hat{x}), h \circ f(\hat{x}))^{\mathrm{T}} = (\hat{x}, -p(k)\hat{x}_1^2(k) - 2X_{f_1} p(k)\hat{x}_1(k) + \hat{x}_2(k))^{\mathrm{T}},$$

system (5.57) is changed to

$$\begin{cases} \hat{z}_1(k+1) = \hat{z}_2(k), \\ \hat{z}_2(k+1) = -p(k)z_2^2(k) - 2X_{f_1}(k)p(k)\hat{z}_2(k) + q(k)\hat{z}_1(k) + u, \\ \hat{y}(k) = \hat{z}_1(k). \end{cases}$$

With the same transformation, system (5.56) is changed to

$$\begin{cases} z_1(k+1) = z_2(k), \\ z_2(k+1) = -p_0 z_2^2(k) - 2X_{f_1} p_0 z_2(k) + q_0 z_1(k), \\ y(k+1) = z_1(k). \end{cases}$$

Setting $e_1(k) = \hat{z}_1(k) - z_1(k)$ and $e_2(k) = \hat{z}_2(k) - z_2(k)$, we get the following error system:

$$\begin{cases} e_1(k+1) = e_1(k), \\ e_2(k+1) = -p(k)z_2^2(k) - 2X_{f_1}(k)p(k)\hat{z}_2(k) + q(k)\hat{z}_1(k) \\ \qquad\qquad\quad +p_0 z_2^2(k) + 2X_{f_1} p_0 z_2(k) - q_0 z_1(k) + u. \end{cases}$$

The controller is designed according to (5.53) with $Q = (0.25, 1)^{\mathrm{T}}$. Choose $p_0 = 1.4$, $q_0 = 0.3$, $p(0) = 1.4$, and $q(0) = 0.3$. Initial values of the drive system and the response system are $(0.3, 0.1)^{\mathrm{T}}$ and $(1.3, 0.5)^{\mathrm{T}}$. Simulation results are shown in Figs. 5.18 and 5.19. Before $k = 1000$ the two systems have the same parameters. Within $0 \le k < 500$ no controller is added on the response system. When $500 \le k < 1000$ the controller is turned on and the two Hénon systems become synchronized very

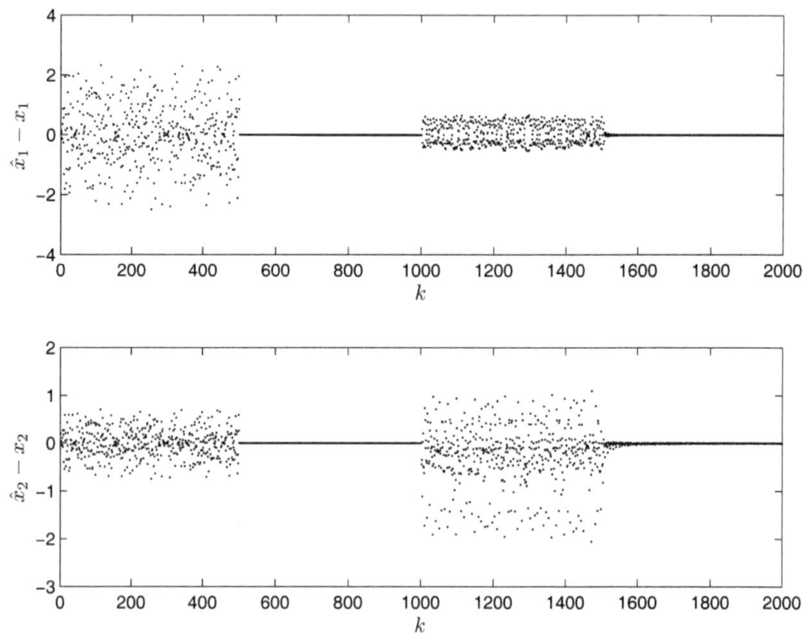

Fig. 5.18 The errors of synchronized Hénon systems

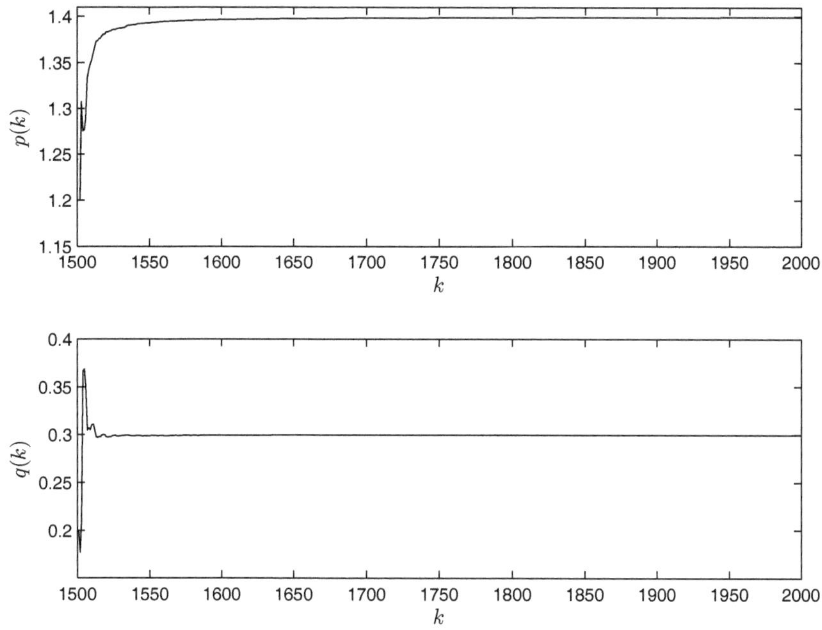

Fig. 5.19 Parameter curves of $p(k)$ and $q(k)$

quickly. When $k = 1000$ the parameters of the response system are disturbed to 1.2 and 0.2, and the previous controller cannot synchronize the Hénon systems any more. At $k = 1500$, parameter update laws are added and the two Hénon systems are synchronized again. From Fig. 5.19 we can see that parameters of the response system converge to those of the drive system when synchronization is achieved.

5.6 Summary

We have studied the problems of synchronization of both continuous-time and discrete-time chaotic systems in this chapter. The methods developed in Sects. 5.2, 5.3, and 5.5 are mainly based on geometric control theory of nonlinear systems. In these methods there are some common steps. First, outputs should be determined such that the response system has a relative degree as large as possible. After that the original system is transformed to normal form so that the controller can easily be designed. The controller designed in Sect. 5.4 is much more complicated than those in other sections. The reason is that the problem considered in this section is also more difficult than those in the other sections. This controller aims for synchronizing general chaotic systems. We do not exclude the special cases in which the controller can be simplified.

References

1. Boccaletti S, Kurths J, Osipov G, Valladares DL, Zhou CS (2002) The synchronization of chaotic systems. Phys Rep 366:1–101
2. Pecora LM, Carroll TL (1990) Synchronization in chaotic systems. Phys Rev Lett 64:821–824
3. Pecora LM, Carroll TL, Johnson GA, Mar DJ (1997) Fundamentals of synchronization in chaotic systems, concepts, and applications. Chaos 7:520–543
4. Chen G, Dong X (1998) *From Chaos to Order: Methodoligies, Perspectives and Applications*. World Scientific, Singapore
5. Cruz C, Nijmeijer H (2000) Synchronization through filtering. Int J Bifurc Chaos 10:763–775
6. Fuh CC, Tsai HH (2002) Control of discrete-time chaotic systems with feedback linearization. Chaos Solitons Fractals 13:285–294
7. Goodwin GC, Sin KS (1984) *Adaptive Filtering, Prediction and Control*. Prentice Hall, New Jersey
8. Hénon M (1976) A two-dimensional mapping with a strange attractor. Commun Math Phys 50:69–77
9. Isidori A (1989) *Nonlinear Control Systems: An Introduction*, 2nd edn. Springer, New York
10. Khalil HK (2002) *Nonlinear Systems*, 3rd edn. Prentice Hall, New Jersey
11. Lee HG, Arapostathis A, Marcus SI (1987) Linearization of discrete-time systems. Int J Control 45:1803–1822
12. Liao TL, Huang NG (1997) Control and synchronization of discrete-time chaotic systems via variable structure control technique. Phys Lett A 234:262–268
13. Liao XX (2000) *Theory and Applications of Stability for Dynamical Systems*. National Defence Industry Press, Beijing
14. Rössler OE (1979) An equation for hyperchaos. Phys Lett A 71:155–157
15. Wang ZL, Zhang HG (2002) Synchronization of high dimensional chaotic systems based on nonlinear feedback control. In: *Proc 15th IFAC Conf*, Barcelona, Spain, pp. 347–351
16. Wang ZL, Zhang HG (2002) Adaptive synchronization of discrete chaotic systems. In: *Proc 1st Int Conf Machine Learning and Cybernetics*, Beijing, China, pp. 336–339
17. Zhang HG, Huang W, Wang ZL, Chai T (2006) Adaptive synchronization between two different chaotic systems with unknown parameters. Phys Lett A 350:363–366

Chapter 6
Synchronizing Chaotic Systems via Impulsive Control

Abstract Impulsive control is an efficient method to deal with dynamical systems which cannot be controlled by continuous control. In addition, in the synchronization process, the response system receives information from the drive system only at discrete time instants, which drastically reduces the amount of synchronization information transmitted from the drive system to the response system and makes this method more efficient in a great number of real-life applications. In this chapter, we will study how to synchronize two identical or different chaotic systems by impulsive control methods. We first study the complete synchronization of a class of chaotic systems, and after that we develop synchronization methods for unified systems with channel time delay in the sense of practical stability. Then, robust synchronization schemes are studied for chaotic systems with parametric uncertainty and parametric mismatch. The aim is to present some practical impulsive control methods for different synchronization schemes.

6.1 Introduction

In this chapter, we will study how to synchronize two identical or different chaotic systems by impulsive control methods. Impulsive control is an efficient method to deal with dynamical systems which cannot be controlled by continuous control [18]. In addition, in the synchronization process, the response system receives information from the drive system only at discrete time instants, which drastically reduces the amount of synchronization information transmitted from the drive system to the response system and makes this method more efficient in a great number of real-life applications.

As is well known, for many applications, parametric uncertainties of systems are inevitable, and the effects of these uncertainties will destroy the synchronization. In the past few years, the analysis of synchronization for chaotic systems with parametric uncertainties has gained much research attention [4, 8, 11]. Most of the researches only concern the synchronization between two identical chaotic systems with known parameters or identical unknown parameters. However, in practice, it is difficult to ensure that the structures of the drive and the response systems are identical. Moreover, the parametric uncertainties of the drive and response systems are always different and time varying. Therefore, it is significant to investigate synchronization between two different chaotic systems in the presence of different time-varying parametric uncertainties. Furthermore, most of the existing literature has only considered chaotic systems described by impulsive ordinary differential equations (ODEs) [11, 20]. Due to the difficulty in investigating the stability of impulsive delayed or functional differential equations (DDEs or FDEs), there is little literature dealing with the synchronization problems of delayed chaotic systems via impulsive control [7, 8].

In this chapter we first study the complete synchronization of a class of chaotic systems, and after that we develop synchronization methods for unified systems with channel time delay in the sense of practical stability. Then, robust synchronization schemes are studied for chaotic systems with parametric uncertainties and parametric mismatch. Our goal is to present some practical impulsive control methods for different synchronization schemes.

6.2 Complete Synchronization of a Class of Chaotic Systems via Impulsive Control

6.2.1 System Description and the Synchronization Problem

Consider the drive system as follows:

$$\dot{x}(t) = Ax(t) + h(x(t)), \tag{6.1}$$

where $x \in \mathbb{R}^n$ is the state vector, $A \in \mathbb{R}^{n\times n}$ is a constant matrix, and $h(x(t)) \in \mathbb{R}^n$ is a continuous nonlinear function. Assume that

$$h(x(t)) - h(\tilde{x}(t)) = N(x(t), \tilde{x}(t))(x(t) - \tilde{x}(t)), \tag{6.2}$$

where $N(x(t), \tilde{x}(t)) \in \mathbb{R}^{n\times n}$ is a bounded matrix with elements depending on $x(t)$ and $\tilde{x}(t)$. Most chaotic systems can be described by (6.1) and (6.2), such as the Lorenz system, the Rössler system, the Chen system, the Lü system, the unified system, several variants of Chua's circuit, etc.

Suppose that a set of discrete instants $\{t_k\}$ satisfies

$$0 < t_1 < t_2 < \cdots < t_k < t_{k+1} < \cdots, \quad \lim_{k\to\infty} t_k = \infty \ (k \in \mathbb{N}), \ 0 \le t_0 < t_1.$$

At discrete time instant t_k, the state variables of the drive system are transmitted to the response system and the state variables of the response system are suddenly changed at these instants. Therefore, the response system can be written in the following form:

$$\begin{cases} \dot{\tilde{x}}(t) = A\tilde{x}(t) + h(\tilde{x}(t)), \ t \neq t_k, \\ \Delta\tilde{x}(t_k) = \tilde{x}(t_k^+) - \tilde{x}(t_k^-) = \tilde{x}(t_k^+) - \tilde{x}(t_k) = -B_k e(t_k), \ t = t_k, \ k \in \mathbb{N}, \\ \tilde{x}(t_0^+) = \tilde{x}_0, \end{cases} \tag{6.3}$$

where

$$e(t) = x(t) - \tilde{x}(t) = (x_1(t) - \tilde{x}_1(t), x_2(t) - \tilde{x}_2(t), \ldots, x_n(t) - \tilde{x}_n(t))^{\mathrm{T}}$$

is the synchronization error. Let $\tilde{x}(t_k^+) = \lim_{t \to t_k^+} \tilde{x}(t)$, $\tilde{x}(t_k^-) = \lim_{t \to t_k^-} \tilde{x}(t)$, i.e., t_k^+ and t_k^- denote the times after and before t_k, respectively. $\tilde{x}(t_k^-) = \tilde{x}(t_k)$ implies that $\tilde{x}(t)$ is left-continuous for $t \geq t_0$. $B_k \in \mathbb{R}^{n\times n}$ are the impulsive control gains. $\Delta x(t_k) = x(t_k^+) - x(t_k^-) = 0$. Then, from (6.1) and (6.3), the following error system is obtained:

$$\begin{cases} \dot{e}(t) = (A + N(x(t), \tilde{x}(t)))e(t), \ t \neq t_k, \\ \Delta e|_{t=t_k} = e(t_k^+) - e(t_k) = B_k e(t_k), \ t = t_k, \ k \in \mathbb{N}, \\ e(t_0^+) = e_0. \end{cases} \tag{6.4}$$

Here, the objective is to find the conditions on the control gains B_k and the impulsive distances $\delta_k = t_k - t_{k-1}$ ($k \in \mathbb{N}$) such that the error system (6.4) is asymptotically stable, which implies that the impulsively controlled response system (6.3) is asymptotically synchronized with the drive system (6.1) for arbitrary initial conditions.

6.2.2 Main Results

Theorem 6.1 ([20]). Let β_k, λ_A, and λ_N be the largest eigenvalues of $(I + B_k)^{\mathrm{T}}(I + B_k)$, $A + A^{\mathrm{T}}$, and $N(x, \tilde{x}) + N^{\mathrm{T}}(x, \tilde{x})$, respectively. If there exists a constant $\rho > 1$ such that

$$\ln(\rho\beta_k) + (\lambda_A + \lambda_N)\bar{\delta}_k \leq 0, \ \ k \in \mathbb{N}, \tag{6.5}$$

then the impulsively controlled response system (6.3) is asymptotically synchronized with the drive system (6.1).

Proof. Let the Lyapunov function be in the form of

$$V(e) = e^{\mathrm{T}} e.$$

For $t \in (t_{k-1}, t_k]$, $k \in \mathbb{N}$, its derivative along the solution of (6.4) is

$$\begin{aligned}\dot{V}(e) &= \dot{e}^{\mathrm{T}}e + e^{\mathrm{T}}\dot{e} \\ &= [(A+N(x,\tilde{x}))e]^{\mathrm{T}}e + e^{\mathrm{T}}[(A+N(x,\tilde{x}))e] \\ &= e^{\mathrm{T}}(A+A^{\mathrm{T}})e + e^{\mathrm{T}}[N(x,\tilde{x}) + N^{\mathrm{T}}(x,\tilde{x})]e \\ &\le (\lambda_A + \lambda_N)V(e).\end{aligned}$$

Therefore,

$$V(e(t)) \le V(e(t_{k-1}^{+}))\exp[(\lambda_A+\lambda_N)(t-t_{k-1})],\ \ t\in(t_{k-1},t_k],\ \ k\in\mathbb{N}. \tag{6.6}$$

On the other hand, when $t = t_k$, we have

$$\begin{aligned}V(e(t_k^{+})) &= [(I+B_k)e(t_k)]^{\mathrm{T}}(I+B_k)e(t_k) \\ &= e^{\mathrm{T}}(t_k)(I+B_k)^{\mathrm{T}}(I+B_k)e(t_k) \\ &\le \beta_k V(e(t_k)),\ \ k\in\mathbb{N}.\end{aligned} \tag{6.7}$$

From inequalities (6.6) and (6.7), we know for any $t\in(t_0,t_1]$,

$$V(e(t)) \le V(e(t_0^{+}))\exp[(\lambda_A+\lambda_N)(t-t_0)],$$

and this leads to

$$V(e(t_1)) \le V(e(t_0^{+}))\exp[(\lambda_A+\lambda_N)(t_1-t_0)]$$

and

$$V(e(t_1^{+})) \le \beta_1 V(e(t_1)) \le V(e(t_0^{+}))\beta_1\exp[(\lambda_A+\lambda_N)(t_1-t_0)].$$

Thus, for any $t\in(t_1,t_2]$,

$$V(e(t)) \le V(e(t_1^{+}))\exp[(\lambda_A+\lambda_N)(t-t_1)] \le V(e(t_0^{+}))\beta_1\exp[(\lambda_A+\lambda_N)(t-t_0)].$$

In general, for $t\in(t_k,t_{k+1}]$,

$$\begin{aligned}V(e(t)) &\le V(e(t_0^{+}))\beta_1\beta_2\cdots\beta_k\exp[(\lambda_A+\lambda_N)(t-t_0)] \\ &= V(e_0)\beta_1\exp[(\lambda_A+\lambda_N)\bar{\delta}_1]\cdots\beta_k\exp[(\lambda_A+\lambda_N)\bar{\delta}_k] \\ &\quad\times\exp[(\lambda_A+\lambda_N)(t-t_k)] \\ &\le V(e_0)\frac{1}{\rho^k}\exp[(\lambda_A+\lambda_N)(t-t_k)] \\ &\le \begin{cases} V(e_0)\dfrac{1}{\rho^k}\exp[(\lambda_A+\lambda_N)\bar{\delta}_{k+1}], & \text{for } \lambda_A+\lambda_N>0, \\ V(e_0)\dfrac{1}{\rho^k}, & \text{for } \lambda_A+\lambda_N\le 0. \end{cases}\end{aligned}$$

This means that $V(e(t))\to 0$ and $e(t)\to 0$ as $t\to\infty$. Thus, the origin of system (6.4) is asymptotically stable, which implies that the impulsively controlled response system (6.3) is asymptotically synchronized with the drive system (6.1). This completes the proof. □

Based on the matrix theory, we can obtain $\lambda_N \leq \left\|N(x,\tilde{x})+N^{\mathrm{T}}(x,\tilde{x})\right\|$, where $\|\cdot\|$ refers to any induced matrix norm in this section. Then, we can derive the following corollary.

Corollary 6.1. Let β_k and λ_A be the largest eigenvalues of $(I+B_k)^{\mathrm{T}}(I+B_k)$ and $A+A^{\mathrm{T}}$, respectively. If there exists a constant $\rho > 1$ such that

$$\ln(\rho\beta_k) + \left(\lambda_A + \left\|N(x,\tilde{x})+N^{\mathrm{T}}(x,\tilde{x})\right\|\right)\bar{\delta}_k \leq 0,\ k\in\mathbb{N}, \tag{6.8}$$

then the impulsively controlled response system (6.3) is asymptotically synchronized with the drive system (6.1). □

Remark 6.1. Since the trajectory of a chaotic system is bounded, inequalities (6.5) and (6.8) hold for suitable values of β_k and $\bar{\delta}_k$ $(k\in\mathbb{N})$. □

Remark 6.2. Generally speaking, the estimate of a matrix norm is easier than that of eigenvalues of a matrix. Therefore, it is more convenient to apply Corollary 6.1 than Theorem 6.1 in practice. □

Remark 6.3. In Theorem 6.1 and Corollary 6.1, if $\bar{\delta}_k = \bar{\delta} > 0$ and $B_k = B$ $(k\in\mathbb{N})$, then the conditions of Theorem 6.1 and Corollary 6.1 can be reduced to

$$\ln(\rho\beta) + (\lambda_A+\lambda_N)\bar{\delta} \leq 0$$

and

$$\ln(\rho\beta) + \left(\lambda_A + \left\|N(x,\tilde{x})+N^{\mathrm{T}}(x,\tilde{x})\right\|\right)\bar{\delta} \leq 0,$$

respectively, where β is the largest eigenvalue of $(I+B)^{\mathrm{T}}(I+B)$. □

Theorem 6.2 ([12]). Let β_k, λ_A, and λ_N be the largest eigenvalues of $(I+B_k)^{\mathrm{T}}(I+B_k)$, $A+A^{\mathrm{T}}$, and $N(x,\tilde{x})+N^{\mathrm{T}}(x,\tilde{x})$, respectively. If one of the following conditions holds:

(i) $\lambda_A+\lambda_N \geq 0$, and there exists a constant $\rho > 1$ such that

$$\ln(\rho\beta_{2k-1}\beta_{2k}) + (\lambda_A+\lambda_N)(t_{2k+1}-t_{2k-1}) \leq 0 \tag{6.9}$$

and

$$\sup_k\{\beta_k \exp[(\lambda_A+\lambda_N)(t_{k+1}-t_k)]\} = \Upsilon < \infty; \tag{6.10}$$

(ii) $\lambda_A+\lambda_N < 0$, and there exists a constant ρ $(0\leq\rho<-(\lambda_A+\lambda_N))$ such that

$$\ln(\beta_k) - \rho(t_k - t_{k-1}) \leq 0,\ k\in\mathbb{N}; \tag{6.11}$$

then the impulsively controlled response system (6.3) is asymptotically synchronized with the drive system (6.1).

Proof. Choose the following Lyapunov function:

$$V(e) = e^{\mathrm{T}} e.$$

For $t \in (t_{k-1}, t_k]$ ($k \in \mathbb{N}$), the time derivative of $V(e)$ along the solution of (6.4) is

$$\begin{aligned}\dot{V}(e(t)) &= \dot{e}^{\mathrm{T}} e + e^{\mathrm{T}} \dot{e}\\ &= [(A + N(x,\tilde{x}))e]^{\mathrm{T}} e + e^{\mathrm{T}}[(A + N(x,\tilde{x}))e]\\ &= e^{\mathrm{T}}(A + A^{\mathrm{T}})e + e^{\mathrm{T}}[N(x,\tilde{x}) + N^{\mathrm{T}}(x,\tilde{x})]e\\ &\leq (\lambda_A + \lambda_N)V(e(t)),\end{aligned}$$

which implies that

$$V(e(t)) \leq V(e(t_{k-1}^{+}))\exp[(\lambda_A + \lambda_N)(t - t_{k-1})]. \tag{6.12}$$

On the other hand, when $t = t_k$, we obtain

$$\begin{aligned}V(e(t_k^{+})) &= [(I + B_k)e(t_k)]^{\mathrm{T}}(I + B_k)e(t_k)\\ &= e^{\mathrm{T}}(t_k)(I + B_k)^{\mathrm{T}}(I + B_k)e(t_k)\\ &\leq \beta_k V(e(t_k)).\end{aligned} \tag{6.13}$$

For $t \in (t_k, t_{k+1}]$, $k \in \mathbb{N}$, from (6.12) and (6.13), we get

$$V(e(t)) \leq V(e(t_0^{+}))\beta_1\beta_2\cdots\beta_k \exp[(\lambda_A + \lambda_N)(t - t_0)]. \tag{6.14}$$

If $\lambda_A + \lambda_N \geq 0$, there exists a constant $\rho > 1$. From (6.9), (6.10), and (6.14), when (i) $t \in (t_{2k-1}, t_{2k}]$, we have

$$\begin{aligned}V(e(t)) &\leq V(e(t_0^{+}))\prod_{i=1}^{2k-1}\beta_i \exp[(\lambda_A + \lambda_N)(t - t_0)]\\ &\leq V(e_0)\prod_{i=1}^{2k-1}\beta_i \exp[(\lambda_A + \lambda_N)(t_{2k} - t_0)]\\ &= V(e_0)\beta_1\beta_2 \exp[(\lambda_A + \lambda_N)(t_3 - t_1)]\cdots\\ &\quad\times \beta_{2k-3}\beta_{2k-2}\exp[(\lambda_A + \lambda_N)(t_{2k-1} - t_{2k-3})]\\ &\quad\times \beta_{2k-1}\exp[(\lambda_A + \lambda_N)(t_{2k} - t_{2k-1})]\\ &\quad\times \exp[(\lambda_A + \lambda_N)(t_1 - t_0)]\\ &\leq \Upsilon\frac{V(e_0)}{\rho^{k-1}}\exp[(\lambda_A + \lambda_N)(t_1 - t_0)],\end{aligned} \tag{6.15}$$

(ii) $t \in (t_{2k}, t_{2k+1}]$, we have

$$
\begin{aligned}
V(e(t)) &\leq V(e(t_0^+)) \prod_{i=1}^{2k} \beta_i \exp[(\lambda_A + \lambda_N)(t - t_0)] \\
&\leq V(e_0) \prod_{i=1}^{2k} \beta_i \exp[(\lambda_A + \lambda_N)(t_{2k+1} - t_0)] \\
&= V(e_0)\beta_1\beta_2 \exp[(\lambda_A + \lambda_N)(t_3 - t_1)] \cdots \\
&\quad \times \beta_{2k-1}\beta_{2k} \exp[(\lambda_A + \lambda_N)(t_{2k+1} - t_{2k-1})] \\
&\quad \times \exp[(\lambda_A + \lambda_N)(t_1 - t_0)] \\
&\leq \frac{V(e_0)}{\rho^k} \exp[(\lambda_A + \lambda_N)(t_1 - t_0)]. \qquad (6.16)
\end{aligned}
$$

If $\lambda_A + \lambda_N < 0$, and there exists a constant ρ $(0 \leq \rho < -(\lambda_A + \lambda_N))$ such that (6.11) holds, i.e., $\beta_k \leq \exp[\rho(t_k - t_{k-1})]$, then

$$
\begin{aligned}
V(e(t)) &\leq V(e(t_0^+))\beta_1\beta_2 \cdots \beta_k \exp[(\lambda_A + \lambda_N)(t - t_0)] \\
&\leq V(e_0) \exp[\rho(t_k - t_0)] \exp[(\lambda_A + \lambda_N)(t - t_0)] \\
&\leq V(e_0) \exp[(\lambda_A + \lambda_N + \rho)(t - t_0)]. \qquad (6.17)
\end{aligned}
$$

Thus, from (6.15)–(6.17), we have $V(e(t)) \to 0$ and $e(t) \to 0$ as $t \to \infty$. Then, the origin of the error system (6.4) is asymptotically stable. This completes the proof. □

Corollary 6.2. Let β_k be the largest eigenvalue of $(I + B_k)^{\mathrm{T}}(I + B_k)$ $(k \in \mathbb{N})$. If there exists a constant $\rho > 1$ such that

$$
\ln(\rho\beta_{2k-1}\beta_{2k}) + \left(\left\|A + A^{\mathrm{T}}\right\| + \left\|N(x,\tilde{x}) + N^{\mathrm{T}}(x,\tilde{x})\right\|\right)(t_{2k+1} - t_{2k-1}) \leq 0,
$$

then the impulsively controlled response system (6.3) is asymptotically synchronized with the drive system (6.1). □

Remark 6.4. From (6.9) in Theorem 6.2, it can be seen that we need only to choose the odd switching sequence $\{t_{2k-1}\}$ instead of the whole switching sequence $\{t_k\}$ as in Theorem 6.1. □

Remark 6.5. The inequality (6.9) can be generalized to the following condition. There exist a finite integer $n_0 > 0$ and a constant $\rho > 1$ such that

$$
\ln(\rho\beta_{n_0(k-1)+1} \cdots \beta_{n_0 k}) + (\lambda_A + \lambda_N)(t_{n_0 k+1} - t_{n_0(k-1)+1}) \leq 0,\ k \in \mathbb{N}. \qquad (6.18)
$$

The choice of n_0 in (6.18) depends on the actual system to be considered. Especially, $n_0 = 1$ corresponds to the case of the whole switching sequence $\{t_k\}$ and $n_0 = 2$ corresponds to the case of the odd switching sequence $\{t_{2k-1}\}$. □

Remark 6.6. In Theorem 6.2 and Corollary 6.2, if $t_{2k+1} - t_{2k-1} = \Delta > 0$ and $B_k = B$ $(k \in \mathbb{N})$, then the conditions of Theorem 6.2 and Corollary 6.2 can be reduced to

$$
\ln(\rho\beta^2) + (\lambda_A + \lambda_N)\Delta \leq 0
$$

and

$$\ln(\rho\beta^2) + \left(\left\|A + A^{\mathrm{T}}\right\| + \left\|N(x,\tilde{x}) + N^{\mathrm{T}}(x,\tilde{x})\right\|\right)\Delta \leq 0,$$

respectively, where β is the largest eigenvalue of $(I+B)^{\mathrm{T}}(I+B)$. □

6.2.3 Simulations

We take both the Lorenz system and the unified system as examples to show how to use the proposed methods to synchronize chaotic systems.

6.2.3.1 Synchronizing the Lorenz System

The Lorenz system is described by the following differential equations:

$$\begin{cases} \dot{x}_1 = \alpha_1(x_2 - x_1), \\ \dot{x}_2 = \alpha_2 x_1 - x_2 - x_1 x_3, \\ \dot{x}_3 = x_1 x_2 - \alpha_3 x_3. \end{cases} \tag{6.19}$$

First, we decompose the linear and nonlinear parts of the Lorenz system (6.19), and rewrite it as follows:

$$\dot{x} = Ax + h(x), \tag{6.20}$$

where $x(t) = (x_1, x_2, x_3)^{\mathrm{T}}$, $A = \begin{pmatrix} -\alpha_1 & \alpha_1 & 0 \\ \alpha_2 & -1 & 0 \\ 0 & 0 & -\alpha_3 \end{pmatrix}$, and $h(x) = \begin{pmatrix} 0 \\ -x_1 x_3 \\ x_1 x_2 \end{pmatrix}$. The impulsively controlled response system is given by

$$\begin{cases} \dot{\tilde{x}} = A\tilde{x} + h(\tilde{x}), \quad t \neq t_k, \\ \Delta\tilde{x} = -B_k e(t_k), \quad t = t_k, \quad k \in \mathbb{N}, \\ \tilde{x}(t_0^+) = \tilde{x}_0, \end{cases} \tag{6.21}$$

where $\tilde{x} = (\tilde{x}_1, \tilde{x}_2, \tilde{x}_3)^{\mathrm{T}}$. From (6.20) and (6.21), we have

$$\begin{cases} \dot{e}(t) = Ae + h(x) - h(\tilde{x}), \ t \neq t_k, \\ \Delta e = B_k e(t_k), \ t = t_k, \ k \in \mathbb{N}, \\ e(t_0^+) = e_0, \end{cases}$$

where $e = (e_1, e_2, e_3)^{\mathrm{T}} = (x_1 - \tilde{x}_1, x_2 - \tilde{x}_2, x_3 - \tilde{x}_3)^{\mathrm{T}}$ is the synchronization error and

$$h(x) - h(\tilde{x}) = \begin{pmatrix} 0 \\ \tilde{x}_1\tilde{x}_3 - x_1 x_3 \\ x_1 x_2 - \tilde{x}_1\tilde{x}_2 \end{pmatrix} = \begin{pmatrix} 0 & 0 & 0 \\ -x_3 & 0 & -\tilde{x}_1 \\ x_2 & \tilde{x}_1 & 0 \end{pmatrix} \begin{pmatrix} e_1 \\ e_2 \\ e_3 \end{pmatrix}.$$

Therefore, we obtain

$$N(x,\tilde{x}) = \begin{pmatrix} 0 & 0 & 0 \\ -x_3 & 0 & -\tilde{x}_1 \\ x_2 & \tilde{x}_1 & 0 \end{pmatrix},$$

$$N(x,\tilde{x}) + N^{\mathrm{T}}(x,\tilde{x}) = \begin{pmatrix} 0 & -x_3 & x_2 \\ -x_3 & 0 & 0 \\ x_2 & 0 & 0 \end{pmatrix},$$

and

$$\left\| N(x,\tilde{x}) + N^{\mathrm{T}}(x,\tilde{x}) \right\|_{\infty} = \max\{|x_3| + |x_2|, |x_3|, |x_2|\}.$$

In this simulation, we take $\alpha_1 = 10$, $\alpha_2 = 28$, and $\alpha_3 = 8/3$. Then, $\lambda_A = 28.0512$. From the chaotic attractor of the Lorenz system, we find $-20 \leq x_1 \leq 20$, $-30 \leq x_2 \leq 30$, and $0 \leq x_3 \leq 50$, which leads to $\left\| N(x,\tilde{x}) + N^{\mathrm{T}}(x,\tilde{x}) \right\|_{\infty} \leq 80$. Let $\delta_k = \delta$ ($k \in \mathbb{N}$) and $B_k = B = \mathrm{diag}\{\sigma, \sigma, \sigma\}$. Then, $\beta_k = \beta = (\sigma+1)^2$. According to Corollary 6.1 and Remark 6.3, the impulsively controlled response system (6.21) is asymptotically synchronized with the drive system (6.20) if the following condition is satisfied:

$$0 < \bar{\delta} \leq -\frac{\ln\rho + \ln(\sigma+1)^2}{108.0512}.$$

Letting $\sigma = -0.7$ and $\rho = 1.1$, we have $0 < \bar{\delta} \leq 0.0214$. The numerical simulation results with $\sigma = -0.7$ and $\delta = 0.02$ are shown in Fig. 6.1. The initial conditions of the drive and response systems are taken as $x_0 = (6,2,4)^{\mathrm{T}}$ and $\tilde{x}_0 = (4,1,5)^{\mathrm{T}}$, respectively.

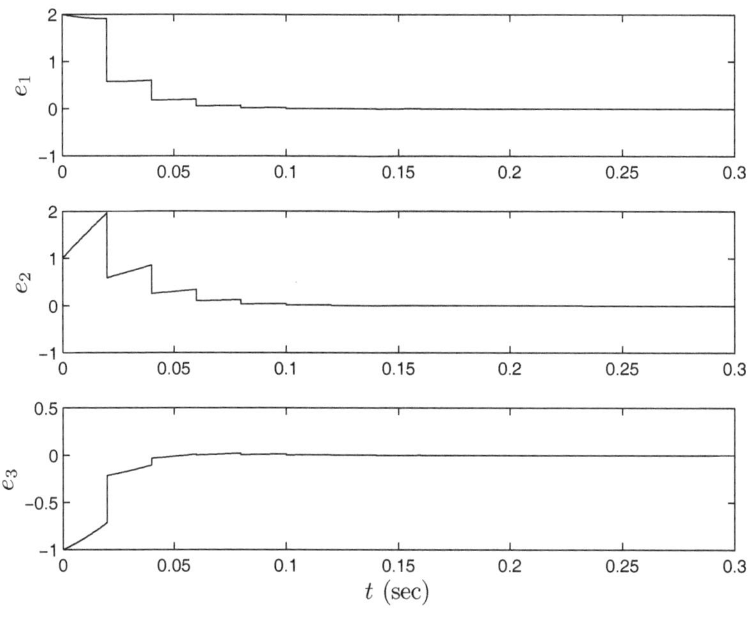

Fig. 6.1 Synchronization error curves of the Lorenz system with $\delta = 0.02$

6.2.3.2 Synchronizing the Unified System

The unified system is described by the following differential equations:

$$\begin{cases} \dot{x}_1 = (25a+10)(x_2-x_1), \\ \dot{x}_2 = (28-35a)x_1 - x_1x_3 + (29a-1)x_2, \\ \dot{x}_3 = x_1x_2 - \frac{1}{3}(a+8)x_3. \end{cases} \tag{6.22}$$

When $a \in [0,1]$, the unified system has a chaotic attractor, which is shown in Fig. 6.2. First, we decompose the linear and nonlinear parts of the unified system (6.22), and rewrite it as follows:

$$\dot{x} = Ax + h(x), \tag{6.23}$$

where $x(t) = (x_1, x_2, x_3)^{\mathrm{T}}$,

$$A = \begin{pmatrix} -(25a+10) & 25a+10 & 0 \\ 28-35a & 29a-1 & 0 \\ 0 & 0 & -\frac{a+8}{3} \end{pmatrix}, \quad \text{and } h(x) = \begin{pmatrix} 0 \\ -x_1x_3 \\ x_1x_2 \end{pmatrix}.$$

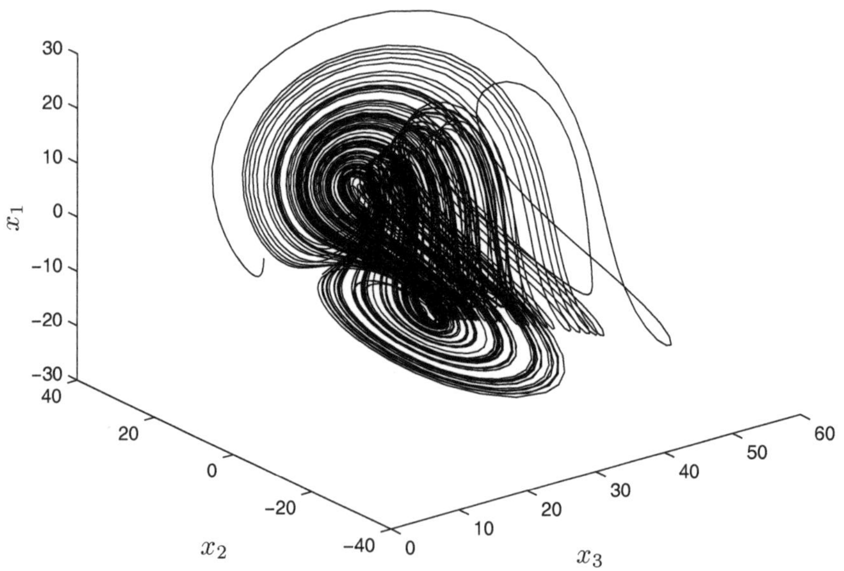

Fig. 6.2 The attractor of the unified system with $a = 1$

The impulsively controlled response system is given by

$$\begin{cases} \dot{\tilde{x}} = A\tilde{x} + h(\tilde{x}), \ \ t \neq t_k, \\ \Delta\tilde{x} = -B_k e(t_k), \ \ t = t_k, \ \ k \in \mathbb{N}, \\ \tilde{x}(t_0^+) = \tilde{x}_0, \end{cases} \tag{6.24}$$

where $\tilde{x} = (\tilde{x}_1, \tilde{x}_2, \tilde{x}_3)^{\mathrm{T}}$. From (6.23) and (6.24), we have

$$\begin{cases} \dot{e}(t) = Ae + h(x) - h(\tilde{x}), \ t \neq t_k, \\ \Delta e = B_k e(t_k), \ t = t_k, \ k \in \mathbb{N}, \\ e(t_0^+) = e_0, \end{cases}$$

where $e = (e_1, e_2, e_3)^{\mathrm{T}} = (x_1 - \tilde{x}_1, x_2 - \tilde{x}_2, x_3 - \tilde{x}_3)^{\mathrm{T}}$ is the synchronization error and

$$h(x) - h(\tilde{x}) = \begin{pmatrix} 0 \\ \tilde{x}_1\tilde{x}_3 - x_1x_3 \\ x_1x_2 - \tilde{x}_1\tilde{x}_2 \end{pmatrix} = \begin{pmatrix} 0 & 0 & 0 \\ -x_3 & 0 & -\tilde{x}_1 \\ x_2 & \tilde{x}_1 & 0 \end{pmatrix} \begin{pmatrix} e_1 \\ e_2 \\ e_3 \end{pmatrix}.$$

Therefore, we obtain

$$N(x, \tilde{x}) = \begin{pmatrix} 0 & 0 & 0 \\ -x_3 & 0 & -\tilde{x}_1 \\ x_2 & \tilde{x}_1 & 0 \end{pmatrix},$$

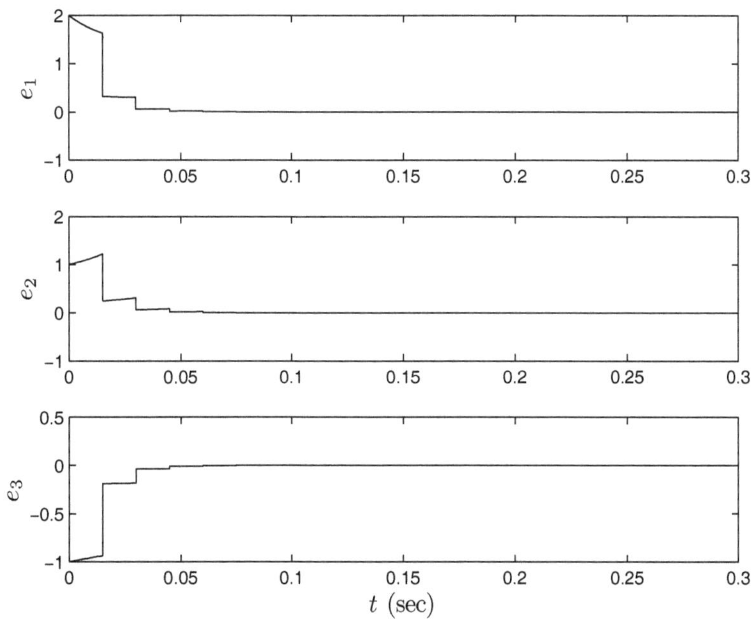

Fig. 6.3 Synchronization error curves of the unified system with $\Delta = 0.03$

$$N(x,\tilde{x}) + N^{\mathrm{T}}(x,\tilde{x}) = \begin{pmatrix} 0 & -x_3 & x_2 \\ -x_3 & 0 & 0 \\ x_2 & 0 & 0 \end{pmatrix},$$

and

$$\left\|N(x,\tilde{x}) + N^{\mathrm{T}}(x,\tilde{x})\right\|_{\infty} = \max\{|x_3| + |x_2|, |x_3|, |x_2|\}.$$

In this simulation, we choose $a = 1$. Then, $\left\|A + A^{\mathrm{T}}\right\|_{\infty} = 98$. From Fig. 6.2, we find $-30 \leq x_1 \leq 30$, $-30 \leq x_2 \leq 35$, and $0 \leq x_3 \leq 55$. Therefore,

$$\left\|N(x,\tilde{x}) + N^{\mathrm{T}}(x,\tilde{x})\right\|_{\infty} \leq 90.$$

Let $t_{2k+1} - t_{2k-1} = \Delta$ $(k \in \mathbb{N})$ and $B_k = B = \mathrm{diag}\{\sigma, \sigma, \sigma\}$. Then, $\beta_k = \beta = (\sigma + 1)^2$. According to Corollary 6.2 and Remark 6.6, the impulsively controlled response system (6.24) is asymptotically synchronized with the drive system (6.23) if the following condition is satisfied:

$$0 < \Delta \leq -\frac{\ln\rho + 2\ln(\sigma + 1)^2}{188}.$$

Let $\sigma = -0.8$ and $\rho = 1.1$. Then, $0 < \Delta \leq 0.0338$. The numerical simulation results with $\sigma = -0.8$ and $\Delta = 0.03$ are shown in Fig. 6.3. The initial conditions of the drive and response systems are taken as $x_0 = (6, 2, 4)^{\mathrm{T}}$ and $\tilde{x}_0 = (4, 1, 5)^{\mathrm{T}}$, respectively.

6.3 Lag-Synchronization of the Unified Systems via Impulsive Control

Strictly speaking, in the practical environment with signal propagation delays, it is not reasonable to require the response system to synchronize with the drive system at exactly the same time. For this reason, Jiang et al. [5] redefined the chaotic synchronization in such a way that the state of the drive system at time $t - \eta$ asymptotically synchronizes with that of the response system at time t, namely,

$$\lim_{t \to \infty} \|x(t - \eta) - \tilde{x}(t)\| = 0,$$

where x and $\tilde{x}$ are the state vectors of the drive system and the response system, respectively, and η is a finite channel time delay, which is often unknown.

In this section, we study how to achieve the synchronization of the unified systems with channel time delay and parametric uncertainties based on the practical stability theory. Meanwhile, a new definition of chaos synchronization is described, namely, the state of the drive system at time $t - \eta$ synchronizes with that of the response system at time t in the sense of practical stability, i.e., for any different initial conditions, $x_0(t - \eta)$ and $\tilde{x}_0(t)$,

$$\lim_{t \to \infty} \|x(t - \eta) - \tilde{x}(t)\| < \zeta,\ 0 < \zeta < +\infty,$$

where $\|\cdot\|$ refers to the Euclidean vector norm or the induced matrix 2-norm.

6.3.1 Preliminaries

Let a general nonautonomous chaotic system be

$$\dot{x} = f(t, x, u(t)),$$

where $f : \mathbb{R}^+ \times \mathbb{R}^n \times \mathbb{R}^m \to \mathbb{R}^n$ is continuous, $x \in \mathbb{R}^n$ is the state vector, and $u : \mathbb{R}^+ \to \mathbb{R}^m$ is the external force which is independent of the system. Suppose that a set of discrete instants $\{t_k\}$ satisfies $0 < t_1 < t_2 < \cdots < t_k < t_{k+1} < \cdots$, $\lim_{k\to\infty} t_k = \infty$ $(k \in \mathbb{N})$, and $0 \le t_0 < t_1$.

Let $U(k,x) = \Delta x|_{t=t_k} = x(t_k^+) - x(t_k^-) = x(t_k^+) - x(t_k)$ be the change of the state vector at the instant t_k. Then, the impulsively controlled chaotic system is given by

$$\begin{cases} \dot{x} = f(t, x, u(t)), & t \neq t_k, \\ \Delta x = U(k, x), & t = t_k, \ k \in \mathbb{N}, \\ x(t_0^+) = x_0, & t_0 \ge 0. \end{cases} \tag{6.25}$$

To study the practical stability of impulsive differential equation given in (6.25), we introduce the following definitions and lemmas.

Definition 6.1 ([17]). Let $V : \mathbb{R}^+ \times \mathbb{R}^n \to \mathbb{R}^+$. Then, V is said to belong to class $\mathscr{V}_0$ if

(i) V is continuous in $(t_{k-1}, t_k] \times \mathbb{R}^n$ and

$$\lim_{(t,y)\to(t_k^+,x)} V(t,y) = V(t_k^+, x).$$

(ii) V is locally Lipschitz in x. □

Definition 6.2 ([17]). Let $V \in \mathscr{V}_0$. For any $(t,x) \in (t_{k-1}, t_k] \times \mathbb{R}^n$, the upper and right (Dini) derivative of $V(t, x(t))$, along the solution $x(t)$ of (6.25), is defined as

$$D^+V(t,x) := \lim_{h\to 0^+} \sup \frac{1}{h}[V(t+h, x+hf(t,x,u(t))) - V(t,x)]. \qquad \square$$

Instead of studying the stability of the nth-order impulsive differential equation given in (6.25), it is convenient to study that of a scalar impulsive differential equation, whose stability properties are related to (6.25).

Definition 6.3 ([19]). *Comparison system.* Let $V \in \mathscr{V}_0$ and assume that

$$D^+V(t,x) \le g(t, V(t,x), v(t)), \ t \neq t_k,$$

$$V(t, x + U(k,x)) \le \psi_k(V(t,x)), \ t = t_k,$$

where $g\colon \mathbb{R}^+ \times \mathbb{R}^+ \times \mathbb{R}^+ \to \mathbb{R}$ is continuous and $\psi_k\colon \mathbb{R}^+ \to \mathbb{R}^+$ is nondecreasing. $\omega, v \in \Theta := \{\sigma\colon \mathbb{R}^+ \to \mathbb{R}^+,\ \sigma(s)$ is continuous for $s \in (t_{k-1}, t_k],\ \lim_{t\to t_k^+} \sigma(t) = \sigma(t_k^+)\}$. Then, the system

$$\begin{cases} \dot{\omega} = g(t, \omega, v(t)),\ t \neq t_k, \\ \omega(t_k^+) = \psi_k(\omega(t_k)),\ t = t_k,\ k \in \mathbb{N}, \\ \omega(t_0^+) = \omega_0 \geq 0 \end{cases} \tag{6.26}$$

is the comparison system of system (6.25). □

Definition 6.4 ([19]). *$PC(\mathbb{R}^+ \times \mathbb{R}^n, \mathbb{R}^+)$* denotes the set of functions $\Phi\colon \mathbb{R}^+ \times \mathbb{R}^n \to \mathbb{R}^+$ which are continuous for $t \in \mathbb{R}^+, t \neq t_k$, and are left-continuous for every $t \in \mathbb{R}^+$. At points t_k, Φ has a discontinuity of the first kind. □

Let a set Ω be

$$\Omega := \{u \in R^m\colon \Gamma(t, u) \leq r(t),\ t \geq t_0\},$$

where $\Gamma \in C(\mathbb{R}^+ \times \mathbb{R}^m, \mathbb{R}^+)$ and $r(t)$ is the maximal solution of the comparison system (6.26).

Definition 6.5 ([6]). *Practical stability*. System (6.25) is said to be practically stable if, given (λ, ζ), $\lambda > 0$, and $\zeta > 0$, we have that $\|x_0\| < \lambda$ implies that $\|x(t)\| < \zeta$, $t \geq t_0$, for some $t_0 \in \mathbb{R}^+$ and every $u \in \Omega$. □

Definition 6.6 ([17]). A function α is said to belong to class $\mathcal{K}$ if $\alpha \in C(\mathbb{R}^+, \mathbb{R}^+)$, $\alpha(0) = 0$, and $\alpha(x)$ is strictly increasing in x. □

Lemma 6.1 ([13]). Assume that

(i) $g \in PC(\mathbb{R}^+ \times \mathbb{R}^+ \times \mathbb{R}^+, \mathbb{R}^+)$ and $g(t, u, v)$ is nondecreasing in u for each (t, v) and nondecreasing in v for each (t, u).
(ii) ψ_k is nondecreasing for each k.
(iii) $0 < \lambda < \zeta$ is given.
(iv) $V \in PC(\mathbb{R}^+ \times \mathbb{R}^n, \mathbb{R}^+)$, $V(t, x)$ is locally Lipschitz in x, and there exist $\alpha, \beta \in \mathcal{K}$ such that

$$\beta(\|x\|) \leq V(t, x) \leq \alpha(\|x\|),\ (t, x) \in \mathbb{R}^+ \times S(\rho),\ \rho > \zeta.$$

(v) For $(t, x) \in (t_k, t_{k+1}] \times S(\rho)$ and $u(t) \in \Omega$,

$$D^+V(t, x) \leq g(t, V(t, x), \Gamma(t, u)).$$

(vi) $x \in S(\zeta)$ implies that $x + U(k, x) \in S(\rho)$ and $V(t_k^+, x + U(k, x)) \leq \psi_k(V(t, x))$.
(vii) $\alpha(\lambda) < \beta(\zeta)$.

Then, the practical stability properties of the comparison system (6.26) with respect to $(\alpha(\lambda), \beta(\zeta))$ imply the corresponding practical stability properties of system (6.25) with respect to (λ, ζ) for every $u(t) \in \Omega$, where $S(\rho) := \{x \in \mathbb{R}^n\colon \|x\| < \rho\}$. □

Remark 6.7. In Lemma 6.1, if the conditions in (iii) and (vii) are changed to $0 < \zeta < \lambda$ and $\alpha(\lambda) > \beta(\zeta)$, respectively, the conclusion is still valid. □

Lemma 6.2 ([19]). Let the comparison system (6.26) be described by the following equations:

$$\begin{cases} g(t,\omega,v(t)) = \varphi\omega + \theta,\ \varphi > 0,\ \theta > 0,\ t \neq t_k, \\ \psi_k(\omega) = \gamma_k\omega,\ \gamma_k > 0,\ t = t_k,\ k \in \mathbb{N}, \\ \omega(t_0^+) = \omega_0 \geq 0. \end{cases} \tag{6.27}$$

Let $\gamma_k = \gamma > 0$ be a constant. For given (λ,ζ), $\lambda > 0$, $\zeta > 0$, if

$$\frac{1}{\delta}\ln(\gamma) + \varphi < 0$$

and

$$\frac{\theta\left|1 - \mathrm{e}^{\varphi\delta}\right|}{\varphi(1 - \gamma\mathrm{e}^{\varphi\delta})} < \zeta,$$

where $\delta = t_{k+1} - t_k$, $0 < \delta < +\infty$, then system (6.25) is practically stable with respect to (λ,ζ) for any $\lambda < +\infty$. □

6.3.2 Main Results

Consider the following unified system with parametric uncertainty:

$$\dot{x}(t) = (A + \Delta A)x(t) + h(x(t)), \tag{6.28}$$

where $x(t) \in \mathbb{R}^n$ is the state vector, $A \in \mathbb{R}^{n\times n}$ is a constant matrix, $h(x(t))\colon \mathbb{R}^n \to \mathbb{R}^n$ is a continuous nonlinear function, and $\Delta A \in \mathbb{R}^{n\times n}$ is a time-varying parametric uncertainty, which satisfies the following assumptions:

(i) $\Delta A = \mu F(t)$, $\mu > 0$, $F(t) = [f_{ij}(t)]_{n\times n}$, $\left|f_{ij}(t)\right| \leq 1$.
(ii) ΔA does not destroy the chaotic behavior of the chaotic system (6.28).

Taking system (6.28) as the drive system, the response system can be described by

$$\dot{\tilde{x}}(t) = (A + \Delta\tilde{A})\tilde{x}(t) + h(\tilde{x}(t)). \tag{6.29}$$

Considering channel time delay η, we replace t with $t-\eta$ in (6.28), and write the drive system at time $t-\eta$ as follows:

$$\dot{x}(t-\eta) = (A + \overline{\Delta A})x(t-\eta) + h(x(t-\eta)). \tag{6.30}$$

In the above, $\Delta\tilde{A}$, $\overline{\Delta A}$, and ΔA satisfy the same assumptions, $\Delta\tilde{A} = \mu\tilde{F}(t) = \mu[\tilde{f}_{ij}(t)]_{n\times n}$, and $\overline{\Delta A} = \mu F(t-\eta) = \mu[f_{ij}(t-\eta)]_{n\times n}$.

If the linear error feedback composed of the state variables of the drive system (6.30) and the response system (6.29) is used as impulsive control signal, the controlled response system can be written in the following form:

$$\begin{cases} \dot{\tilde{x}}(t) = (A+\Delta\tilde{A})\tilde{x}(t) + h(\tilde{x}(t)), \ t \neq t_k, \\ \Delta\tilde{x}(t) = B_k(\tilde{x}(t) - x(t-\eta)), \ t = t_k, \ k \in \mathbb{N}, \\ \tilde{x}(t_0^+) = \tilde{x}_0. \end{cases} \tag{6.31}$$

From (6.30) and (6.31), the following error system can be obtained:

$$\begin{cases} \dot{e}(t) = Ae(t) + h(x(t-\eta)) - h(\tilde{x}(t)) \\ \qquad\qquad + \Delta\bar{A}x(t-\eta) - \Delta\tilde{A}\tilde{x}(t), \ t \neq t_k, \\ \Delta e|_{t=t_k} = B_k e(t_k), \ t = t_k, \ k \in \mathbb{N}, \\ e(t_0^+) = e_0, \end{cases} \tag{6.32}$$

where $e(t) = x(t-\eta) - \tilde{x}(t) = (x_1(t-\eta) - \tilde{x}_1(t), x_2(t-\eta) - \tilde{x}_2(t), \ldots, x_n(t-\eta) - \tilde{x}_n(t))^{\mathrm{T}}$ is the synchronization error. Here, the objective is to design the impulsive control gains $B_k \in \mathbb{R}^{n\times n}$ ($k \in \mathbb{N}$) such that the error system (6.32) is practically stable, which implies that the impulsively controlled system (6.31) is synchronized with the drive system (6.30) for any initial conditions.

Let $B_k = \mathrm{diag}\{\gamma_k - 1, \gamma_k - 1, \gamma_k - 1\}$ ($k \in \mathbb{N}$). From system (6.22), system (6.32) can be described by

$$\begin{cases} \dot{e}_1 = -(25a+10)e_1 + (25a+10)e_2 + \mu\left(\sum\limits_{i=1}^{3} f_{1i}x_i - \sum\limits_{i=1}^{3} \tilde{f}_{1i}\tilde{x}_i\right), \\ \dot{e}_2 = (28-35a)e_1 + (29a-1)e_2 - x_3e_1 - \tilde{x}_1e_3 + \mu\left(\sum\limits_{i=1}^{3} f_{2i}x_i - \sum\limits_{i=1}^{3} \tilde{f}_{2i}\tilde{x}_i\right), \\ \dot{e}_3 = -\dfrac{1}{3}(a+8)e_3 + x_2e_1 + \tilde{x}_1e_2 + \mu\left(\sum\limits_{i=1}^{3} f_{3i}x_i - \sum\limits_{i=1}^{3} \tilde{f}_{3i}\tilde{x}_i\right), \ t \neq t_k, \\ \Delta e|_{t=t_k} = B_k e, \ t = t_k, \ k \in \mathbb{N}, \\ e(t_0^+) = e_0. \end{cases} \tag{6.33}$$

For simplicity, in (6.33) and the rest of this section, $e(t)$, $\tilde{x}(t)$, $x(t-\eta)$, $\tilde{F}(t)$, $F(t-\eta)$, $\tilde{f}_{ij}(t)$, and $f_{ij}(t-\eta)$ are denoted by e, $\tilde{x}$, x, $\tilde{F}$, F, $\tilde{f}_{ij}$, and f_{ij}, respectively.

Remark 6.8. Due to the boundedness of the chaotic signals, there exist constants M_j such that $|\tilde{x}_j| \leq M_j$, $|x_j| \leq M_j$, $j = 1,2,3$. □

Theorem 6.3 ([11]). The practical stability of the synchronization error system (6.32) with respect to (λ, ζ) is equivalent to that of the comparison system (6.27) with respect to $(\sqrt{3}\lambda, \zeta)$, where $\varphi = H_{\max} = \max\{H_1, H_2, H_3\}$, $\theta = H_f = 6\mu \sum\limits_{j=1}^{3} M_j$, $H_1 = |28-35a| - (25a+10) + M_2 + M_3 + 3\mu$, $H_2 = |25a+10| + (29a-1) + M_1 + 3\mu$, and $H_3 = -\dfrac{1}{3}(a+8) + M_1 + 3\mu$.

Proof. Choose a Lyapunov function as

$$V(t,e) = |e_1| + |e_2| + |e_3|.$$

Then, we have

$$\begin{aligned}
D^+V(t,e) &= \dot{e}_1\mathrm{sgn}(e_1) + \dot{e}_2\mathrm{sgn}(e_2) + \dot{e}_3\mathrm{sgn}(e_3)\\
&= \left[-(25a+10)e_1 + (25a+10)e_2 + \mu\left(\sum_{i=1}^{3} f_{1i}x_i - \sum_{i=1}^{3}\tilde{f}_{1i}\tilde{x}_i\right)\right]\mathrm{sgn}(e_1)\\
&\quad + \left[(28-35a)e_1 + (29a-1)e_2 - x_3e_1 - \tilde{x}_1e_3\right.\\
&\quad \left.+ \mu\left(\sum_{i=1}^{3} f_{2i}x_i - \sum_{i=1}^{3}\tilde{f}_{2i}\tilde{x}_i\right)\right]\mathrm{sgn}(e_2)\\
&\quad + \left[-\frac{1}{3}(a+8)e_3 + x_2e_1 + \tilde{x}_1e_2 + \mu\left(\sum_{i=1}^{3} f_{3i}x_i - \sum_{i=1}^{3}\tilde{f}_{3i}\tilde{x}_i\right)\right]\mathrm{sgn}(e_3)\\
&\le \left[|28-35a| - (25a+10) + |x_2| + |x_3| + \mu\sum_{i=1}^{3}|f_{i1}|\right]|e_1|\\
&\quad + \left[|25a+10| + (29a-1) + |\tilde{x}_1| + \mu\sum_{i=1}^{3}|f_{i2}|\right]|e_2|\\
&\quad + \left[-\frac{1}{3}(a+8) + |\tilde{x}_1| + \mu\sum_{i=1}^{3}|f_{i3}|\right]|e_3|\\
&\quad + \mu\sum_{j=1}^{3}\sum_{i=1}^{3}|f_{ij}-\tilde{f}_{ij}|\,|\tilde{x}_j|\\
&\le H_1|e_1| + H_2|e_2| + H_3|e_3| + \mu\sum_{j=1}^{3}\sum_{i=1}^{3}|f_{ij}-\tilde{f}_{ij}|M_j\\
&\le \max\{H_1,H_2,H_3\}V(t,e) + 6\mu\sum_{j=1}^{3}M_j.
\end{aligned}$$

Hence, $g(t,\omega,v) = \varphi\omega + \theta$, where $\varphi = H_{\max} = \max\{H_1,H_2,H_3\}$ and $\theta = H_f = 6\mu\sum_{j=1}^{3}M_j$.

Since $V(t_k^+, e(t_k^+)) = V(t_k, (I+B_k)e(t_k)) = \gamma_k V(t_k, e(t_k))$, we have $\psi_k(\omega) = \gamma_k\omega$. As is well known, $|e_1| + |e_2| + |e_3| \ge \sqrt{e_1^2 + e_2^2 + e_3^2}$, and from

$$\begin{aligned}(|e_1|+|e_2|+|e_3|)^2 &= e_1^2+e_2^2+e_3^2+2|e_1||e_2|+2|e_2||e_3|+2|e_1||e_3| \\ &\le e_1^2+e_2^2+e_3^2+(e_1^2+e_2^2)+(e_2^2+e_3^2)+(e_1^2+e_3^2) \\ &= 3(e_1^2+e_2^2+e_3^2),\end{aligned}$$

we get $|e_1|+|e_2|+|e_3| \le \sqrt{3}\sqrt{e_1^2+e_2^2+e_3^2}$, i.e., $\alpha(s)=\sqrt{3}s$ and $\beta(s)=s$.

From $e \in S(\zeta)$, or $\|e\| < \zeta$, we have

$$\|e+U(k,e)\| = \|e+B_k e\| = \|\gamma_k e\| < \gamma_k \zeta < \rho.$$

If there exists $\gamma_k \in (0,1)$ such that the last inequality holds, then $e+U(k,e) \in S(\rho)$. Other conditions in Theorem 6.3 can easily be proved, which are omitted here. This completes the proof of the theorem. □

Based on the results of Lemma 6.2 and Theorem 6.3, we will give the following practical stability criterion of the error system (6.32).

Theorem 6.4 ([11]). Let $\gamma_k = \gamma > 0$ be a constant, $\delta = t_{k+1} - t_k$, $0 < \delta < +\infty$, $k \in \mathbb{N}$, (λ, ζ) be given, $\lambda > 0$, and $\zeta > 0$. If

$$\frac{1}{\delta}\ln(\gamma) + H_{\max} < 0, \tag{6.34}$$

$$\frac{H_f\left|1-\mathrm{e}^{H_{\max}\delta}\right|}{H_{\max}(1-\gamma\mathrm{e}^{H_{\max}\delta})} < \zeta, \tag{6.35}$$

then the synchronization error system (6.32) is practically stable with respect to (λ, ζ) for any $\lambda < +\infty$, where $H_{\max}$ and H_f are defined in Theorem 6.3.

Proof. Replace φ and θ in Lemma 6.2 with $H_{\max}$ and H_f in Theorem 6.3, respectively. Then, the conclusion can easily be obtained. For details, which are omitted here, please refer to [19]. This completes the proof. □

6.3.3 Simulations

Let $a = 1$. From the attractor of the unified system, we get $-30 \le x_1 \le 30$, $-30 \le x_2 \le 35$, $0 \le x_3 \le 55$, and then $M_1 = 30$, $M_2 = 35$, $M_3 = 55$. Assume that the parametric uncertainty $\Delta A = 0.03\begin{pmatrix}\sin t & \cos t & \sin t \\ \cos t & \cos t & \cos t \\ \sin t & \cos t & \sin t\end{pmatrix}$, i.e., $\mu = 0.03$ and $F(t) = [f_{ij}(t)]_{3\times 3} = \begin{pmatrix}\sin t & \cos t & \sin t \\ \cos t & \cos t & \cos t \\ \sin t & \cos t & \sin t\end{pmatrix}$. By computation, we have $H_1 = 62.09$, $H_2 = 93.09$, $H_3 = 27.09$, and $H_f = 21.6$. So, $H_{\max} = 93.09$. The initial conditions

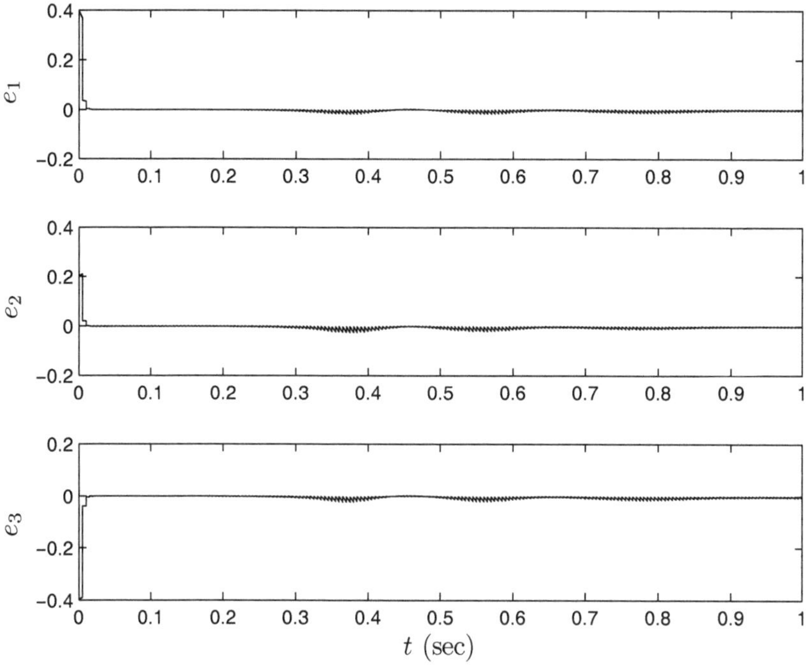

Fig. 6.4 Synchronization error curves with $\delta = 0.005$, $\gamma = 0.1$

of the drive and response systems are $x_0 = (4,1,5)^{\mathrm{T}}$ and $\tilde{x}_0 = (3.6,0.8,5.4)^{\mathrm{T}}$, respectively. Let $\delta = 0.005$. From (6.34) it gives $\gamma < 0.6279$. We choose $\gamma = 0.1$, and from (6.35) it gives $\zeta \approx 0.1636$. It means that under this condition, the maximum amplitudes of e_1, e_2, and e_3 are less than 0.1636. We choose the 'unknown' channel time delay $\eta = 3$ sec. The simulation result is shown in Fig. 6.4. Obviously, the synchronization errors fluctuate around zero with small amplitudes that are estimated by Theorem 6.4.

Remark 6.9. ζ will decrease as the parametric uncertainty ΔA decreases. Specifically, ζ will converge to zero when $\Delta A = 0$; that is, the response system (6.31) is asymptotically synchronized with the drive system (6.30). □

6.4 Impulsive Synchronization of Different Chaotic Systems

In practice, it is difficult to ensure the structures of the drive and the response systems to be identical. Moreover, the parametric uncertainties of the drive and the response systems are always different and time varying. Therefore, it is significant to investigate synchronization schemes between two different chaotic systems in the presence of different time-varying parametric uncertainties.

6.4.1 System Description and Synchronization Problem

Consider the drive system as follows:

$$\dot{x}(t) = (A + \Delta A(t))x(t) + f(x(t)), \tag{6.36}$$

where $x \in \mathbb{R}^n$ is the state vector, $A \in \mathbb{R}^{n\times n}$ is a known constant matrix, and $f(x(t)) \in \mathbb{R}^n$ is a continuous nonlinear function. Assume that

$$f(x(t)) - f(\tilde{x}(t)) = N(x(t), \tilde{x}(t))(x(t) - \tilde{x}(t)), \tag{6.37}$$

where $N(x(t), \tilde{x}(t)) \in \mathbb{R}^{n\times n}$ is a matrix with bounded elements depending on $x(t) \in \mathbb{R}^n$ and $\tilde{x}(t) \in \mathbb{R}^n$.

The parametric uncertainty $\Delta A(t)$ is said to be admissible if $\Delta A(t)$ satisfies the following assumptions:

(i) $\Delta A(t) = EF(t)H$, where E and H are known real constant matrices with appropriate dimensions, and the uncertain matrix $F(t) \in \mathbb{R}^{n\times n}$ satisfies $\|F(t)\| \le 1$.
(ii) The parametric uncertainty $\Delta A(t)$ does not destroy the chaotic behavior of the chaotic system (6.36).

Remark 6.10. The parametric uncertainty $\Delta A(t)$ is a time-varying perturbation, which is often used in other papers to deal with the scheme of robust stabilization for uncertain systems [2, 14].

Consider the channel time delay η; we replace t with $t - \eta$ in (6.36), and write the drive system at time $t - \eta$ as follows:

$$\dot{x}(t-\eta) = (A + \Delta A(t-\eta))x(t-\eta) + f(x(t-\eta)). \tag{6.38}$$

The response system is as follows:

$$\dot{\tilde{x}}(t) = (\tilde{A} + \Delta\tilde{A}(t))\tilde{x}(t) + g(\tilde{x}(t)), \tag{6.39}$$

where $\tilde{x} \in \mathbb{R}^n$ is the state vector, $\tilde{A} \in \mathbb{R}^{n\times n}$ is a known constant matrix, and $g(x(t)) \in \mathbb{R}^n$ is a continuous nonlinear function. The parametric uncertainty $\Delta\tilde{A}(t) = \tilde{E}\tilde{F}(t)\tilde{H}$ satisfies the same assumptions as that for $\Delta A(t)$. Suppose that the following condition holds:

$$f(\tilde{x}(t)) - g(\tilde{x}(t)) = U(\tilde{x}(t))\tilde{x}(t), \tag{6.40}$$

where $U(\tilde{x}(t)) \in \mathbb{R}^{n\times n}$ is a matrix with each element bounded depending on $\tilde{x}(t)$.

Suppose that a set of discrete instants $\{t_k\}$ satisfies

$$0 < t_1 < t_2 < \cdots < t_k < t_{k+1} < \cdots, \ \lim_{k\to\infty} t_k = \infty \ (k \in \mathbb{N}), \ 0 \le t_0 < t_1.$$

At the discrete time t_k, the state variables of the drive system are transmitted to the response system as part of the control inputs such that the state variables of the response system are suddenly changed at these instants. Therefore, the response system can be written in the following form:

$$\begin{cases} \dot{\tilde{x}}(t) = (\tilde{A} + \Delta\tilde{A}(t))\tilde{x}(t) + g(\tilde{x}(t)), \ t \neq t_k, \\ \Delta\tilde{x}(t) = \tilde{x}(t_k^+) - \tilde{x}(t_k^-) = \tilde{x}(t_k^+) - \tilde{x}(t_k) = -B_k e(t), \ t = t_k, \ k \in \mathbb{N}, \\ \tilde{x}(t_0^+) = \tilde{x}_0, \end{cases} \tag{6.41}$$

where

$$e(t) = x(t-\eta) - \tilde{x}(t) = [x_1(t-\eta) - \tilde{x}_1(t), x_2(t-\eta) - \tilde{x}_2(t), \ldots, x_n(t-\eta) - \tilde{x}_n(t)]^{\mathrm{T}}$$

is the synchronization error. $\tilde{x}(t)$ is left-continuous at $t = t_k$ and $B_k \in \mathbb{R}^{n\times n}$ are the impulsive control gains. Then, from (6.38) and (6.41), the following error system is obtained:

$$\begin{cases} \dot{e}(t) = [A + \Delta A(t-\eta) + N(x(t-\eta), \tilde{x}(t))]\, e(t) \\ \qquad + [A - \tilde{A} + \Delta A(t-\eta) - \Delta\tilde{A}(t) + U(\tilde{x}(t))]\, \tilde{x}(t), \ t \neq t_k, \\ \Delta e|_{t=t_k} = e(t_k^+) - e(t_k) = B_k e(t_k), \ \ t = t_k, \ k \in \mathbb{N}, \\ e(t_0^+) = e_0. \end{cases} \tag{6.42}$$

For simplicity, in the rest of the section, $\tilde{x}(t), x(t-\eta), N(x(t-\eta), \tilde{x}(t)), U(\tilde{x}(t))$, $\Delta A(t-\eta)$, and $\Delta\tilde{A}(t)$ are denoted by $\tilde{x}$, x, N, U, ΔA, and $\Delta\tilde{A}$, respectively.

Remark 6.11. Due to the boundedness of the chaotic signals, there exist positive constants ς and $\tilde{\varsigma}$ such that $\|x\| \leq \varsigma$ and $\|\tilde{x}\| \leq \tilde{\varsigma}$. □

Definition 6.7. The synchronization of systems (6.38) and (6.41) is said to have been achieved if, for arbitrary initial conditions x_0 and $\tilde{x}_0$, the trivial solution of the error system (6.42) converges to a predetermined neighborhood of the origin for any admissible parametric uncertainties. □

Here, the objective is to find the conditions on the control gains B_k and the impulsive distances $\delta_k = t_k - t_{k-1}$ $(k \in \mathbb{N})$ such that the error magnitude, i.e., $\|e\|$, converges to below some constant ξ, which implies that the impulsively controlled response system (6.41) is synchronized with the drive system (6.38) for arbitrary initial conditions.

6.4.2 Main Results

The following theorem gives sufficient conditions for robust stability of the error system (6.42), which imply that the impulsively controlled response system (6.41) is synchronized with the drive system (6.38).

Theorem 6.5. Let β_k, λ_A, and λ_N be the largest eigenvalues of $(I+B_k)^{\mathrm{T}}(I+B_k)$, $A+A^{\mathrm{T}}$, and $N+N^{\mathrm{T}}$, respectively. The response system (6.41) is synchronized with the drive system (6.38) if one of the following conditions holds:

(i) when $\theta_1 \geq 0$, there exists a constant $\rho > 1$ such that

$$\ln(\rho\beta_{2k-1}\beta_{2k}) + \theta_1(t_{2k+1} - t_{2k-1}) \leq 0 \tag{6.43}$$

and

$$\sup_k\{\beta_k \exp[\theta_1(t_{k+1} - t_k)]\} = \Upsilon < \infty; \tag{6.44}$$

(ii) when $\theta_1 < 0$, there exists a constant ρ $(0 \leq \rho < -\theta_1)$ such that

$$\ln(\beta_k) - \rho(t_k - t_{k-1}) \leq 0, \quad k \in \mathbb{N}, \tag{6.45}$$

where $\xi > 0$ is the bound of the error magnitude $\|e\|$ and can be chosen small enough, where

$$\theta_1 = \lambda_A + \lambda_N + 2\|E\|\,\|H\| + \frac{\varsigma}{\xi}\left(2\left\|A - \tilde{A}\right\| + 2\|E\|\,\|H\| + 2\left\|\tilde{E}\right\|\left\|\tilde{H}\right\| + \left\|U + U^{\mathrm{T}}\right\|\right).$$

Proof. Choose the following Lyapunov function:

$$V(e) = e^{\mathrm{T}}e.$$

For $t \in (t_{k-1}, t_k]$, $k \in \mathbb{N}$, the time derivative of $V(e)$ along the solution of (6.42) is

$$\begin{aligned}
\dot{V}(e(t)) &= \dot{e}^{\mathrm{T}}e + e^{\mathrm{T}}\dot{e} \\
&= \left[(A + \Delta A + N)e + (A - \tilde{A} + \Delta A - \Delta\tilde{A} + U)\tilde{x}\right]^{\mathrm{T}} e \\
&\quad + e^{\mathrm{T}}\left[(A + \Delta A + N)e + (A - \tilde{A} + \Delta A - \Delta\tilde{A} + U)\tilde{x}\right] \\
&= e^{\mathrm{T}}(A + A^{\mathrm{T}})e + e^{\mathrm{T}}(N + N^{\mathrm{T}})e + e^{\mathrm{T}}(\Delta A + \Delta A^{\mathrm{T}})e \\
&\quad + 2e^{\mathrm{T}}(A - \tilde{A} + \Delta A - \Delta\tilde{A} + U)\tilde{x} \\
&\leq (\lambda_A + \lambda_N + 2\|E\|\,\|H\|)\, e^{\mathrm{T}}e \\
&\quad + \left(2\left\|A - \tilde{A}\right\| + 2\|E\|\,\|H\| + 2\left\|\tilde{E}\right\|\left\|\tilde{H}\right\| + \left\|U + U^{\mathrm{T}}\right\|\right)\varsigma\,\|e\|.
\end{aligned}$$

For $\|e\| \geq \xi$ one can conclude that $\|e\| \leq \|e\|^2/\xi$ and, therefore,

$$\begin{aligned}\dot{V}(e(t)) &\leq (\lambda_A + \lambda_N + 2\|E\|\,\|H\|)e^{\mathrm{T}}e \\ &\quad + \frac{\tilde{\varsigma}}{\varsigma}\left(2\left\|A-\tilde{A}\right\| + 2\|E\|\,\|H\| + 2\left\|\tilde{E}\right\|\left\|\tilde{H}\right\| + \left\|U+U^{\mathrm{T}}\right\|\right)\|e\|^2 \\ &\leq \Big[\lambda_A + \lambda_N + 2\|E\|\,\|H\| \\ &\quad + \frac{\tilde{\varsigma}}{\varsigma}\left(2\left\|A-\tilde{A}\right\| + 2\|E\|\,\|H\| + 2\left\|\tilde{E}\right\|\left\|\tilde{H}\right\| + \left\|U+U^{\mathrm{T}}\right\|\right)\Big]V(e(t)) \\ &= \theta_1 V(e(t)), \end{aligned} \tag{6.46}$$

which implies that

$$V(e(t)) \leq V(e(t_{k-1}^{+}))\exp[\theta_1(t-t_{k-1})],\ t \in (t_{k-1}, t_k],\ k \in \mathbb{N}. \tag{6.47}$$

On the other hand, when $t = t_k$, we obtain

$$\begin{aligned} V(e(t_k^{+})) &= [(I+B_k)e(t_k)]^{\mathrm{T}}(I+B_k)e(t_k) \\ &= e^{\mathrm{T}}(t_k)(I+B_k)^{\mathrm{T}}(I+B_k)e(t_k) \\ &\leq \beta_k V(e(t_k)). \end{aligned}$$

Let $k=1$ in (6.47). Then, we have

$$V(e(t)) \leq V(e(t_0^{+}))\exp[\theta_1(t-t_0)],\ \ t \in (t_0, t_1],$$

which leads to

$$V(e(t_1)) \leq V(e(t_0^{+}))\exp[\theta_1(t_1-t_0)]$$

and

$$V(e(t_1^{+})) \leq \beta_1 V(e(t_1)) \leq V(e(t_0^{+}))\beta_1\exp[\theta_1(t_1-t_0)].$$

Similarly, for $t \in (t_1, t_2]$, we have

$$V(e(t)) \leq V(e(t_1^{+}))\exp[\theta_1(t-t_1)] \leq V(e(t_0^{+}))\beta_1\exp[\theta_1(t-t_0)].$$

In general, for $t \in (t_k, t_{k+1}]$,

$$V(e(t)) \leq V(e(t_0^{+}))\beta_1\beta_2\cdots\beta_k\exp[\theta_1(t-t_0)]. \tag{6.48}$$

If $\theta_1 \geq 0$, there exists a constant $\rho > 1$ and, from (6.43), (6.44), and (6.48), when (i) $t \in (t_{2k-1}, t_{2k}]$, we have

$$
\begin{aligned}
V(e(t)) &\le V(e(t_0^+)) \prod_{i=1}^{2k-1} \beta_i \exp[\theta_1 (t - t_0)] \\
&\le V(e_0) \prod_{i=1}^{2k-1} \beta_i \exp[\theta_1 (t_{2k} - t_0)] \\
&= V(e_0)\beta_1\beta_2 \exp[\theta_1 (t_3 - t_1)] \cdots \beta_{2k-3}\beta_{2k-2} \exp[\theta_1 (t_{2k-1} - t_{2k-3})] \\
&\quad \times \beta_{2k-1} \exp[\theta_1 (t_{2k} - t_{2k-1})] \exp[\theta_1 (t_1 - t_0)] \\
&\le \Upsilon \frac{V(e_0)}{\rho^{k-1}} \exp[\theta_1 (t_1 - t_0)];
\end{aligned}
\tag{6.49}
$$

(ii) $t \in (t_{2k}, t_{2k+1}]$, we have

$$
\begin{aligned}
V(e(t)) &\le V(e(t_0^+)) \prod_{i=1}^{2k} \beta_i \exp[\theta_1 (t - t_0)] \\
&\le V(e_0) \prod_{i=1}^{2k} \beta_i \exp[\theta_1 (t_{2k+1} - t_0)] \\
&= V(e_0)\beta_1\beta_2 \exp[\theta_1 (t_3 - t_1)] \cdots \beta_{2k-1}\beta_{2k} \exp[\theta_1 (t_{2k+1} - t_{2k-1})] \\
&\quad \times \exp[\theta_1 (t_1 - t_0)] \\
&\le \frac{V(e_0)}{\rho^{k}} \exp[\theta_1 (t_1 - t_0)].
\end{aligned}
\tag{6.50}
$$

If $\theta_1 < 0$, and there exists a constant ρ $(0 \le \rho < -\theta_1)$ such that (6.45) holds, i.e., $\beta_k \le \exp[\rho(t_k - t_{k-1})]$, then

$$
\begin{aligned}
V(e(t)) &\le V(e(t_0^+))\beta_1\beta_2 \cdots \beta_k \exp[\theta_1 (t - t_0)] \\
&\le V(e_0) \exp[\rho(t_k - t_0)] \exp[\theta_1 (t - t_0)] \\
&\le V(e_0) \exp[(\theta_1 + \rho)(t - t_0)].
\end{aligned}
\tag{6.51}
$$

It follows from (6.49)–(6.51) that the error magnitude $\|e\|$ will converge to below the constant ξ if the error started from $\|e\| > \xi$. This completes the proof. □

Remark 6.12. Most typical chaotic systems with parametric uncertainties can be described by (6.36) or (6.39), and the continuous nonlinear functions can satisfy (6.37) and (6.40), such as the Lorenz system, the Chen system, the Rössler system, the unified system, the Lü system, several variants of Chua's circuit, etc. □

Remark 6.13. If $A = \tilde{A}$ and $f(\cdot) = g(\cdot)$, then the structures of system (6.38) and system (6.39) are identical. Therefore, Theorem 6.5 is also applicable to the impulsive synchronization between two identical chaotic systems with or without parametric uncertainties. □

Remark 6.14. In (6.41) or (6.42), the value of the channel time delay η is not required to be known for executing the impulsive control, since we can obtain the time-delay signal $x(t-\eta)$ at the discrete time t_k easily and use it in the response system blindly when it is received. □

Based on the matrix theory and the boundedness of chaotic signals, we obtain $\lambda_A \leq \|A+A^{\mathrm{T}}\|$, $\lambda_N \leq \|N+N^{\mathrm{T}}\| \leq \phi$, and $\|U+U^{\mathrm{T}}\| \leq \varphi$, where ϕ and φ are positive real constants that can be obtained for different chaotic systems. Then, we have the following corollary.

Corollary 6.3. Let β_k be the largest eigenvalue of $(I+B_k)^{\mathrm{T}}(I+B_k)$. If there exists a constant $\rho > 1$ such that

$$\ln(\rho\beta_{2k-1}\beta_{2k}) + \theta_2(t_{2k+1}-t_{2k-1}) \leq 0,\ k \in \mathbb{N}, \tag{6.52}$$

and

$$\sup_k\{\beta_k \exp[\theta_2(t_{k+1}-t_k)]\} = \Upsilon < \infty,$$

then the response system (6.41) is synchronized with the drive system (6.38), where $\xi > 0$ is the bound of the error magnitude $\|e\|$ and can be chosen small enough, where

$$\theta_2 = \|A+A^{\mathrm{T}}\| + \phi + 2\|E\|\,\|H\| + \frac{\tilde{\varsigma}}{\xi}\left(2\|A-\tilde{A}\| + 2\|E\|\,\|H\| + 2\|\tilde{E}\|\,\|\tilde{H}\| + \varphi\right).$$

□

Remark 6.15. Though Corollary 6.3 will be more conservative than Theorem 6.5, the computation of $\|A+A^{\mathrm{T}}\|$ and $\|N+N^{\mathrm{T}}\|$ will be easier than that of λ_A and λ_N in general. Therefore, the result in Corollary 6.3 will be more convenient for use in practice. □

Remark 6.16. From the boundedness of chaotic systems, we know that λ_A, $\|A+A^{\mathrm{T}}\|$, λ_N, ϕ, $\|A-\tilde{A}\|$, and φ are bounded, and the inequalities (6.43) and (6.52) can be satisfied by choosing appropriate β_k and $\Delta_k = t_{2k+1}-t_{2k-1}$ $(k \in \mathbb{N})$. □

In practice, for purpose of convenience, the gains B_k are always selected as a constant matrix and the impulsive distances Δ_k are set to be a positive constant. Then, we have the following corollary.

Corollary 6.4. Assume that $t_{2k+1}-t_{2k-1} = \Delta > 0$ and $B_k = B$ $(k \in \mathbb{N})$. If there exists a constant $\rho > 1$ such that

$$\ln(\rho\beta^2) + \Delta\theta_2 \leq 0 \ \text{ and } \ \beta\exp(\Delta\theta_2) = \Upsilon < \infty,$$

then the response system (6.41) is synchronized with the drive system (6.38), where $\xi > 0$ is the bound of the error magnitude $\|e\|$ and can be chosen small enough, β is the largest eigenvalue of $(I+B)^{\mathrm{T}}(I+B)$, and θ_2 is defined in Corollary 6.3. □

Remark 6.17. It is worth pointing out that the topic of synchronization between different chaotic systems with different parametric uncertainties has attracted a great deal of attention, and the scheme considered here is more general in a practical environment. In this section, to compensate for the different structures of the drive and the response systems, stronger impulsive strength and higher impulsive frequency are necessary. On the other hand, to save the energy of impulsive control and improve its performance, one can introduce an additional feedback controller to compensate for the different structures, and this will lead to a more complicated controller. Hence, to achieve synchronization, we can choose one of the two schemes mentioned above by considering implementation simplicity and performance improvement, which depend on the practical situation.

6.4.3 Simulations

In this simulation, the impulsive distances Δ_k are set to be a positive constant Δ, and the gains B_k are selected as a constant matrix, i.e., $B_k = B = \mathrm{diag}\{d,d,d\}$. Then, $\beta = (d+1)^2$.

In order to observe the synchronization behavior of different chaotic systems, we assume that the Lorenz system drives the Chen system. Therefore, the drive and the response systems are as follows:

$$\begin{cases} \dot{x}_1 = \alpha_1(x_2 - x_1), \\ \dot{x}_2 = \alpha_2 x_1 - x_1 x_3 - x_2, \\ \dot{x}_3 = x_1 x_2 - \alpha_3 x_3 \end{cases} \tag{6.53}$$

and

$$\begin{cases} \dot{\tilde{x}}_1 = \tilde{\alpha}_1(\tilde{x}_2 - \tilde{x}_1), \\ \dot{\tilde{x}}_2 = (\tilde{\alpha}_2 - \tilde{\alpha}_1)\tilde{x}_1 - \tilde{x}_1\tilde{x}_3 + \tilde{\alpha}_2\tilde{x}_2, \\ \dot{\tilde{x}}_3 = \tilde{x}_1\tilde{x}_2 - \tilde{\alpha}_3\tilde{x}_3. \end{cases} \tag{6.54}$$

From (6.42), (6.53), and (6.54), the following compact form of the error system is obtained:

$$\begin{cases} \dot{e}(t) = (A + \Delta A + N)e(t) + (A - \tilde{A} + \Delta A - \Delta\tilde{A} + U)\tilde{x}(t),\ t \neq t_k, \\ \Delta e|_{t=t_k} = e(t_k^+) - e(t_k) = B_k e(t_k),\ t = t_k,\ k \in \mathbb{N}, \\ e(t_0^+) = e_0. \end{cases}$$

Let $\alpha_1 = 10$, $\alpha_2 = 28$, $\alpha_3 = 8/3$, $\tilde{\alpha}_1 = 35$, $\tilde{\alpha}_2 = 28$, and $\tilde{\alpha}_3 = 3$. Systems (6.53) and (6.54) are chaotic with the above parameters. Then,

$$A = \begin{pmatrix} -\alpha_1 & \alpha_1 & 0 \\ \alpha_2 & -1 & 0 \\ 0 & 0 & -\alpha_3 \end{pmatrix} = \begin{pmatrix} -10 & 10 & 0 \\ 28 & -1 & 0 \\ 0 & 0 & -8/3 \end{pmatrix},\quad A + A^{\mathrm{T}} = \begin{pmatrix} -20 & 38 & 0 \\ 38 & -2 & 0 \\ 0 & 0 & -16/3 \end{pmatrix},$$

$$\tilde{A} = \begin{pmatrix} -\tilde{\alpha}_1 & \tilde{\alpha}_1 & 0 \\ \tilde{\alpha}_2 - \tilde{\alpha}_1 & \tilde{\alpha}_2 & 0 \\ 0 & 0 & -\tilde{\alpha}_3 \end{pmatrix} = \begin{pmatrix} -35 & 35 & 0 \\ -7 & 28 & 0 \\ 0 & 0 & -3 \end{pmatrix},\ \text{and}\ A - \tilde{A} = \begin{pmatrix} 25 & -25 & 0 \\ 35 & -29 & 0 \\ 0 & 0 & 1/3 \end{pmatrix}.$$

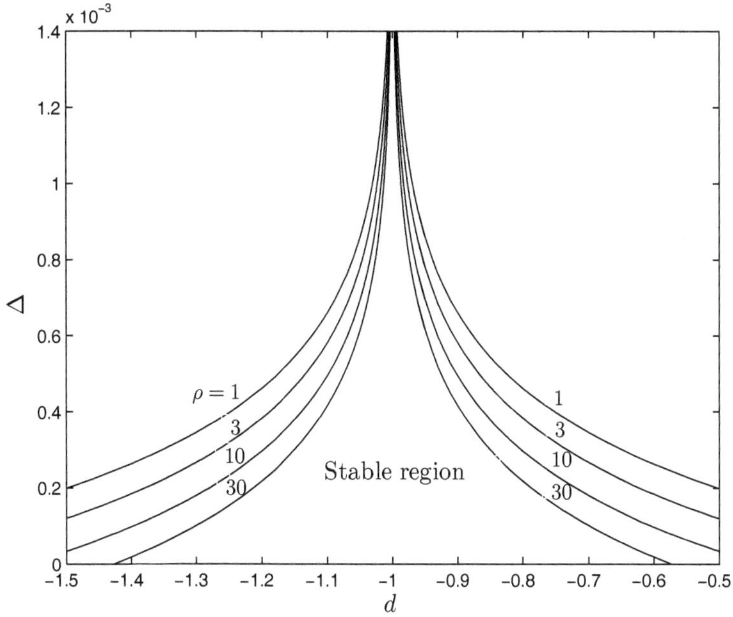

Fig. 6.5 The boundaries of stable region with different ρ

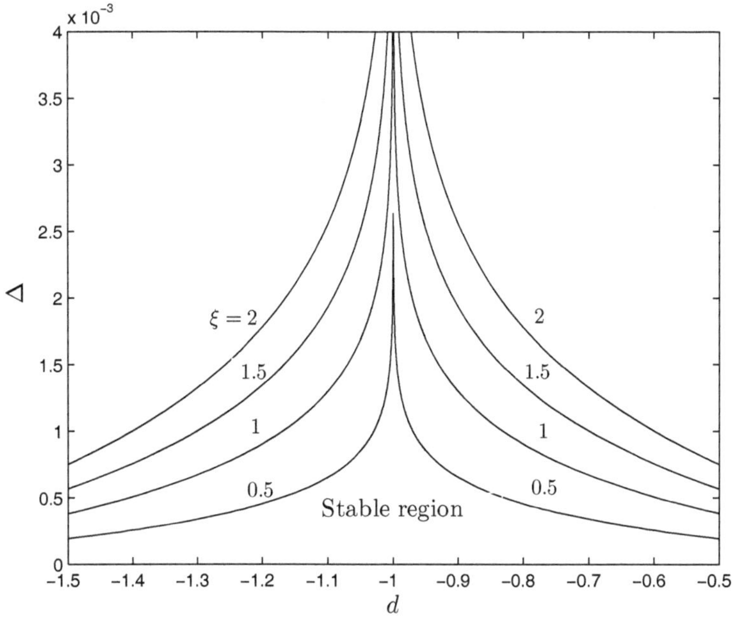

Fig. 6.6 The boundaries of stable region with different ξ

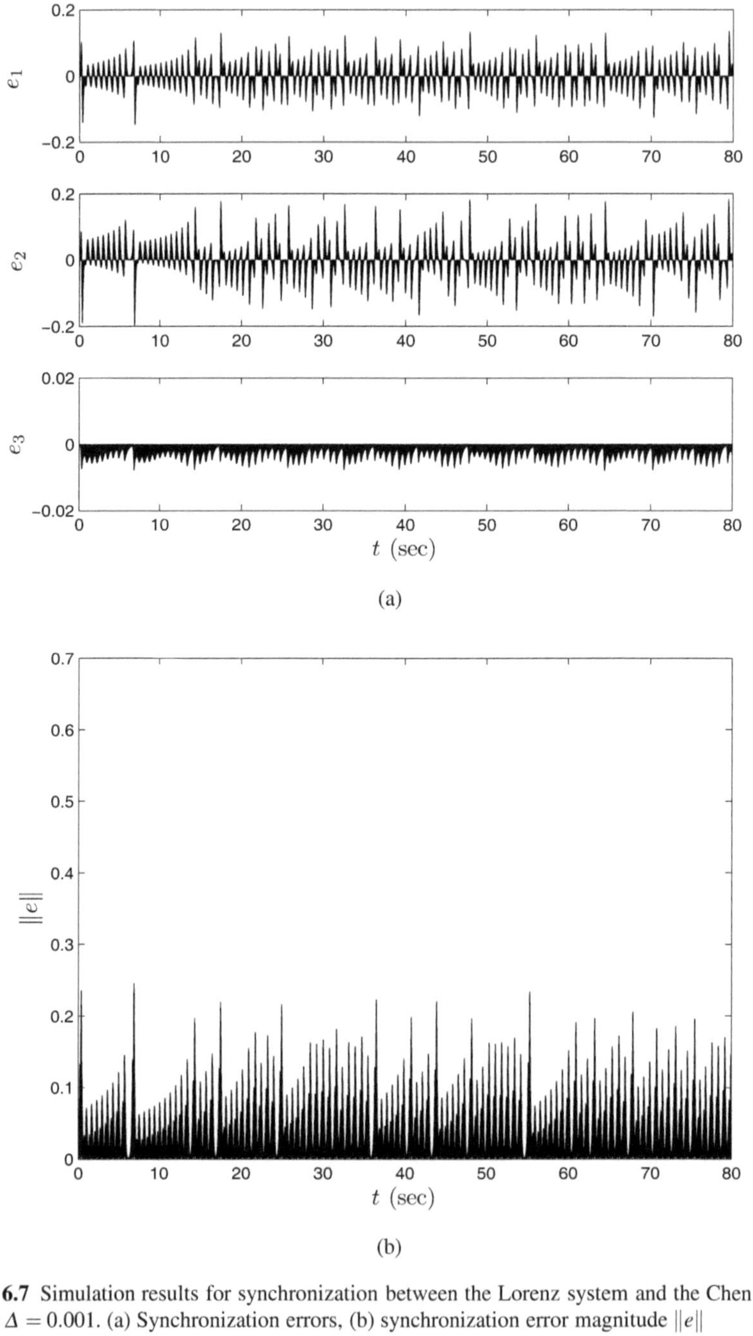

Fig. 6.7 Simulation results for synchronization between the Lorenz system and the Chen system with $\Delta = 0.001$. (a) Synchronization errors, (b) synchronization error magnitude $||e||$

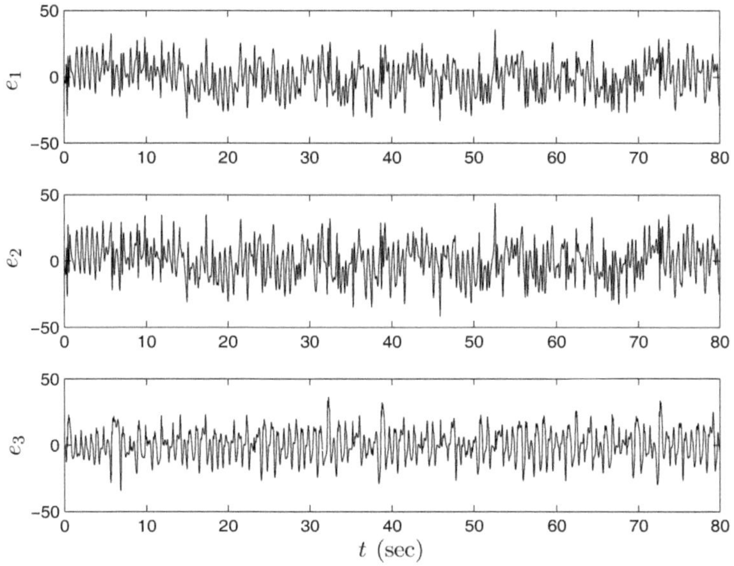

Fig. 6.8 Synchronization errors without impulsive control between the Lorenz system and the Chen system with $\Delta = 0.001$

Let $\Delta A = -\Delta\tilde{A} = 0.04\begin{pmatrix} \sin t & 0 & 0 \\ 0 & \cos t & 0 \\ 0 & 0 & \sin t \end{pmatrix}$, where $E = H = \tilde{E} = \tilde{H} = 0.2I$ and $F(t) = -\tilde{F}(t) = \begin{pmatrix} \sin t & 0 & 0 \\ 0 & \cos t & 0 \\ 0 & 0 & \sin t \end{pmatrix}$. We have

$$Ne = f(x) - f(\tilde{x}) = \begin{pmatrix} 0 \\ -x_1x_3 \\ x_1x_2 \end{pmatrix} - \begin{pmatrix} 0 \\ -\tilde{x}_1\tilde{x}_3 \\ \tilde{x}_1\tilde{x}_2 \end{pmatrix} = \begin{pmatrix} 0 & 0 & 0 \\ -x_3 & 0 & -\tilde{x}_1 \\ x_2 & \tilde{x}_1 & 0 \end{pmatrix}\begin{pmatrix} e_1 \\ e_2 \\ e_3 \end{pmatrix},$$

i.e., $N = \begin{pmatrix} 0 & 0 & 0 \\ -x_3 & 0 & -\tilde{x}_1 \\ x_2 & \tilde{x}_1 & 0 \end{pmatrix}$. Then, $N + N^{\mathrm{T}} = \begin{pmatrix} 0 & -x_3 & x_2 \\ -x_3 & 0 & 0 \\ x_2 & 0 & 0 \end{pmatrix}$ and

$$\|N + N^{\mathrm{T}}\| = \sqrt{x_2^2 + x_3^2} \leq \sqrt{x_1^2 + x_2^2 + x_3^2} = \|x\| \leq \varsigma = \phi.$$

From $f(\tilde{x}) = g(\tilde{x}) = \begin{pmatrix} 0 \\ -\tilde{x}_1\tilde{x}_3 \\ \tilde{x}_1\tilde{x}_2 \end{pmatrix}$, we have $U = 0$ and $\varphi = 0$.

In this simulation, we choose $\varsigma = 55$ and $\tilde{\varsigma} = 60$. By computation, we get $\|A+A^{\mathrm{T}}\| = 50.0512$, $\|A-\tilde{A}\| = 57.5257$, $\|E\|\,\|H\| = \|\tilde{E}\|\,\|\tilde{H}\| = 0.04$, and $\phi = \varsigma = 55$. Fig. 6.5 shows the stable region for different ρ. The whole region under the curve of $\rho = 1$ is the stable region. When $\rho \to \infty$, the stable region approaches a vertical line $d = -1$. Similarly, Fig. 6.6 shows the stable region for different ξ.

We choose $\rho = 1.1$, $\xi = 0.5$, $d = -0.98$, and the 'unknown' channel time delay $\eta = 3$ sec. The initial conditions of the drive and response systems are taken as $x_0 = (4,1,5)^{\mathrm{T}}$ and $\tilde{x}_0 = (3.6,0.8,5.4)^{\mathrm{T}}$, respectively. Hence, from Corollary 6.4, the estimate of the bound of the stable region is given by

$$0 < \Delta \le -\frac{\ln\rho + \ln\beta^2}{13930} = 0.0011.$$

Using these parameters, conditions in Corollary 6.4 are satisfied for $\Delta \le 0.0011$. Robust impulsive synchronization between the Lorenz system and the Chen system with impulsive distance $\Delta = 0.001$ is given in Fig. 6.7. We can see that the synchronization has been achieved practically and $\|e\|$ is smaller than $\xi = 0.5$. Fig. 6.8 shows the synchronization errors without impulsive control.

6.5 Impulsive Synchronization of a Class of Chaotic Delayed Neural Networks

Recently, there have been many efforts for the study of dynamical properties of delayed neural networks (DNNs) [1, 9]. Furthermore, it has been shown that these networks can exhibit some complicated dynamics and even chaotic behavior if the parameters and time delays are appropriately chosen [3, 10, 23]. Thus, the problem of stabilization and synchronization of chaotic DNNs has received extensive consideration [15, 21, 22]. This section addresses a practical issue of using an impulsive control method to synchronize a class of chaotic DNNs with different time-varying parametric uncertainties. This class of chaotic DNNs includes several well-known chaotic DNNs, such as chaotic delayed Hopfield neural networks (DHNNs) and chaotic delayed cellular neural networks (DCNNs), etc. Based on the theory of Lyapunov stability and impulsive FDEs, some new sufficient conditions are derived, which are expressed in terms of matrix norm inequalities. Furthermore, the synchronization error magnitude can be reduced arbitrarily as long as some specific conditions hold.

6.5.1 Preliminaries

A class of chaotic DNNs considered in this section is described by the following delayed differential equations:

$$\dot{u}_i(t) = -d_i \left[c_i u_i(t) - \sum_{j=1}^{n} a_{ij} \hat{f}_j(u_j(t)) - \sum_{j=1}^{n} b_{ij} \hat{f}_j(u_j(t-\tau)) + J_i \right], \quad i = 1,2,\ldots,n, \tag{6.55}$$

or

$$\dot{u}(t) = -D\left[Cu(t) - A\hat{f}(u(t)) - B\hat{f}(u(t-\tau)) + J\right], \tag{6.56}$$

where $n \geq 2$ denotes the number of neurons, $u(t) = (u_1(t), u_2(t), \ldots, u_n(t))^{\mathrm{T}}$ denotes the state vector associated with the neurons, $D = \mathrm{diag}\{d_1, d_2, \ldots, d_n\} > 0$ represents the amplification gain matrix, and $C = \mathrm{diag}\{c_1, c_2, \ldots, c_n\} > 0$ is an appropriate coefficient matrix such that the solution of the model given in (6.55) or (6.56) remains bounded. The feedback matrix $A = (a_{ij})_{n\times n}$ indicates the strength of neuron interconnections, while the delayed feedback matrix $B = (b_{ij})_{n\times n}$ indicates the strength of neuron interconnections with constant delay parameter τ. The activation function, $\hat{f}(u) = (\hat{f}_1(u_1), \hat{f}_2(u_2), \ldots, \hat{f}_n(u_n))^{\mathrm{T}}$, describes the behavior of the neurons' response, and $J = (J_1, J_2, \ldots, J_n)^{\mathrm{T}}$ is a real constant input vector.

The activation function, $\hat{f}_i(x_i)$, satisfies the following assumption.

Assumption 6.1. Each function $\hat{f}_i \colon \mathbb{R} \to \mathbb{R}$ $(i = 1,2,\ldots,n)$ is bounded, and satisfies the Lipschitz condition with a Lipschitz constant $\sigma_i > 0$, i.e., $\left|\hat{f}_i(u) - \hat{f}_i(v)\right| \leq \sigma_i |u - v|$ for all $u, v \in \mathbb{R}$. □

Remark 6.18. Equation (6.55) unifies several well-known neural networks such as Hopfield neural networks with or without delays and cellular neural networks with or without delays. In particular, if $d_i = 1$ and the activation function $\hat{f}_i(u_i)$ is sigmoid, then (6.55) describes the dynamics of Hopfield neural networks. Similarly, if $d_i = 1$ and the activation function $\hat{f}_i(u_i) = (|u_i + 1| - |u_i - 1|)/2$, then (6.55) describes the dynamics of cellular neural networks. □

Under Assumption 6.1, it is not difficult to prove the existence of an equilibrium point of system (6.56) by using Brouwer's fixed point theorem. Let $u^* = (u_1^*, u_2^*, \ldots, u_n^*)^{\mathrm{T}}$ be the equilibrium point of system (6.56). For convenience, we shift the equilibrium point u^* of (6.56) to the origin by letting $x(t) = u(t) - u^*$, and then system (6.56) can be transformed into

$$\dot{x}(t) = -D\left[Cx(t) - Af(x(t)) - Bf(x(t-\tau))\right], \tag{6.57}$$

where $x(t) = (x_1(t), x_2(t), \ldots, x_n(t))^{\mathrm{T}}$ is the state vector of the new system (6.57), and $f(x(t)) = (f_1(x_1), f_2(x_2), \ldots, f_n(x_n))^{\mathrm{T}}$ with $f_i(x_i) = \hat{f}_i(x_i + u_i^*) - \hat{f}_i(x_i)$, $i = 1,2,\ldots,n$. It is easy to see that by Assumption 6.1, $f_i(x_i)$ satisfies the following assumption.

Assumption 6.2. Each function $f_i \colon \mathbb{R} \to \mathbb{R}$ $(i = 1,2,\ldots,n)$ is bounded, and satisfies the Lipschitz condition with a Lipschitz constant $\sigma_i > 0$, i.e., $|f_i(u) - f_i(v)| \leq \sigma_i |u - v|$ for all $u, v \in \mathbb{R}$. □

To derive our main results, the following lemma is necessary.

First, some notation about impulsive FDEs is given. We denote

$$\mathcal{K}^* := \{\psi \in \mathcal{K} : \psi(s) < s, \ \forall s > 0\}$$

and

$$\Xi := \{H \in C(\mathbb{R}^+, \mathbb{R}^+) : H(0) = 0, \ H(s) > 0, \ \forall s > 0\}.$$

Consider the following impulsive FDEs:

$$\begin{cases} \dot{x}(t) = f(t, x_t), \ t \geq t_0, \\ x(t_k^+) = J_k(x(t_k^-)), \ k \in \mathbb{N}, \end{cases} \tag{6.58}$$

where $f\colon [t_0, \infty) \times PC([-\tau, 0], \mathbb{R}^n) \to \mathbb{R}^n$ and $J_k(x)\colon S(\rho) \to \mathbb{R}^n$ for each $k \in \mathbb{N}$. For any $t \geq t_0$, $x_t \in PC([-\tau, 0], \mathbb{R}^n)$ is defined by $x_t(s) = x(t+s)$, $-\tau \leq s \leq 0$. Suppose that the set of discrete instants $\{t_k\}$ satisfies

$$0 < t_1 < t_2 < \cdots < t_k < t_{k+1} < \cdots, \ \lim_{k \to \infty} t_k = \infty, \ 0 \leq t_0 < t_1, \ k \in \mathbb{N}.$$

We denote $\lim_{t \to t_k^+} x(t) = x(t_k^+)$ and $\lim_{t \to t_k^-} x(t) = x(t_k^-)$. Then, $x(t_k) = x(t_k^-)$ implies that $x(t)$ is left-continuous for $t \geq t_0$.

Lemma 6.3 (Lyapunov-like Stability Theorem [7, 16]). The zero solution of (6.58) is uniformly asymptotically stable if there exist $V \in \mathcal{V}_0$, $\omega_1, \omega_2 \in \mathcal{K}$, $\psi \in \mathcal{K}^*$ and $H \in \Xi$ such that

(i) $\omega_1(\|x\|) \leq V(t, x) \leq \omega_2(\|x\|)$ for $(t, x) \in [t_0, \infty) \times S(\rho)$.
(ii) For all $x \in S(\rho_1)$, $0 < \rho_1 \leq \rho$, and $k \in \mathbb{N}$, $V(t_k, J_k(x)) \leq \psi(V(t_k^-, x))$.
(iii) For any solution $x(t)$ of (6.58), $V(t+s, x(t+s)) \leq \psi^{-1}(V(t, x))$, $-\tau \leq s \leq 0$, implies that $D^+V(t, x(t)) \leq g(t)H(V(t, x(t)))$, where $g\colon [t_0, \infty) \to \mathbb{R}^+$ is locally integrable and ψ^{-1} is the inverse function of ψ.
(iv) H is nondecreasing and there exist constants $\lambda_2 \geq \lambda_1 > 0$ and $A_0 > 0$ such that, for all $k \in \mathbb{N}$ and $\mu > 0$, $\lambda_1 \leq t_k - t_{k-1} \leq \lambda_2$ and

$$\int_{\psi(\mu)}^{\mu} \frac{\mathrm{d}u}{H(u)} - \int_{t_{k-1}}^{t_k} g(s)\mathrm{d}s \geq A_0. \qquad \square$$

6.5.2 Main Results

In this section, we study the robust synchronization scheme for two identical chaotic DNNs with parametric uncertainties via impulsive control. Consider the following drive system described by:

$$\dot{x}(t) = -D\left[Cx(t) - (A + \Delta A(t))f(x(t)) - (B + \Delta B(t))f(x(t-\tau))\right], \tag{6.59}$$

where $x(t) = (x_1(t), x_2(t), \ldots, x_n(t))^{\mathrm{T}}$ is the state vector, $f(x) = (f_1(x_1), f_2(x_2), \ldots, f_n(x_n))^{\mathrm{T}}$ is the nonlinear vector-value function, $D = \mathrm{diag}\{d_1, d_2, \ldots, d_n\} > 0$, $C = \mathrm{diag}\{c_1, c_2, \ldots, c_n\} > 0$, and $A = (a_{ij}) \in \mathbb{R}^{n\times n}$ and $B = (b_{ij}) \in \mathbb{R}^{n\times n}$ are two real constant matrices.

The parametric uncertainties $\Delta A(t) \in \mathbb{R}^{n\times n}$ and $\Delta B(t) \in \mathbb{R}^{n\times n}$ satisfy the following assumptions.

(i) $\Delta A(t) = E_1 F_1(t) H_1$, $\Delta B(t) = E_2 F_2(t) H_2$, where E_i and H_i $(i = 1,2)$ are known real constant matrices with appropriate dimensions, and the uncertain matrices $F_i(t)$ satisfy $\|F_i(t)\| \le 1$.

(ii) The parametric uncertainties $\Delta A(t)$ and $\Delta B(t)$ do not destroy the chaotic behavior of the chaotic neural networks (6.59).

The response system can be described by

$$\dot{\tilde{x}}(t) = -D\Big[C\tilde{x}(t) - (A + \Delta\tilde{A}(t))f(\tilde{x}(t)) - (B + \Delta\tilde{B}(t))f(\tilde{x}(t-\tau))\Big],$$

where $\tilde{x}(t)$ is the state vector of the response system $\Delta\tilde{A}(t) = \tilde{E}_1\tilde{F}_1(t)\tilde{H}_1$ and $\Delta\tilde{B}(t) = \tilde{E}_2\tilde{F}_2(t)\tilde{H}_2$ satisfy the same assumptions as $\Delta A(t)$ and $\Delta B(t)$.

At the discrete time t_k, the state variables of the drive system are transmitted to the response system as the control inputs such that the state variables of the response system are suddenly changed at these instants. Therefore, the response system can be written in the following form:

$$\begin{cases} \dot{\tilde{x}}(t) = -D\left[C\tilde{x}(t) - (A + \Delta\tilde{A}(t))f(\tilde{x}(t))\right. \\ \qquad\qquad \left. -(B + \Delta\tilde{B}(t))f(\tilde{x}(t-\tau))\right],\ t \ge 0,\ t \ne t_k, \\ \tilde{x}(t_k^+) = \tilde{x}(t_k) + (W_k - I)e(t_k),\ t = t_k,\ k \in \mathbb{N}, \end{cases} \tag{6.60}$$

where $W_k \in \mathbb{R}^{n\times n}$, I is an $n\times n$ identity matrix, and $e(t) = \tilde{x}(t) - x(t)$ is the synchronization error. Then, from (6.59) and (6.60), the following error system is obtained:

$$\begin{cases} \dot{e}(t) = -D\left[Ce(t) - (A + \Delta\tilde{A}(t))h(e(t), x(t))\right. \\ \qquad -(B + \Delta\tilde{B}(t))h(e(t-\tau), x(t-\tau)) - (\Delta\tilde{A}(t) - \Delta A(t))f(x(t)) \\ \qquad \left. -(\Delta\tilde{B}(t) - \Delta B(t))f(x(t-\tau))\right],\ t \ge 0,\ t \ne t_k, \\ e(t_k^+) = W_k e(t_k),\ t = t_k,\ k \in \mathbb{N}, \end{cases} \tag{6.61}$$

where $h(e(t), x(t)) = f(e(t) + x(t)) - f(x(t)) = f(\tilde{x}(t)) - f(x(t))$. Using Assumption 6.2, we have

$$\|h(e(t), x(t))\|^2 = \sum_{i=1}^{n} h_i^2(e_i(t), x_i(t)) \le \sum_{i=1}^{n} \sigma_i^2(e_i^2(t)) \le \sigma^2 \|e(t)\|^2$$

and

$$\|h(e(t-\tau), x(t-\tau))\|^2 \le \sigma^2 \|e(t-\tau)\|^2,$$

where $\sigma = \max\limits_{1\le i\le n}\{\sigma_i\}$.

Remark 6.19. Due to the boundedness of the chaotic signals, there exists a positive constant χ such that $\|x\| \leq \chi$ and $\|\tilde{x}\| \leq \chi$. □

Remark 6.20. According to Lemma 6.3, for investigating the stability of the zero solution of the error system (6.61), we construct such an appropriate Lyapunov-like function $V \in \mathscr{V}_0$ that the first two conditions of Lemma 6.3 hold. Then, by calculating and estimating the Dini derivative of V along the trajectory of the error system, one can obtain the candidates of g and H. Then, based on condition (iv), we derive a sufficient condition to ensure robust stability of the error system. Furthermore, to obtain an appropriate construction of V, the estimate of $D^+V(t,x(t))$ often requires some necessary inequality techniques such as the matrix norm inequality. □

Definition 6.8. The synchronization of systems (6.59) and (6.60) is said to have been achieved if, for arbitrary initial conditions x_0 and $\tilde{x}_0$, the trivial solution of the error system (6.61) converges to a predetermined neighborhood of the origin for any admissible parametric uncertainty. □

Here, the objective is to find conditions on the control gains W_k and the impulsive distances $\delta_k = t_k - t_{k-1}$ $(k \in \mathbb{N})$ such that the error magnitude, i.e., $\|e\|$, converges to below some constant ξ, which implies that the impulsively controlled response system (6.60) is synchronized with the drive system (6.59) for arbitrary initial conditions. Similarly, for convenience of analysis, we choose a constant control gain and the same impulsive distance, i.e., $W_k = wI$ and $\delta_k = \delta$.

Theorem 6.6. Assume that $\bar{\delta}_k = \bar{\delta}$ and $W_k = wI$ for $k \in \mathbb{N}$. If the following conditions hold:

(i) $|w| < 1$,
(ii) $2\ln|w| + v\bar{\delta} < 0$,

then the response system (6.60) is synchronized with the drive system (6.59), where $\xi > 0$ is the bound of the error magnitude $\|e\|$ and can be chosen small enough,

$$\begin{aligned} v = & -2\min_{1\leq i\leq n}\{d_i c_i\} + \sigma\bar{d}\Big[2\|\tilde{E}_1\|\,\|\tilde{H}_1\| + (1+w^{-2})\left(\|B\| + \|\tilde{E}_2\|\,\|\tilde{H}_2\|\right) \\ & + 2\|A\| + \frac{2\chi}{\xi}\left(\|E_1\|\,\|H_1\| + \|\tilde{E}_1\|\,\|\tilde{H}_1\| + \|E_2\|\,\|H_2\| + \|\tilde{E}_2\|\,\|\tilde{H}_2\|\right)\Big] \\ & > 0, \end{aligned}$$

and $\bar{d} = \max\limits_{1\leq i\leq n}\{d_i\}$.

Proof. Choose the following Lyapunov-like function:

$$V(e) = e^{\mathrm{T}}e. \tag{6.62}$$

For all $e \in S(\rho_1)$, $0 < \rho_1 \leq \rho$, and $k \in \mathbb{N}$, we have

$$
\begin{aligned}
V(t_k, e(t_k^+)) &= [W_k e(t_k)]^{\mathrm{T}} W_k e(t_k) \\
&= e^{\mathrm{T}}(t_k) W_k^{\mathrm{T}} W_k e(t_k) \\
&= w^2 e^{\mathrm{T}}(t_k) e(t_k) \\
&= w^2 V(t_k, e(t_k)).
\end{aligned}
$$

Let $\psi(s) = w^2 s$. Then, $\psi \in \mathscr{K}^*$. For any solution of (6.61), if

$$
V(t-\tau, e(t-\tau)) \leq \psi^{-1}(V(t, e(t))),
$$

then we have

$$
e^{\mathrm{T}}(t-\tau) e(t-\tau) \leq w^{-2} e^{\mathrm{T}}(t) e(t).
$$

For $\|e\| \geq \xi$ one can conclude that $\|e\| \leq \|e\|^2/\xi$. Subsequently, calculating the Dini derivative of $V(e)$ along the trajectory of (6.61) yields

$$
\begin{aligned}
D^+ V(t, e(t)) = {} & 2e^{\mathrm{T}}(t) D \big[-Ce(t) + (A + \Delta\tilde{A}(t)) h(e(t), x(t)) \\
& + (B + \Delta\tilde{B}(t)) h(e(t-\tau), x(t-\tau)) \\
& + (\Delta\tilde{A}(t) - \Delta A(t)) f(x(t)) + (\Delta\tilde{B}(t) - \Delta B(t)) f(x(t-\tau)) \big] \\
\leq {} & -2 \textstyle\sum_{i=1}^{n} d_i c_i e_i^2(t) + 2\sigma \|D\| \left(\|A\| + \|\tilde{E}_1\| \, \|\tilde{H}_1\| \right) \|e(t)\|^2 \\
& + 2\sigma \|D\| \left(\|B\| + \|\tilde{E}_2\| \, \|\tilde{H}_2\| \right) \|e(t)\| \, \|e(t-\tau)\| \\
& + 2\sigma \|D\| \left(\|E_1\| \, \|H_1\| + \|\tilde{E}_1\| \, \|\tilde{H}_1\| \right) \|x(t)\| \, \|e(t)\| \\
& + 2\sigma \|D\| \left(\|E_2\| \, \|H_2\| + \|\tilde{E}_2\| \, \|\tilde{H}_2\| \right) \|x(t-\tau)\| \, \|e(t)\| \\
\leq {} & -2 \min_{1 \leq i \leq n} \{d_i c_i\} \|e(t)\|^2 + 2\sigma \bar{d} \left(\|A\| + \|\tilde{E}_1\| \, \|\tilde{H}_1\| \right) \|e(t)\|^2 \\
& + \sigma \bar{d} \left(\|B\| + \|\tilde{E}_2\| \, \|\tilde{H}_2\| \right) \left(\|e(t)\|^2 + \|e(t-\tau)\|^2 \right) \\
& + \frac{2\sigma\chi\bar{d}}{\xi} \left(\|E_1\| \, \|H_1\| + \|\tilde{E}_1\| \, \|\tilde{H}_1\| \right) \|e(t)\|^2 \\
& + \frac{2\sigma\chi\bar{d}}{\xi} \left(\|E_2\| \, \|H_2\| + \|\tilde{E}_2\| \, \|\tilde{H}_2\| \right) \|e(t)\|^2 \\
\leq {} & \Bigg[-2 \min_{1 \leq i \leq n} \{d_i c_i\} + 2\sigma \bar{d} \left(\|A\| + \|\tilde{E}_1\| \, \|\tilde{H}_1\| \right) \\
& + \sigma \bar{d} (1 + w^{-2}) \left(\|B\| + \|\tilde{E}_2\| \, \|\tilde{H}_2\| \right) \\
& + \frac{2\sigma\chi\bar{d}}{\xi} \big(\|E_1\| \, \|H_1\| + \|\tilde{E}_1\| \, \|\tilde{H}_1\| \\
& + \|E_2\| \, \|H_2\| + \|\tilde{E}_2\| \, \|\tilde{H}_2\| \big) \Bigg] \|e(t)\|^2 \\
= {} & vV(t, e(t)).
\end{aligned}
$$

Let $g(t) = 1$ and $H(s) = vs$. Condition (iv) of Lemma 6.3 implies that

$$\begin{aligned}\int_{\psi(\mu)}^{\mu}\frac{du}{H(u)}-\int_{t_{k-1}}^{t_k}g(s)ds&=\int_{w^2\mu}^{\mu}\frac{du}{vu}-(t_k-t_{k-1})\\&=-\frac{2}{v}\ln|w|-(t_k-t_{k-1})\\&=-\frac{1}{v}(2\ln|w|+v\bar{\delta})\\&=A_0\\&>0.\end{aligned}$$

It is easy to see that (6.62) satisfies condition (i) of Lemma 6.3. Thus, the conditions of Lemma 6.3 are all satisfied. Therefore, we can conclude that the error magnitude $\|e\|$ can converge to below the constant ξ if the error started from $\|e\|>\xi$, which implies that the response system (6.60) is synchronized with the drive system (6.59). This completes the proof. □

6.5.3 Simulations

In order to observe the synchronization behavior in two chaotic DNNs, we define the drive system and the response system as

$$\dot{x}(t)=-D\left[Cx(t)-(A+\Delta A(t))f(x(t))-(B+\Delta B(t))f(x(t-\tau))\right]$$

and

$$\dot{\tilde{x}}(t)=-D\left[C\tilde{x}(t)-(A+\Delta\tilde{A}(t))f(\tilde{x}(t))-(B+\Delta\tilde{B}(t))f(\tilde{x}(t-\tau))\right],$$

where

$$A=\begin{pmatrix}1+\pi/4 & 20\\ 0.1 & 1+\pi/4\end{pmatrix},\quad B=\begin{pmatrix}-1.3\sqrt{2}\pi/4 & 0.1\\ 0.1 & -1.3\sqrt{2}\pi/4\end{pmatrix},$$

$D=C=I$, $f(x)=(f_1(x_1),f_2(x_2))^{\mathrm{T}}$, and $f_i(x_i)=(|x_i+1|-|x_i-1|)/2$. Let $F_1(t)=-\tilde{F}_1(t)=\begin{pmatrix}\sin t & 0\\ 0 & \sin t\end{pmatrix}$, $F_2(t)=-\tilde{F}_2(t)=\begin{pmatrix}\cos t & 0\\ 0 & \cos t\end{pmatrix}$, and $E_i=\tilde{E}_i=H_i=\tilde{H}_i=0.2I$ $(i=1,2)$.

From (6.59), (6.60), and (6.61), we get the error system as follows:

$$\begin{cases}\dot{e}(t)=-D\left[Ce(t)-(A+\Delta\tilde{A}(t))h(e(t),x(t))\right.\\ \qquad-(B+\Delta\tilde{B}(t))h(e(t-\tau),x(t-\tau))-(\Delta\tilde{A}(t)-\Delta A(t))f(x(t))\\ \qquad\left.-(\Delta\tilde{B}(t)-\Delta B(t))f(x(t-\tau))\right],\ t\geq 0,\ t\neq t_k,\\ e(t_k^+)=W_ke(t_k),\ t=t_k,\ k\in\mathbb{N}.\end{cases}$$

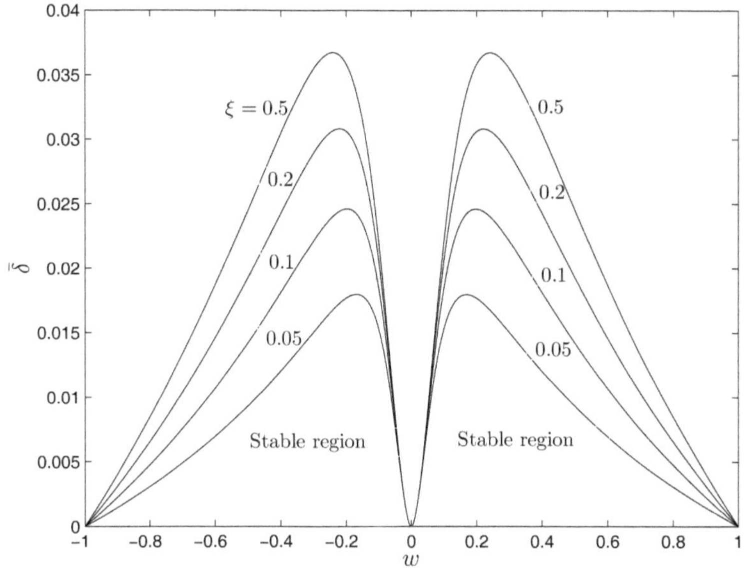

Fig. 6.9 The boundaries of stable region with different ξ

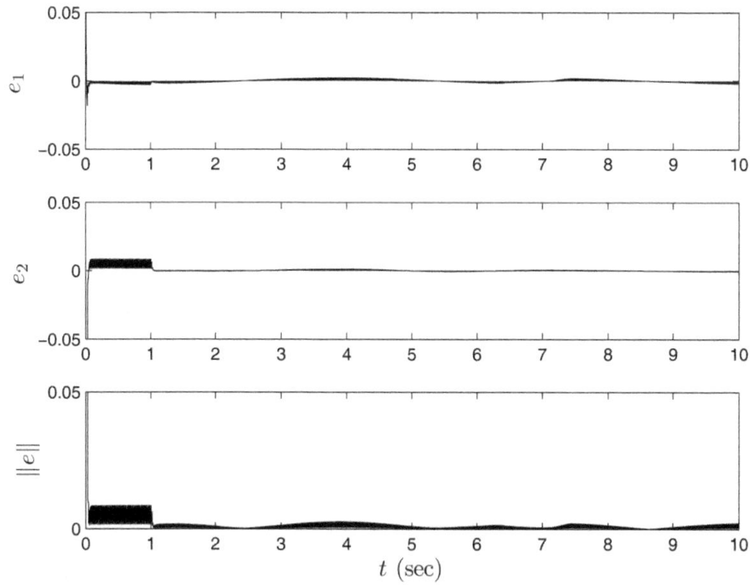

Fig. 6.10 Synchronization errors and synchronization error magnitude with $\bar{\delta} = 0.015$

We choose $\chi = 16$. From Theorem 6.6, we get the stable region for different ξ, which is shown in Fig. 6.9. Let $w = 0.2$, $\xi = 0.05$, and $\sigma_1 = \sigma_2 = 1$. Then, the estimate of the bound of the stable region is given by

$$\bar{\delta} < -\frac{2}{v}\ln|w| = -\frac{2}{181.9792}\ln|0.2| = 0.0177.$$

Using these parameters, conditions in Theorem 6.6 are satisfied for $\bar{\delta} < 0.0177$. Synchronization errors and synchronization error magnitude with impulsive distance $\bar{\delta} = 0.015$ are given in Fig. 6.10. As we can see, synchronization has been achieved practically and $\|e\|$ is smaller than $\xi = 0.05$. The initial conditions of the drive and response systems are taken as $x(s) = (0.1, 0.1)^{\mathrm{T}}$ and $\tilde{x}(s) = (0.2, -0.2)^{\mathrm{T}}$, respectively, for $-1 \leq s \leq 0$.

6.6 Summary

In this chapter we have studied synchronization schemes for both identical and different chaotic systems. Sect. 6.2 studies the complete synchronization of a class of chaotic systems. Sects. 6.3–6.5 mainly focus on robust synchronization, and the synchronization error will converge to a predetermined level, which can be reduced arbitrarily as long as some specific conditions hold. It should be pointed out that the conditions we give for synchronization are only sufficient ones, and continued research is desirable since less conservative conditions still need to be developed, which are interesting and important for secure communications.

References

1. Cao J (2001) Global stability conditions for delayed CNNs. IEEE Trans Circuits Syst I 48:1330–1333
2. Gau RS, Lien CH, Hsieh JG (2007) Global exponential stability for uncertain cellular neural networks with multiple time-varying delays via LMI approach. Chaos Solitons Fractals 32:1258–1267
3. Gilli M (1993) Strange attractors in delayed cellular neural networks. IEEE Trans Circuits Syst I 40:849–853
4. Huang H, Li HX, Zhong J (2006) Master–slave synchronization of general Lur'e systems with time-varying delay and parameter uncertainty. Int J Bifurc Chaos 16:281–294
5. Jiang GP, Zheng WX, Chen GR (2004) Global chaos synchronization with channel time-delay. Chaos Solitons Fractals 20:267–275
6. Lakshmikantham V, Leela S, Martynyuk AA (1990) *Practical Stability of Nonlinear Systems*. World Scientific, Singapore
7. Li CD, Liao XF, Yang XF (2005) Impulsive stabilization and synchronization of a class of chaotic delay systems. Chaos 15:043103
8. Li P, Cao JD, Wang ZD (2007) Robust impulsive synchronization of coupled delayed neural networks with uncertainties. Physica A 373:261–272

9. Liu QS, Cao JD, Xia YS (2005) A delayed neural network for solving linear projection equations and its analysis. IEEE Trans Neural Networks 16:834–843
10. Lu HT (2002) Chaotic attractors in delayed neural networks. Phys Lett A 298:109–116
11. Ma TD, Zhang HG, Wang ZL (2007) Impulsive synchronization for unified chaotic systems with channel time-delay and parameter uncertainty. Acta Phys Sinica 56:3796–3802
12. Ma TD, Zhang HG (2008) Impulsive synchronization of a class of unified chaotic systems with parameter uncertainty. J Syst Simu , 18:4923–4926
13. McRae FA (1994) Practical stability of impulsive control systems. J Math Anal Appl 181:656–672
14. Moon YS, Park P, Kwon WH, Lee YS (2001) Delay-dependent robust stabilization of uncertain state-delayed systems. Int J Control 74:1447–1455
15. Wang ZS, Zhang HG (2006) Global synchronization of a class of chaotic neural networks. Acta Phys Sinica 55:2687–2693
16. Yan JR, Shen JH (1999) Impulsive stabilization and synchronization of functional differential equations by Lyapunov-Razumikhin functions. Nonlinear Anal 37:245–255
17. Yang T (1999) Impulsive control. IEEE Trans Autom Control 44:1081–1083
18. Yang T (2001) *Impulsive Control Theory*. Springer, New York
19. Yang T, Chua LO (2000) Practical stability of impulsive synchronization between two nonautonomous chaotic systems. Int J Bifurc Chaos 10:859-867
20. Zhang HG, Huang W, Wang ZL (2005) Impulsive control for synchronization of a class of chaotic systems. Dyn Contin Discrete Impuls Syst B 12:153–161
21. Zhang HG, Xie YH, Liu D (2006) Synchronization of a class of delayed chaotic neural networks with fully unknown parameters. Dyn Contin Discrete Impuls Syst B 13:297–308
22. Zhang HG, Xie YH, Wang ZL, Zheng CD (2007) Adaptive synchronization between two different chaotic neural networks with time delay. IEEE Trans Neural Networks 18:1841–1845
23. Zou F, Nossek JA (1993) Bifurcation and chaos in cellular neural networks. IEEE Trans Circuits Syst I 40:166–173

Chapter 7
Synchronization of Chaotic Systems with Time Delay

Abstract In many physical, industrial, and engineering systems, delays occur due to the finite capabilities of information processing and data transmission among various parts of the system. Delays could arise as well from inherent physical phenomena like mass transport flow or recycling. Also, they could be by-products of computational delays or could intentionally be introduced for some design considerations. Such delays could be constant or time varying, known or unknown, deterministic or stochastic depending on the system under consideration. In all of these cases, the time-delay factors have counteracting effects on the system behavior and most of the time lead to poor performance. Therefore, the subject of time-delay systems has been investigated as functional differential equations over the past three decades. In this chapter, we study how to synchronize chaotic systems when time delay exists and the synchronized systems have different structures. We first develop synchronization methods for a class of delayed chaotic systems when the drive system and the response system have the same structure but different parameters. After that, the problem of synchronizing different chaotic systems is studied. Some concrete examples are presented to show how to design the controller. Based on that, a more general case, synchronizing two different delayed chaotic neural networks with known and unknown parameters, is considered.

7.1 Introduction

In the last two chapters, we have studied the synchronization of chaotic systems which are described by ordinary differential equations. However, in many physical, industrial, and engineering systems, delays occur due to the finite capabilities of information processing and data transmission among various parts of the system. Delays could arise as well from inherent physical phenomena like mass transport flow or recycling. Also, they could be by-products of computational delays or could intentionally be introduced for some design considerations. Such delays could be constant or time varying, known or unknown, deterministic or stochastic depending on the system under consideration. In all of these cases, the time-delay factors have, by and large, counteracting effects on the system behavior and most of the time lead to poor performance. Therefore, the subject of time-delay systems has been investigated as functional differential equations over the past three decades. This has occupied a separate discipline in mathematical sciences falling between differential and difference equations.

It is well known that chaos can easily be generated in a time-delay system. When synchronizing two time-delay chaotic systems, the values of delays and parameters of the systems are often unknown in advance. In some special cases, the structure or parameters of the drive system are even unknown in advance. Therefore, in this chapter, we study the synchronization of chaotic systems with time delays and with different structures. In Sect. 7.2, we first develop the synchronization methods for a class of delayed chaotic systems when the drive system and the response system have the same structure but different parameters. After that, in Sect. 7.3, the problem of synchronizing different chaotic systems is studied. Some examples are presented to show how to design the controller. Based on that, in Sect. 7.4, a more general case, synchronizing two different delayed chaotic neural networks with known and unknown parameters, is considered.

7.2 Adaptive Synchronization of a Class of Delayed Chaotic Systems

In this section, the synchronization problem of two identical delayed chaotic systems with unknown parameters is studied. We take the delayed Chen system and the delayed Lorenz system as examples to show the effectiveness of the proposed method.

7.2.1 Adaptive Synchronization of the Delayed Chen Chaotic Systems

The drive delayed Chen chaotic system [14] in this section is described by the following differential equations:

$$\begin{cases} \dot{x}_1(t) = a(x_2(t) - x_1(t)), \\ \dot{x}_2(t) = (c-a)x_1(t) - x_1(t)x_3(t) + (c-d)x_2(t) + dx_2(t-\tau), \\ \dot{x}_3(t) = x_1(t)x_2(t) - bx_3(t), \end{cases} \tag{7.1}$$

where a, b, c, and d are the unknown parameters of system (7.1), $d < \eta$, and η is a positive constant. The chaotic behavior of the drive system (7.1) is shown in Fig. 7.1, when the initial values of the parameters of the drive system (7.1) are selected as $a = 35$, $b = 3$, $c = 28$, $d = 3$, and $\tau = 0.1$.

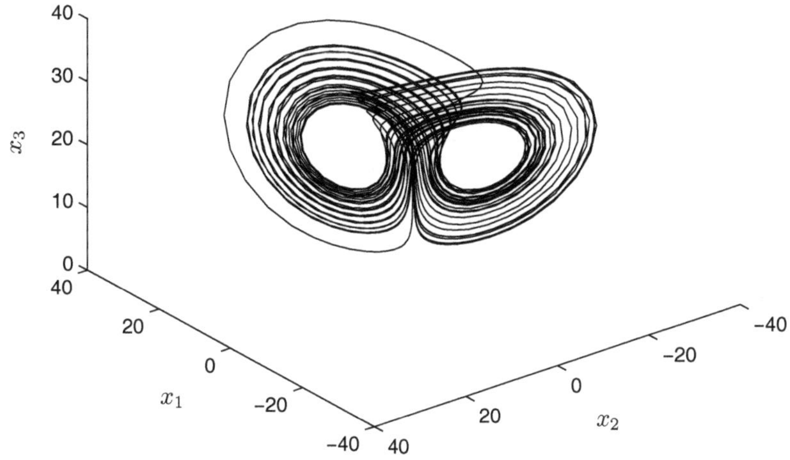

Fig. 7.1 Chaotic behavior of the delayed Chen chaotic system

The response delayed Chen chaotic system is described by the following equations:

$$\begin{cases} \dot{y}_1(t) = a(y_2(t) - y_1(t)) + u_1, \\ \dot{y}_2(t) = (c-a)y_1(t) - y_1(t)y_3(t) + (c-d)\,y_2(t) + dy_2(t-\tau) + u_2, \\ \dot{y}_3(t) = y_1(t)y_2(t) - by_3(t) + u_3, \end{cases} \tag{7.2}$$

where u_1, u_2, and u_3 denote the designed controllers that will realize the synchronization of system (7.1) and system (7.2).

Let us define the synchronization error signal as $e_i(t) = y_i(t) - x_i(t)$ and $e_{i\tau}(t) = y_i(t-\tau) - x_i(t-\tau)$ $(i = 1,2,3)$. The fact that $e_i(t) \to 0$ as $t \to \infty$ means that the drive system and the response system are synchronized. The goal of the controllers' design is to obtain $u_i(t)$ so that $\lim_{t\to\infty} \|e_i(t)\| = \lim_{t\to\infty} \|y_i(t) - x_i(t)\| = 0$, where $\|\cdot\|$ denotes the Euclidean norm.

Therefore, the error system between (7.1) and (7.2) can be expressed as

$$\begin{cases} \dot{e}_1 = a(e_2 - e_1) + u_1, \\ \dot{e}_2 = (c-a)e_1 + (c-d)\,e_2 - e_1e_3 - e_3x_1 - e_1x_3 + de_{2\tau} + u_2, \\ \dot{e}_3 = e_1e_2 + e_1x_2 + e_2x_1 - be_3 + u_3. \end{cases} \tag{7.3}$$

Then, we have the following theorem.

Theorem 7.1 ([16]). By the following controller:

$$\begin{cases} u_1 = \hat{a}(e_1 - e_2) - e_1, \\ u_2 = e_1 x_3 - (\hat{c} - \hat{a}) e_1 + \left(\hat{d} - \hat{c} - k\right) e_2, \\ u_3 = -e_1 x_2 + \left(\hat{b} - 1\right) e_3 \end{cases} \tag{7.4}$$

and the parameter update law

$$\begin{cases} \dot{\hat{a}} = -e_1^2, \\ \dot{\hat{b}} = -e_3^2, \\ \dot{\hat{c}} = e_1 e_2 + e_2^2, \\ \dot{\hat{d}} = -e_2^2, \end{cases} \tag{7.5}$$

the drive system (7.1) and the response system (7.2) will be globally asymptotically synchronized. Here, $\hat{a}$, $\hat{b}$, $\hat{c}$, and $\hat{d}$ are the estimated values of unknown parameters a, b, c, and d, respectively, and $k > (\eta^2 + 1)/2$.

Proof. Substituting (7.4) into (7.3), the error system (7.3) can be rewritten as

$$\begin{cases} \dot{e}_1 = (a - \hat{a})(e_2 - e_1) - e_1, \\ \dot{e}_2 = (c - \hat{c} - a + \hat{a}) e_1 + \left(c - \hat{c} - d + \hat{d} - k\right) e_2 - e_1 e_3 - e_3 x_1 + d e_{2\tau}, \\ \dot{e}_3 = e_1 e_2 + e_2 x_1 - \left(b - \hat{b}\right) e_3 - e_3. \end{cases} \tag{7.6}$$

Let $e_a = a - \hat{a}$, $e_b = b - \hat{b}$, $e_c = c - \hat{c}$, and $e_d = d - \hat{d}$ be the estimated errors of the unknown parameters a, b, c, and d, respectively. Now construct a Lyapunov function in the form of

$$V(e) = \frac{1}{2}\left(e_1^2 + e_2^2 + e_3^2 + e_a^2 + e_b^2 + e_c^2 + e_d^2 + \int_{t-\tau}^{t} e_2^2(s)\,\mathrm{d}s\right).$$

Then, the time derivative of $V(e)$ along the trajectory of the error dynamical system (7.6) is as follows:

$$\begin{aligned} \dot{V}(e) &= e_1\dot{e}_1 + e_2\dot{e}_2 + e_3\dot{e}_3 + e_a\dot{e}_a + e_b\dot{e}_b + e_c\dot{e}_c + e_d\dot{e}_d + \frac{1}{2}e_2^2 - \frac{1}{2}e_{2\tau}^2 \\ &= e_1\left(e_a(e_2 - e_1) - e_1\right) + e_3\left(e_1 e_2 + e_2 x_1 - e_b e_3 - e_3\right) \\ &\quad + e_2\left((e_c - e_a)e_1 + (e_c - e_d - k)e_2 - e_1 e_3 - e_3 x_1 + d e_{2\tau}\right) \\ &\quad - \left(e_a\dot{\hat{a}} + e_b\dot{\hat{b}} + e_c\dot{\hat{c}} + e_d\dot{\hat{d}}\right) + \frac{1}{2}e_2^2 - \frac{1}{2}e_{2\tau}^2. \end{aligned} \tag{7.7}$$

Substituting (7.5) into (7.7), we obtain

$$\begin{aligned} \dot{V}(e) &= -e_1^2 - e_3^2 - k e_2^2 + e_2 d e_{2\tau} + \frac{1}{2}e_2^2 - \frac{1}{2}e_{2\tau}^2 \\ &\le -e_1^2 - e_3^2 - k e_2^2 + \frac{d^2}{2}e_2^2 + \frac{1}{2}e_{2\tau}^2 + \frac{1}{2}e_2^2 - \frac{1}{2}e_{2\tau}^2 \\ &\le -e_1^2 - e_3^2 - \left(k - \frac{\eta^2}{2} - \frac{1}{2}\right) e_2^2. \end{aligned}$$

Setting $Q = k - (\eta^2 + 1)/2$, we obtain $\dot{V}(e) \leq -\left(e_1^2 + e_3^2 + Qe_2^2\right)$.

When $k > (\eta^2 + 1)/2$, i.e., $Q > 0$, we get $\dot{V}(e) \leq 0$, which implies that $\dot{V}(e) < 0$ for all $e(t) \neq 0$. Since $V(e)$ is positive definite and $\dot{V}(e)$ is negative semidefinite, it follows that $e_1 \in L_\infty$, $e_2 \in L_\infty$, and $e_3 \in L_\infty$. From the fact that

$$\int_0^t -(e_1^2 + Qe_2^2 + e_3^2)\mathrm{d}s \leq -\int_0^t \dot{V}\mathrm{d}s \leq V(e_1(0), e_2(0), e_3(0)),$$

we can easily show that $e_1 \in L_2$, $e_2 \in L_2$, and $e_3 \in L_2$. From (7.6) we have $\dot{e}_1 \in L_\infty$, $\dot{e}_2 \in L_\infty$, and $\dot{e}_3 \in L_\infty$. Thus, by Barbalat's lemma, we have

$$\lim_{t \to \infty} \|e_i(t)\| = \lim_{t \to \infty} \|y_i(t) - x_i(t)\| = 0.$$

This means that the proposed controller (7.4) and (7.5) can globally asymptotically synchronize the system (7.1) and the system (7.2). This completes the proof. □

The following illustrative example is used to demonstrate the effectiveness of the above method.

According to Theorem 7.1, the initial condition of the drive delayed Chen chaotic system is $x_0 = (5, 5, 15)^{\mathrm{T}}$, and the initial condition of the response delayed Chen chaotic system is $y_0 = (3, 3, 12)^{\mathrm{T}}$. The initial values for the estimation of parameters are selected as follows: $\hat{a}(0) = 32$, $\hat{b}(0) = 4$, $\hat{c}(0) = 25$, and $\hat{d}(0) = 4$. The synchro-

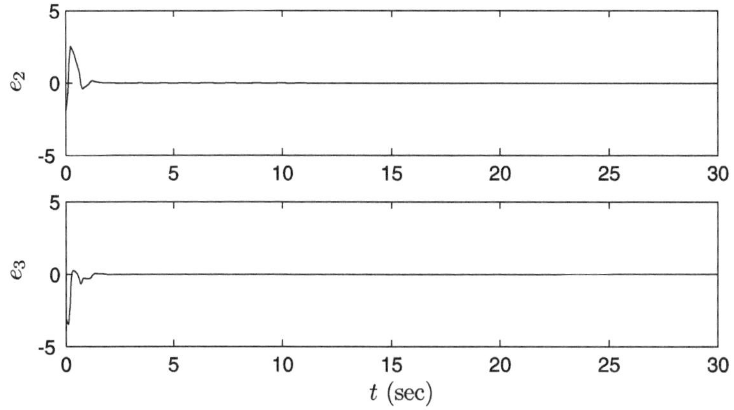

Fig. 7.2 The synchronization error curves of the drive system (7.1) and the response system (7.2)

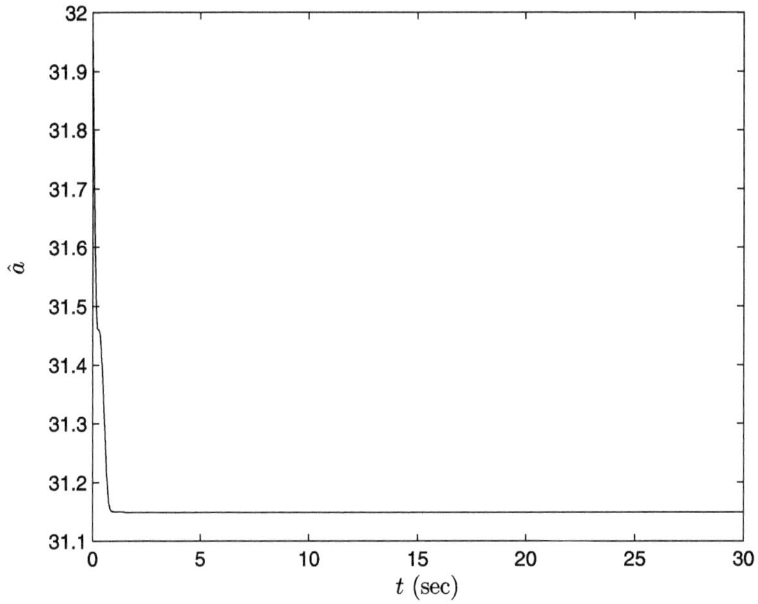

Fig. 7.3 The curve of the adaptive parameter $\hat{a}(t)$

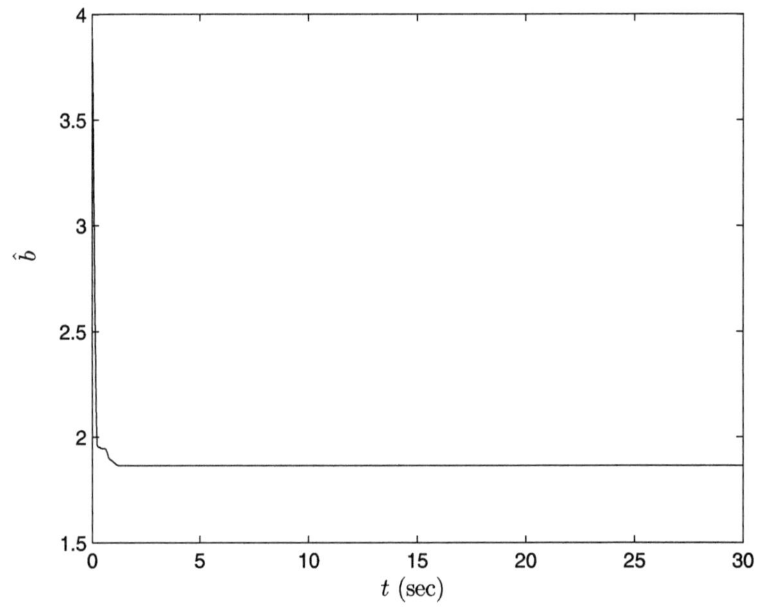

Fig. 7.4 The curve of the adaptive parameter $\hat{b}(t)$

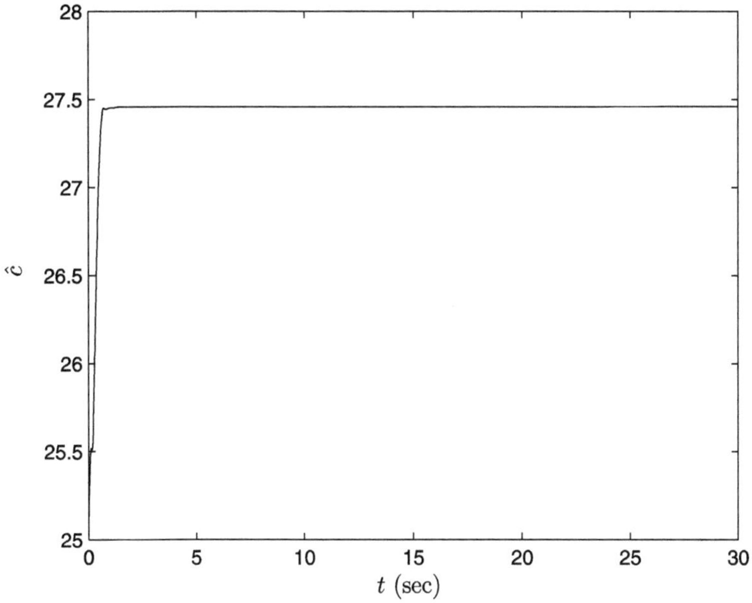

Fig. 7.5 The curve of the adaptive parameter $\hat{c}(t)$

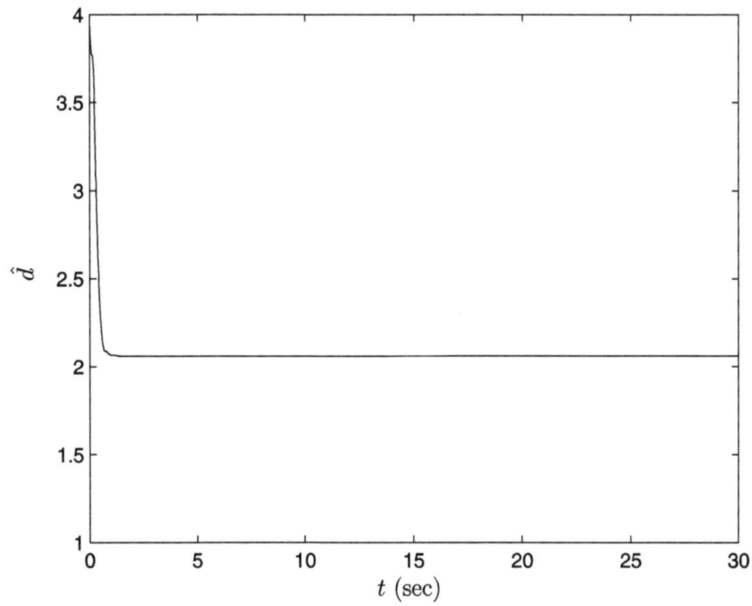

Fig. 7.6 The curve of the adaptive parameter $\hat{d}(t)$

nization error curves of the drive system (7.1) and the response system (7.2) with the controller $u(t) = (u_1(t), u_2(t), u_3(t))^{\mathrm{T}}$ are shown in Fig. 7.2. The curves of the estimated parameters $\hat{a}$, $\hat{b}$, $\hat{c}$, and $\hat{d}$ are shown in Figs. 7.3–7.6, respectively. From simulation results we can see that the synchronization errors converge asymptotically to zero, i.e., the trajectory of the response delayed Chen chaotic system can synchronize asymptotically with the trajectory of the drive delayed Chen chaotic system. The estimated parameters converge asymptotically to some constants, which shows the effectiveness of the adaptive synchronization scheme.

7.2.2 Adaptive Synchronization of the Delayed Lorenz Systems

The drive delayed Lorenz chaotic system [13] is described by the following differential equations:

$$\begin{cases} \dot{x}_1(t) = a(x_2(t) - x_1(t)), \\ \dot{x}_2(t) = cx_1(t) - x_1(t)x_3(t) + (d-1)x_2(t) - dx_2(t-\tau), \\ \dot{x}_3(t) = x_1(t)x_2(t) - bx_3(t), \end{cases} \tag{7.8}$$

where a, b, c, and d are the unknown parameters of the system (7.8), and $d < \eta$. The chaotic behavior of the drive system (7.8) is shown in Fig. 7.7, when the initial values of the parameters of the drive system (7.8) are selected as $a = 10$, $b = 8/3$, $c = 5$, $d = 6.5$, and $\tau = 0.5$.

The response delayed Lorenz chaotic system is described by the following equations:

$$\begin{cases} \dot{y}_1(t) = a(y_2(t) - y_1(t)) + u_1, \\ \dot{y}_2(t) = cy_1(t) - y_1(t)y_3(t) + (d-1)y_2(t) - dy_2(t-\tau) + u_2, \\ \dot{y}_3(t) = y_1(t)y_2(t) - by_3(t) + u_3, \end{cases} \tag{7.9}$$

where u_1, u_2, and u_3 denote the designed controllers that will realize the synchronization between system (7.1) and system (7.2).

Therefore, the error system between (7.8) and (7.9) can be expressed as

$$\begin{cases} \dot{e}_1 = a(e_2 - e_1) + u_1, \\ \dot{e}_2 = ce_1 + (d-1)e_2 - e_1e_3 - e_3x_1 - e_1x_3 - de_{2\tau} + u_2, \\ \dot{e}_3 = e_1e_2 + e_1x_2 + e_2x_1 - be_3 + u_3. \end{cases} \tag{7.10}$$

Then, we have the following theorem.

Theorem 7.2 ([16]). By the controller

$$\begin{cases} u_1 = \hat{a}(e_1 - e_2) - e_1, \\ u_2 = e_1x_3 - \hat{c}e_1 - \left(\hat{d} + k\right)e_2, \\ u_3 = -e_1x_2 + \left(\hat{b} - 1\right)e_3 \end{cases} \tag{7.11}$$

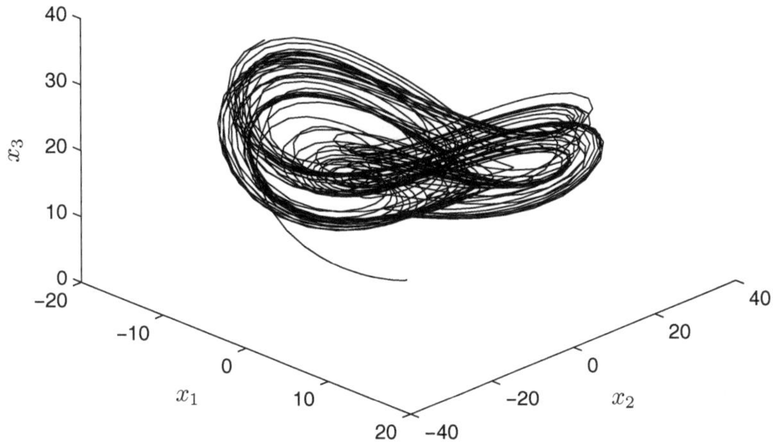

Fig. 7.7 Chaotic behavior of the delayed Lorenz chaotic system

and the parameter update law

$$\begin{cases} \dot{\hat{a}} = e_1 e_2 - e_1^2, \\ \dot{\hat{b}} = -e_3^2, \\ \dot{\hat{c}} = e_1 e_2, \\ \dot{\hat{d}} = e_2^2, \end{cases} \tag{7.12}$$

the drive system (7.8) and the response system (7.9) will be globally asymptotically synchronized. Here, $\hat{a}$, $\hat{b}$, $\hat{c}$, and $\hat{d}$ are the estimated values of unknown parameters a, b, c, and d, respectively, and $k > (\eta^2 - 1)/2$.

Proof. Substituting (7.11) into (7.10), the error system (7.10) can be rewritten by

$$\begin{cases} \dot{e}_1 = (a - \hat{a})(e_2 - e_1) - e_1, \\ \dot{e}_2 = (c - \hat{c})e_1 + \left(d - \hat{d} - k - 1\right) e_2 - e_1 e_3 - e_3 x_1 - d e_{2\tau}, \\ \dot{e}_3 = e_1 e_2 + e_2 x_1 - \left(b - \hat{b}\right) e_3 - e_3. \end{cases} \tag{7.13}$$

Let $e_a = a - \hat{a}$, $e_b = b - \hat{b}$, $e_c = c - \hat{c}$, and $e_d = d - \hat{d}$ be the estimated errors of unknown parameters a, b, c, and d, respectively. Now construct a Lyapunov function in the form of

$$V(e)=\frac{1}{2}\left(e_1^2+e_2^2+e_3^2+e_a^2+e_b^2+e_c^2+e_d^2+\int_{t-\tau}^{t}e_2^2(s)\,\mathrm{d}s\right).$$

Then, the time derivative of $V(e)$ along the trajectory of the error dynamical system (7.13) is as follows:

$$\begin{aligned}\dot V(e)&=e_1\dot e_1+e_2\dot e_2+e_3\dot e_3+e_a\dot e_a+e_b\dot e_b+e_c\dot e_c+e_d\dot e_d+\frac{1}{2}e_2^2-\frac{1}{2}e_{2\tau}^2\\&=e_1\left(e_a(e_2-e_1)-e_1\right)+e_3\left(e_1e_2+e_2x_1-e_be_3-e_3\right)\\&\quad+e_2\left(e_ce_1+(e_d-k-1)\,e_2-e_1e_3-e_3x_1-de_{2\tau}\right)\\&\quad-\left(e_a\dot{\hat a}+e_b\dot{\hat b}+e_c\dot{\hat c}+e_d\dot{\hat d}\right)+\frac{1}{2}e_2^2-\frac{1}{2}e_{2\tau}^2.\end{aligned}\tag{7.14}$$

Substituting (7.12) into (7.14), we obtain

$$\begin{aligned}\dot V(e)&=-e_1^2-e_3^2-ke_2^2-e_2de_{2\tau}-\frac{1}{2}e_2^2-\frac{1}{2}e_{2\tau}^2\\&\le-e_1^2-e_3^2-ke_2^2+\frac{d^2}{2}e_2^2+\frac{1}{2}e_{2\tau}^2-\frac{1}{2}e_2^2-\frac{1}{2}e_{2\tau}^2\\&\le-e_1^2-e_3^2-\left(k-\frac{\eta^2}{2}+\frac{1}{2}\right)e_2^2.\end{aligned}\tag{7.15}$$

Setting $Q=k-(\eta^2-1)/2$, we obtain $\dot V(e)\le-\left(e_1^2+e_3^2+Qe_2^2\right)$.

When $k>(\eta^2-1)/2$, i.e., $Q>0$, we get $\dot V(e)\le 0$, which implies that $\dot V(e)<0$ for all $e(t)\ne 0$. Since $V(e)$ is positive definite and $\dot V(e)$ is negative semidefinite, it follows that $e_1\in L_\infty$, $e_2\in L_\infty$, and $e_3\in L_\infty$. From the fact that

$$\int_0^t-(e_1^2+Qe_2^2+e_3^2)\mathrm{d}s\le-\int_0^t\dot V\mathrm{d}s\le V(e_1(0),e_2(0),e_3(0)),$$

we can easily show that $e_1\in L_2$, $e_2\in L_2$, and $e_3\in L_2$. From (7.10) we have $\dot e_1\in L_\infty$, $\dot e_2\in L_\infty$, and $\dot e_3\in L_\infty$. Thus, by Barbalat's lemma, we have $\lim_{t\to\infty}\|e_i(t)\|=\lim_{t\to\infty}\|y_i(t)-x_i(t)\|=0$. This means that the proposed controller (7.11) with parameter update law (7.12) can globally asymptotically synchronize the system (7.8) and the system (7.9). This completes the proof. □

The following illustrative example is used to demonstrate the effectiveness of the above method.

According to Theorem 7.2, the initial condition of the drive delayed Lorenz chaotic system is $x_0=(-0.1,-0.1,0.2)^{\mathrm T}$, and the initial condition of the response delayed Lorenz chaotic system is $y_0=(3.1,-0.2,0.3)^{\mathrm T}$. The initial values for the estimation of parameters are selected as follows: $\hat a(0)=9$, $\hat b(0)=3$, $\hat c(0)=6$, and $\hat d(0)=7$. The synchronization error curves of the drive system (7.8) and the response system (7.9) with the controller $u(t)=(u_1(t),u_2(t),u_3(t))^{\mathrm T}$ are shown in Fig. 7.8. The curves of the estimated parameters $\hat a$, $\hat b$, $\hat c$, and $\hat d$ are shown in Figs. 7.9–7.12, respectively. From simulation results we can see that the synchronization

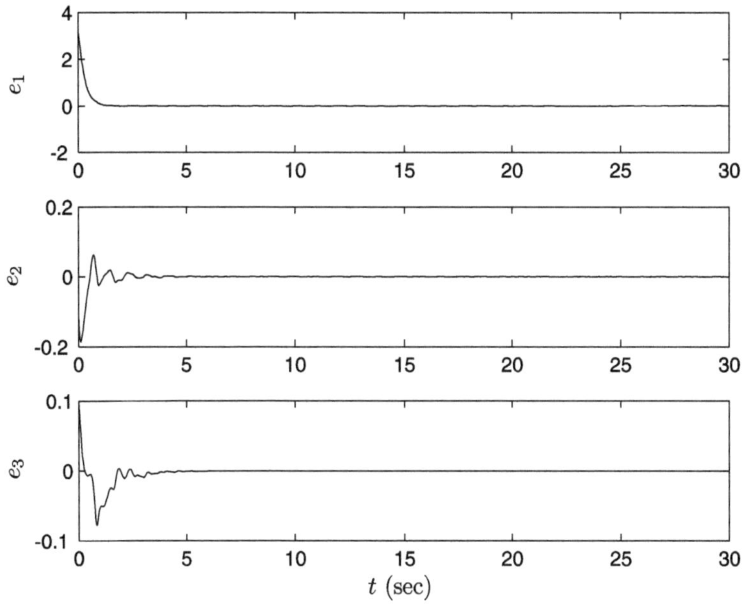

Fig. 7.8 The synchronization error curves of the drive system (7.8) and the response system (7.9)

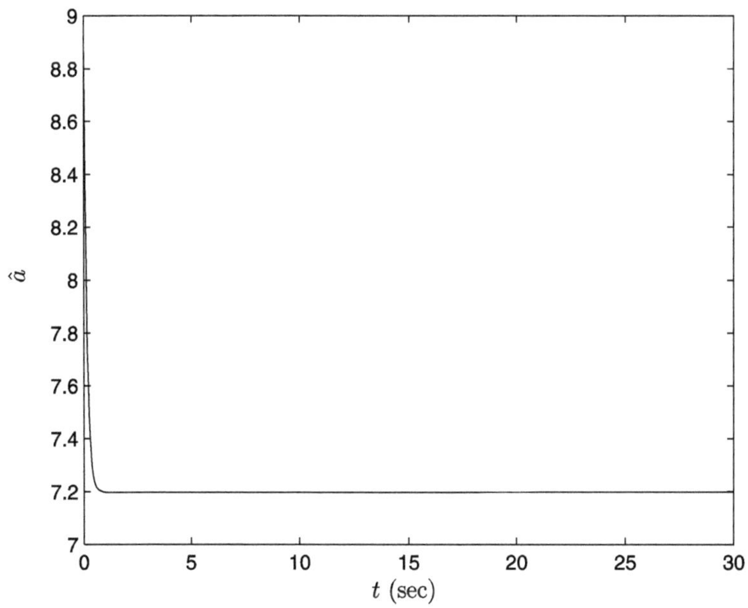

Fig. 7.9 The curve of the adaptive parameter $\hat{a}(t)$

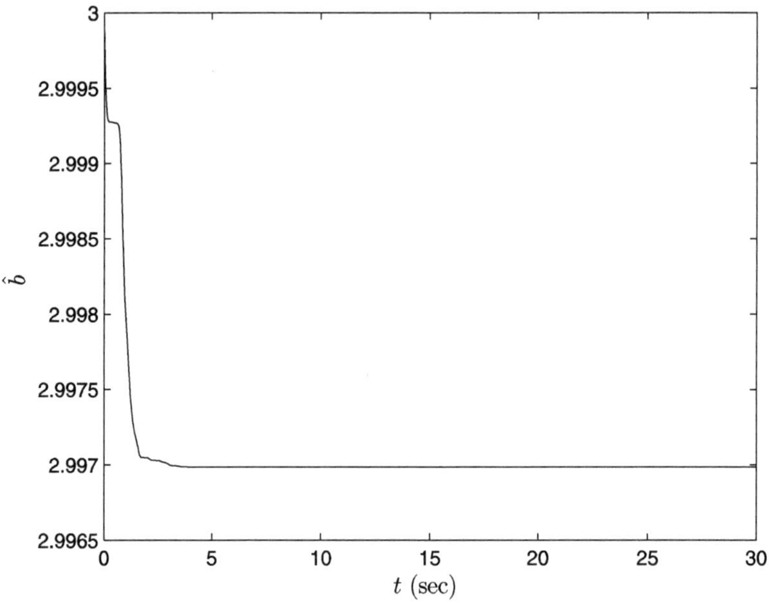

Fig. 7.10 The curve of the adaptive parameter $\hat{b}(t)$

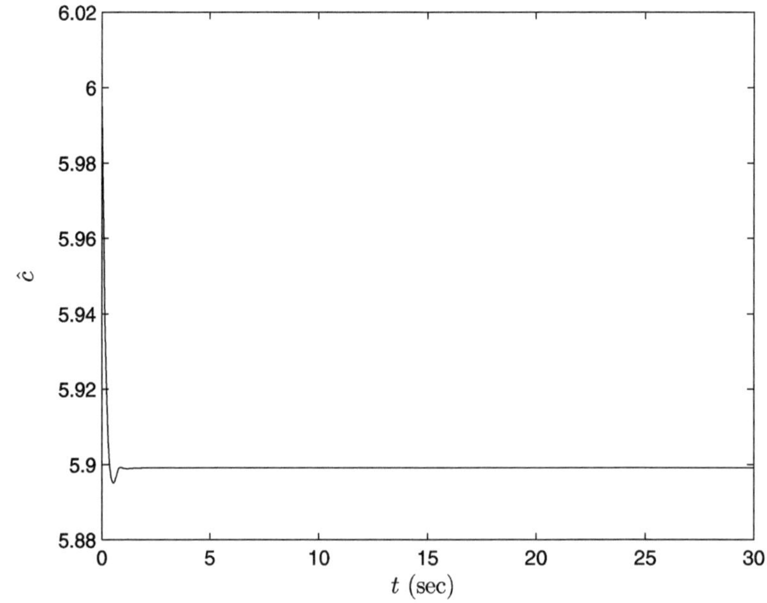

Fig. 7.11 The curve of the adaptive parameter $\hat{c}(t)$

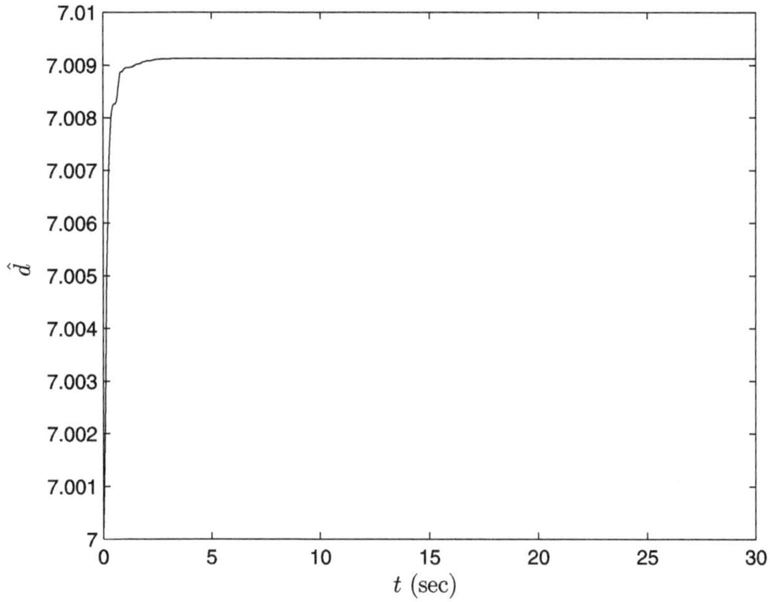

Fig. 7.12 The curve of the adaptive parameter $\hat{d}(t)$

errors converge asymptotically to zero, i.e., the trajectory of the response delayed Lorenz chaotic system can synchronize asymptotically with the trajectory of the drive delayed Lorenz chaotic system. The estimated parameters converge asymptotically to some constants, which shows the effectiveness of the adaptive synchronization scheme.

7.3 Adaptive Synchronization Between Two Different Delayed Chaotic Systems

In the last section, the underlying assumption is that the drive system and the response system have identical dynamic structures. In this section, we consider the problem of synchronizing two different delayed chaotic systems with unknown parameters. Three examples are used to illustrate the design procedure of the proposed synchronization methods.

7.3.1 *Adaptive Synchronization Between Delayed Chen Chaotic Systems and Delayed Lorenz Chaotic Systems*

The drive delayed Chen system in this section is described by differential equations (7.1).

The response delayed Lorenz system is described by the following equations:

$$\begin{cases} \dot{y}_1(t) = \alpha(y_2(t) - y_1(t)) + u_1, \\ \dot{y}_2(t) = \gamma y_1(t) - y_1(t)y_3(t) + (\theta - 1)\,y_2(t) - \theta y_2(t-\tau) + u_2, \\ \dot{y}_3(t) = y_1(t)y_2(t) - \beta y_3(t) + u_3, \end{cases} \tag{7.16}$$

where α, β, γ, and θ are the unknown parameters of system (7.16) and u_1, u_2, and u_3 denote the designed controllers that will realize the synchronization of system (7.1) and system (7.16).

Therefore, the error system between (7.1) and (7.16) can be expressed as

$$\begin{cases} \dot{e}_1 = \alpha(y_2 - y_1) - a(x_2 - x_1) + u_1, \\ \dot{e}_2 = \gamma y_1 - (c-a)x_1 - e_1e_3 - e_3x_1 - e_1x_3 \\ \qquad + (\theta - 1)\,y_2 - (c-d)\,x_2 - \theta y_{2\tau} - dx_{2\tau} + u_2, \\ \dot{e}_3 = e_1e_2 + e_1x_2 + e_2x_1 - \beta y_3 + bx_3 + u_3. \end{cases} \tag{7.17}$$

Thus, we have the following theorem.

Theorem 7.3 ([16]). By the following controller:

$$\begin{cases} u_1 = \hat{\alpha}(e_1 - e_2) + (\hat{a} - \hat{\alpha})(x_2 - x_1) - e_1, \\ u_2 = (\hat{c} - \hat{a})x_1 + (\hat{c} - \hat{d})x_2 - \hat{\gamma}y_1 + (1 - \hat{\theta})y_2 \\ \qquad + e_1x_3 + \hat{\theta}y_{2\tau} + \hat{d}x_{2\tau} - e_2, \\ u_3 = (\hat{\beta} - 1)e_3 + (\hat{\beta} - \hat{b})x_3 - e_1x_2 \end{cases} \tag{7.18}$$

and the parameter update law

$$\begin{cases} \dot{\hat{a}} = e_2x_1 - e_1(x_2 - x_1), \\ \dot{\hat{b}} = e_3x_3, \\ \dot{\hat{c}} = -e_2x_1 - e_2x_2, \\ \dot{\hat{d}} = e_2x_2 - e_2x_{2\tau}, \\ \dot{\hat{\alpha}} = e_1(y_2 - y_1), \\ \dot{\hat{\beta}} = -e_3y_3, \\ \dot{\hat{\gamma}} = e_2y_1, \\ \dot{\hat{\theta}} = e_2y_2 - e_2y_{2\tau}, \end{cases} \tag{7.19}$$

the drive system (7.1) and the response system (7.16) will be globally asymptotically synchronized. Here, $\hat{a}$, $\hat{b}$, $\hat{c}$, $\hat{d}$, $\hat{\alpha}$, $\hat{\beta}$, $\hat{\gamma}$, and $\hat{\theta}$ are the estimated values of unknown parameters a, b, c, d, α, β, γ, and θ, respectively.

Proof. Substituting (7.18) into (7.17), the error system (7.17) can be rewritten as

$$\begin{cases} \dot{e}_1 = (\alpha-\hat{\alpha})(y_2-y_1)-(a-\hat{a})(x_2-x_1)-e_1, \\ \dot{e}_2 = (\gamma-\hat{\gamma})y_1-(c-\hat{c}-a+\hat{a})x_1-\left(c-\hat{c}-d+\hat{d}\right)x_2 \\ \qquad +\left(\theta-\hat{\theta}\right)y_2-\left(\theta-\hat{\theta}\right)y_{2\tau}-\left(d-\hat{d}\right)x_{2\tau}-e_2-e_1e_3-e_3x_1, \\ \dot{e}_3 = e_1e_2+e_2x_1-(\beta-\hat{\beta})y_3+\left(b-\hat{b}\right)x_3-e_3. \end{cases} \tag{7.20}$$

Let $e_a = a-\hat{a}$, $e_b = b-\hat{b}$, $e_c = c-\hat{c}$, $e_d = d-\hat{d}$, $e_\alpha = \alpha-\hat{\alpha}$, $e_\beta = \beta-\hat{\beta}$, $e_\gamma = \gamma-\hat{\gamma}$, and $e_\theta = \theta-\hat{\theta}$ be the estimated errors of unknown parameters a, b, c, d, α, β, γ, and θ, respectively. Now construct a Lyapunov function in the form of

$$V(e) = \frac{1}{2}\left(e_1^2+e_2^2+e_3^2+e_a^2+e_b^2+e_c^2+e_d^2+e_\alpha^2+e_\beta^2+e_\gamma^2+e_\theta^2\right).$$

Then, the time derivative of $V(e)$ along the trajectory of the error dynamical system (7.20) is as follows:

$$\begin{aligned} \dot{V}(e) &= e_1\dot{e}_1+e_2\dot{e}_2+e_3\dot{e}_3+e_a\dot{e}_a+e_b\dot{e}_b+e_c\dot{e}_c \\ &\quad +e_d\dot{e}_d+e_\alpha\dot{e}_\alpha+e_\beta\dot{e}_\beta+e_\gamma\dot{e}_\gamma+e_\theta\dot{e}_\theta \\ &= e_1\left(e_\alpha(y_2-y_1)-e_a(x_2-x_1)-e_1\right) \\ &\quad +e_3\left(e_1e_2+e_2x_1-e_\beta y_3+e_bx_3-e_3\right) \\ &\quad +e_2\left(e_\gamma y_1-(e_c-e_a)x_1-(e_c-e_d)x_2\right) \\ &\quad +e_2\left(e_\theta y_2-e_\theta y_{2\tau}-e_dx_{2\tau}-e_2-e_1e_3-e_3x_1\right) \\ &\quad -\left(e_a\dot{\hat{a}}+e_b\dot{\hat{b}}+e_c\dot{\hat{c}}+e_d\dot{\hat{d}}+e_\alpha\dot{\hat{\alpha}}+e_\beta\dot{\hat{\beta}}+e_\gamma\dot{\hat{\gamma}}+e_\theta\dot{\hat{\theta}}\right) \\ &= e_1e_\alpha(y_2-y_1)-e_ae_1(x_2-x_1)-e_1^2 \\ &\quad +e_3e_1e_2+e_3e_2x_1-e_\beta e_3y_3+e_be_3x_3-e_3^2 \\ &\quad +e_2e_\gamma y_1-e_2e_cx_1+e_2e_ax_1-e_ce_2x_2 \\ &\quad +e_de_2x_2+e_\theta e_2y_2-e_\theta e_2y_{2\tau}-e_de_2x_{2\tau} \\ &\quad -\left(e_a\dot{\hat{a}}+e_b\dot{\hat{b}}+e_c\dot{\hat{c}}+e_d\dot{\hat{d}}+e_\alpha\dot{\hat{\alpha}}+e_\beta\dot{\hat{\beta}}+e_\gamma\dot{\hat{\gamma}}+e_\theta\dot{\hat{\theta}}\right) \\ &\quad -e_2^2-e_1e_3e_2-e_2e_3x_1. \end{aligned} \tag{7.21}$$

Substituting (7.19) into (7.21), we obtain

$$\dot{V}(e) = -\left(e_1^2+e_2^2+e_3^2\right) \le 0,$$

which implies that $\dot{V}(e) < 0$ for all $e(t) \neq 0$. It is clear that $e_1 \in L_\infty$, $e_2 \in L_\infty$, and $e_3 \in L_\infty$. From the fact

$$\int_0^t -(e_1^2+e_2^2+e_3^2)\mathrm{d}s \le -\int_0^t \dot{V}\mathrm{d}s \le V(e_1(0),e_2(0),e_3(0)),$$

we can easily show that $e_1 \in L_2$, $e_2 \in L_2$, and $e_3 \in L_2$. From (7.17) we have $\dot{e}_1 \in L_\infty$, $\dot{e}_2 \in L_\infty$, and $\dot{e}_3 \in L_\infty$. Thus, by Barbalat's lemma, we have $\lim_{t\to\infty} \|e_i(t)\| = \lim_{t\to\infty} \|y_i(t) - x_i(t)\| = 0$. This means that the proposed controller (7.18) and (7.19) can globally asymptotically synchronize the drive system (7.1) and the response system (7.16). This completes the proof. □

The following illustrative example is used to demonstrate the effectiveness of the above method.

According to Theorem 7.3, the initial condition of the drive delayed Chen chaotic system is $x_0 = (5, 5, 15)^{\mathrm{T}}$, and the initial condition of the response delayed Lorenz chaotic system is $y_0 = (-0.1, -0.1, 0.2)^{\mathrm{T}}$. The 'unknown' parameters are chosen as $\alpha = 10$, $\beta = 8/3$, $\gamma = 5$, and $\theta = 6.5$. The initial values of the estimation of parameters are selected as follows: $\hat{a}(0) = 32$, $\hat{b}(0) = 4$, $\hat{c}(0) = 25$, $\hat{d}(0) = 2$, $\hat{\alpha}(0) = 9$, $\hat{\beta}(0) = 3$, $\hat{\gamma}(0) = 4$, and $\hat{\theta}(0) = 5$. The synchronization error curves of the drive system (7.1) and the response system (7.16) are shown in Fig. 7.13. The curves of the estimated parameters $\hat{a}$, $\hat{b}$, $\hat{c}$, $\hat{d}$, $\hat{\alpha}$, $\hat{\beta}$, $\hat{\gamma}$, and $\hat{\theta}$ are shown in Figs. 7.14–7.21, respectively.

From simulation results we can see that the synchronization errors converge asymptotically to zero, i.e., the trajectory of the response delayed Lorenz chaotic system can synchronize asymptotically with the trajectory of the drive delayed Chen

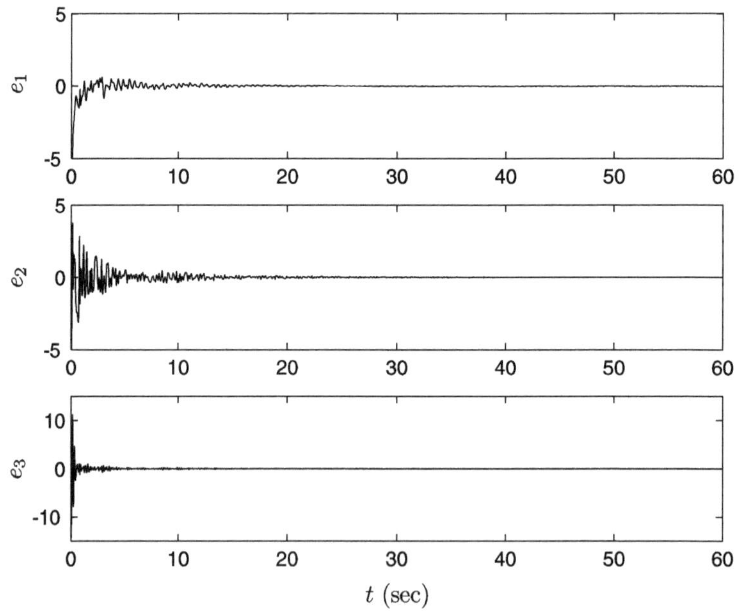

Fig. 7.13 The synchronization error curves of the drive system (7.1) and the response system (7.16)

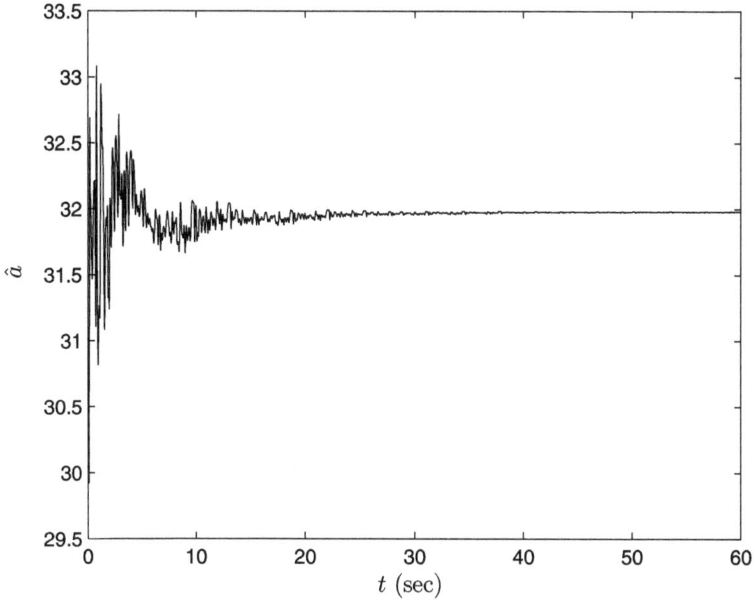

Fig. 7.14 The curve of the adaptive parameter $\hat{a}(t)$

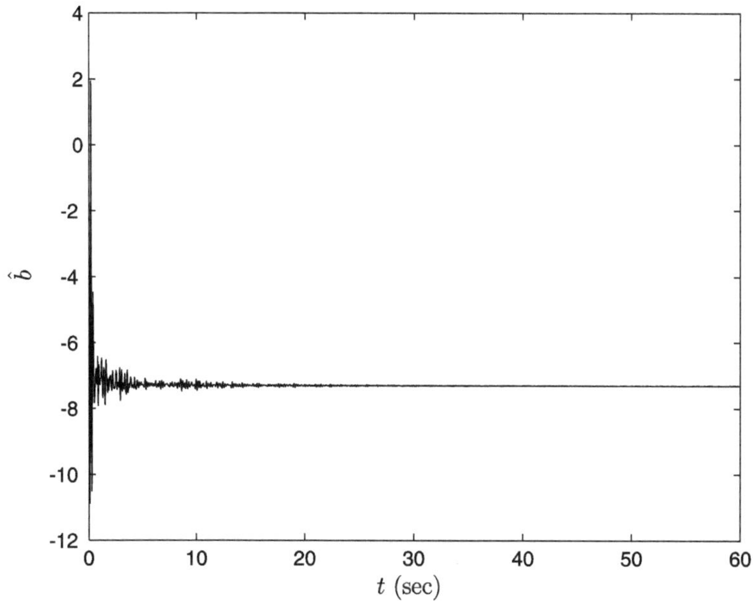

Fig. 7.15 The curve of the adaptive parameter $\hat{b}(t)$

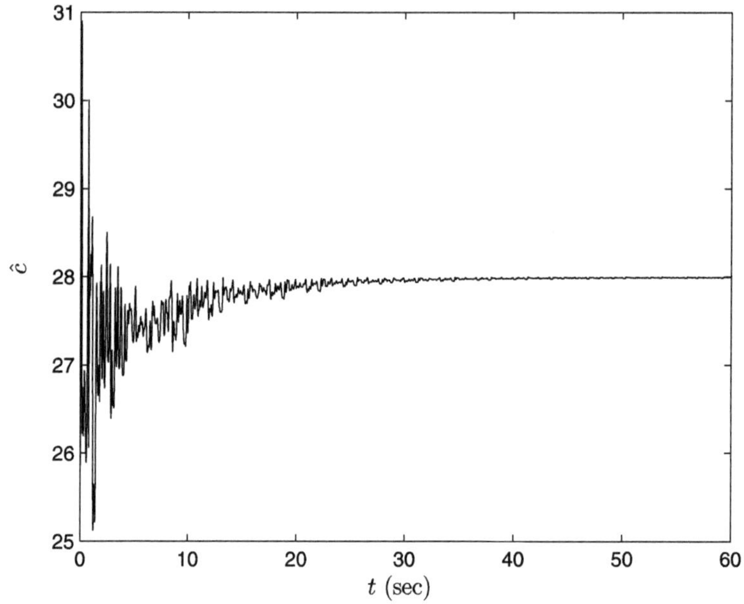

Fig. 7.16 The curve of the adaptive parameter $\hat{c}(t)$

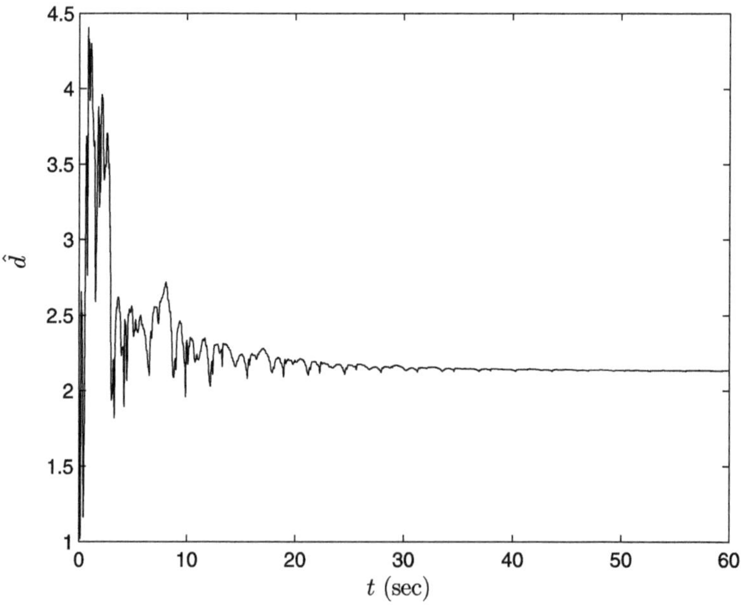

Fig. 7.17 The curve of the adaptive parameter $\hat{d}(t)$

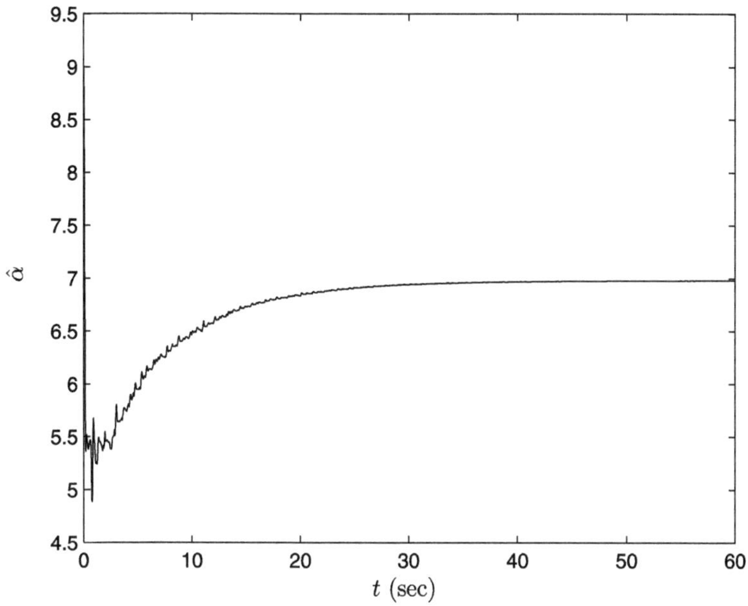

Fig. 7.18 The curve of the adaptive parameter $\hat{\alpha}(t)$

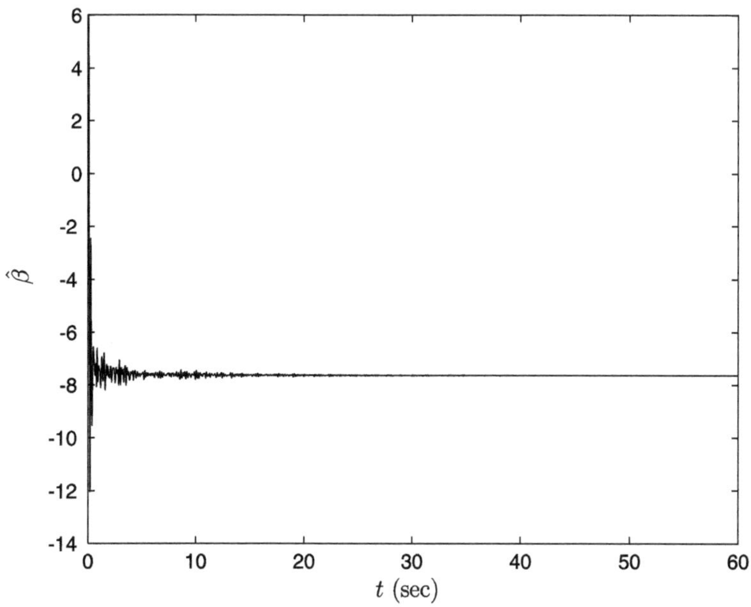

Fig. 7.19 The curve of the adaptive parameter $\hat{\beta}(t)$

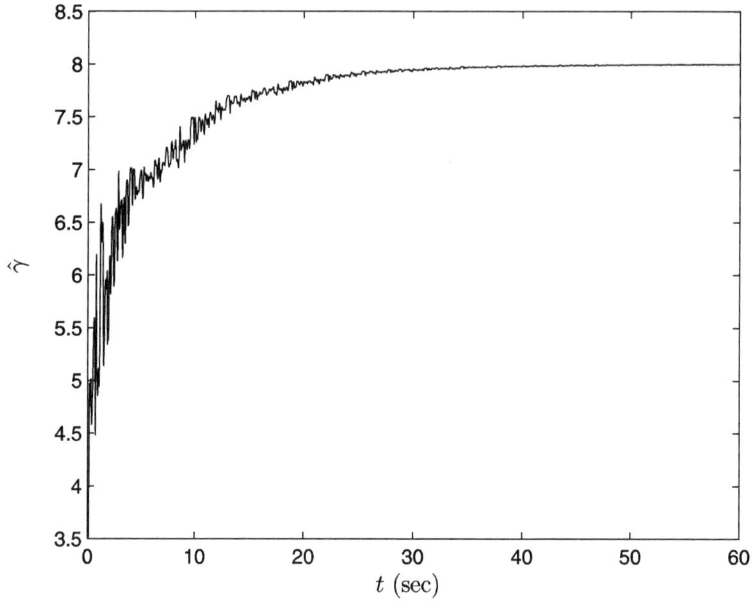

Fig. 7.20 The curve of the adaptive parameter $\hat{\gamma}(t)$

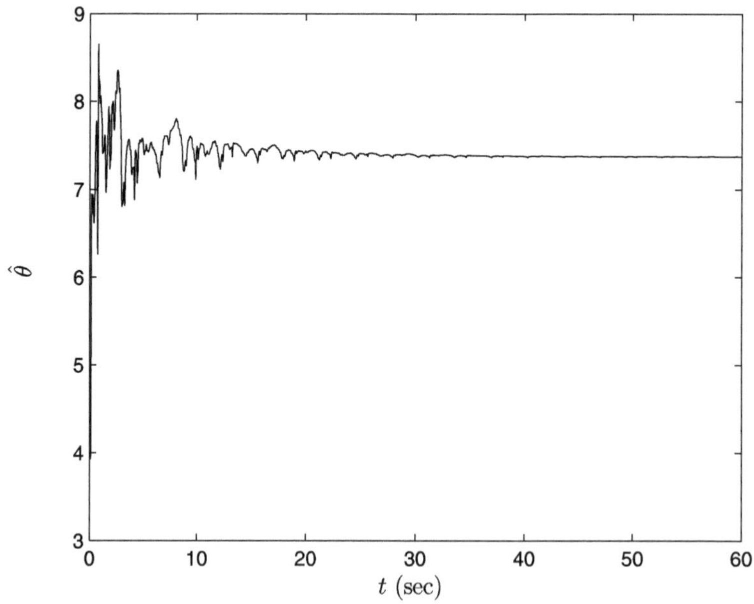

Fig. 7.21 The curve of the adaptive parameter $\hat{\theta}(t)$

chaotic system. The estimated parameters converge asymptotically to some constants, which shows the effectiveness of the adaptive synchronization scheme.

7.3.2 Adaptive Synchronization Between Two Different Delayed Ikeda Chaotic Systems

The drive delayed Ikeda chaotic system [4] in this section is described by the following differential equation:

$$\dot{x}(t) = -cx(t) + af(x(t-\tau)), \tag{7.22}$$

where $\tau > 0$ denotes the bounded delay, $a > 0$, and $c > 0$. We assume that c is the unknown parameter of system (7.22), and a is the known parameter of system (7.22). The initial condition of system (7.22) is given by $x(t) = \varphi_x(t) \in C([-\tau,0],\ \mathbb{R})$, where $C([-\tau,0],\mathbb{R})$ denotes the set of all continuous functions from $[-\tau,0]$ to $\mathbb{R}$. The function $f(x)$ is a one-dimensional and continuous nonlinear function.

The response delayed Ikeda chaotic system is described by the following equation:

$$\dot{y}(t) = -dy(t) + bg(y(t-\tau)) + u(t), \tag{7.23}$$

where $\tau > 0$ denotes the bounded delay, $b > 0$, and $d > 0$. We assume that d is the unknown parameters of the system (7.23), and b is the known parameter of the system (7.23). The initial conditions of system (7.23) are given by $y(t) = \varphi_y(t) \in C([-\tau,0],\ \mathbb{R})$, where $C([-\tau,0],\mathbb{R})$ denotes the set of all continuous functions from $[-\tau,0]$ to $\mathbb{R}$. The function $g(y)$ is a one-dimensional and continuous nonlinear function, and $u(t)$ denotes the designed controller that will achieve the goal of synchronizing system (7.22) and system (7.23), i.e., $\lim_{t\to\infty}\|e(t)\| = \lim_{t\to\infty}\|y(t) - x(t)\| = 0$.

For this problem, we have the following theorem.

Theorem 7.4 ([16]). By the following controller:

$$u(t) = -\hat{c}x(t) + af(x(t-\tau)) + \hat{d}y(t) - bg(y(t-\tau)) - ke(t) \tag{7.24}$$

and the parameter update law

$$\begin{cases} \dot{\hat{c}} = e^{\mathrm{T}}(t)x(t), \\ \dot{\hat{d}} = -e^{\mathrm{T}}(t)y(t), \end{cases} \tag{7.25}$$

the drive system (7.22) and the response system (7.23) will be globally asymptotically synchronized. Here, $k > 0$ and $\hat{c}$ and $\hat{d}$ are the estimated values of unknown parameters c and d, respectively.

Proof. From (7.22)–(7.24), we get the error system as follows:

$$\dot{e}(t) = -\left(d - \hat{d}\right)y(t) + (c - \hat{c})x(t) - ke(t). \tag{7.26}$$

Define $\varepsilon := \left(e^{\mathrm{T}}(t), \hat{c}, \hat{d}\right)^{\mathrm{T}}$. If a Lyapunov function candidate is chosen as

$$V(\varepsilon) = \frac{1}{2}\left(e^{\mathrm{T}}(t)\,e(t) + (c-\hat{c})^2 + \left(d-\hat{d}\right)^2\right),$$

then the time derivative of $V(\varepsilon)$ along the trajectory of the error system (7.26) is as follows:

$$\begin{aligned}\dot{V}(\varepsilon) &= e^{\mathrm{T}}(t)\,\dot{e}(t) + (c-\hat{c})\left(-\dot{\hat{c}}\right) + \left(d-\hat{d}\right)\left(-\dot{\hat{d}}\right)\\ &= e^{\mathrm{T}}(t)\left(-\left(d-\hat{d}\right)y(t) + (c-\hat{c})\,x(t)\right)\\ &\quad -(c-\hat{c})\,\dot{\hat{c}} - \left(d-\hat{d}\right)\dot{\hat{d}} - ke^{\mathrm{T}}(t)\,e(t). \end{aligned} \tag{7.27}$$

Substituting (7.25) into (7.27), we obtain

$$\dot{V}(\varepsilon) = -ke^{\mathrm{T}}(t)\,e(t) \le 0,$$

which implies that $\dot{V}(\varepsilon) < 0$ for all $e(t) \ne 0$. It is clear that $e(t) \in L_\infty$, $\hat{c} \in L_\infty$, and $\hat{d} \in L_\infty$. From the fact that

$$\int_0^t \| e(t)\|^2 \mathrm{d}s = \frac{[V(e(0)) - V(e(t))]}{k} \le \frac{V(e(0))}{k},$$

we can easily show that $e(t) \in L_2$. From (7.26) we have $\dot{e}(t) \in L_\infty$. Thus, by Barbalat's lemma, we have $\lim\limits_{t\to\infty} e(t) = 0$. This means that the proposed controller (7.24) can globally asymptotically synchronize the system (7.22) and the system (7.23). This completes the proof. □

The following illustrative example is used to demonstrate the effectiveness of the above method.

Assume that the following delayed Ikeda chaotic system [7] is the drive system:

$$\dot{x}(t) = -cx(t) + af(x(t-\tau)), \tag{7.28}$$

where the activation function $f(x(t-\tau)) = \sin(x(t-\tau))$, and the following delayed Ikeda chaotic system [6] is the response system:

$$\dot{y}(t) = -dy(t) + bg(y(t-\tau)), \tag{7.29}$$

where the activation function $g(y(t-\tau)) = \cos(my(t-\tau))$.

The initial conditions of the drive system (7.28) and the response system (7.29) are taken as $x(s) = 1$ and $y(s) = 1.5$ for $-2 \le s \le 0$, respectively. The parameters of the drive system (7.28) and the response system (7.29) are selected as $c = 1$, $a = 20$, $d = 1$, $b = 2$, $m = 2$, and $\tau = 2$. The state curve of the drive system (7.28) is shown in Fig. 7.22, and the state curve of the response system (7.29) is shown in Fig. 7.23.

According to Theorem 7.4, the initial values of the drive system and the response system are $x(0) = 0.5$ and $y(0) = 1$. We choose $\hat{c}(0) = 1.5$, $\hat{d}(0) = 1.1$, and $k = 1$. The synchronization error curve of the drive system (7.28) and the response sys-

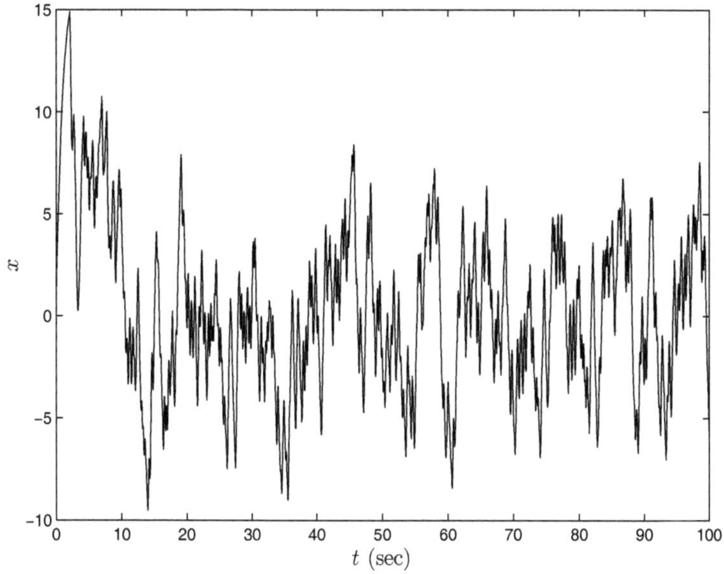

Fig. 7.22 The state curve of the drive chaotic system (7.28)

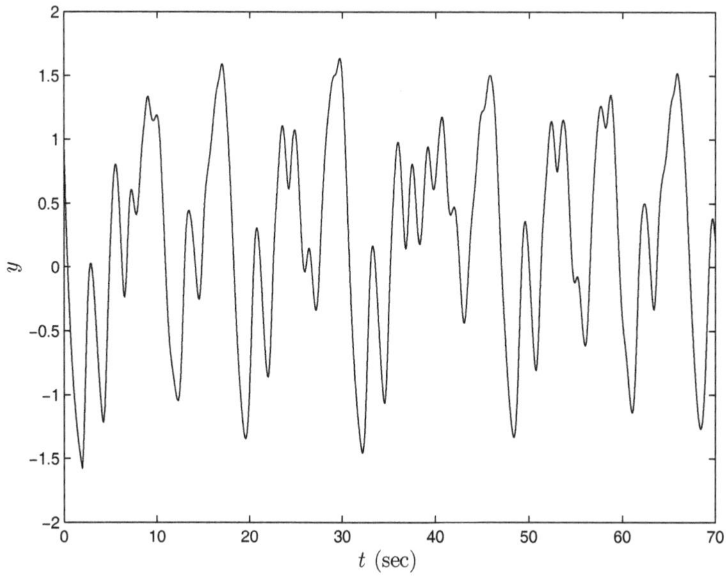

Fig. 7.23 The state curve of the response chaotic system (7.29)

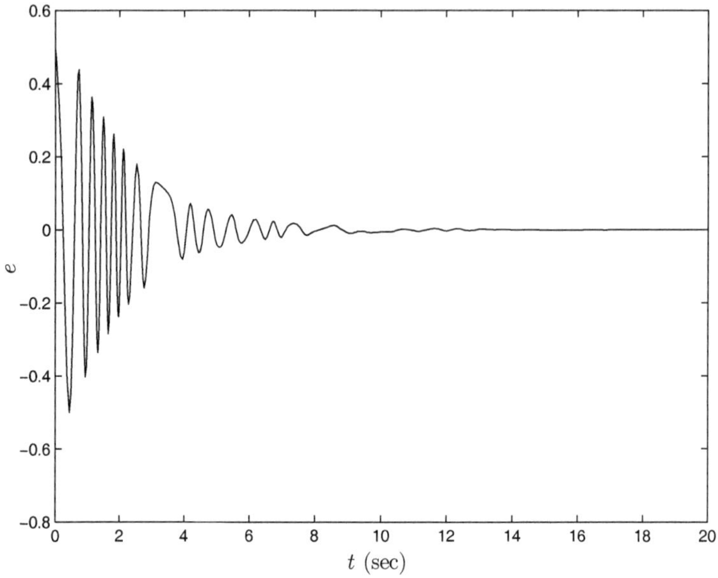

Fig. 7.24 The synchronization error curve of the drive system (7.28) and the response system (7.29)

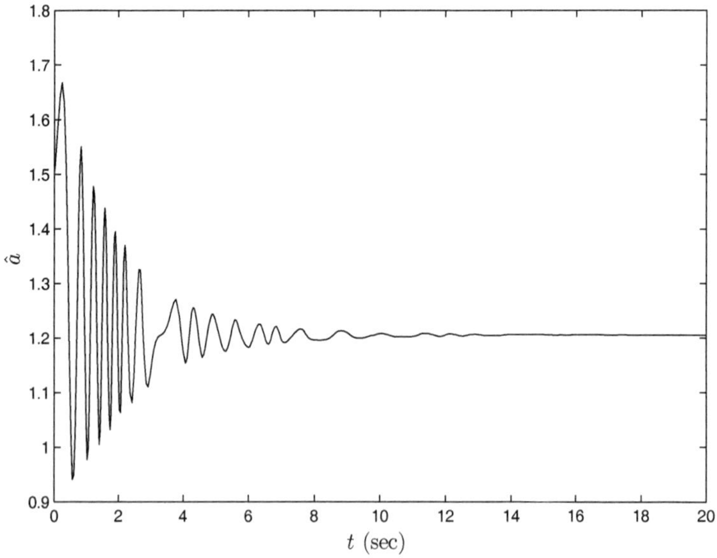

Fig. 7.25 The curve of the adaptive parameter $\hat{c}(t)$

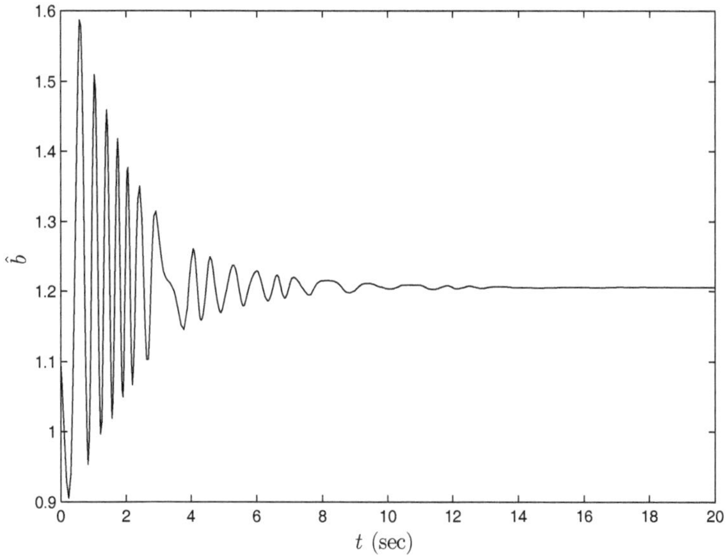

Fig. 7.26 The curve of the adaptive parameter $\hat{d}(t)$

tem (7.29) is shown in Fig. 7.24. The curves of the estimated parameters $\hat{c}$ and $\hat{d}$ are shown in Figs. 7.25 and 7.26, respectively. From simulation results we can see that the synchronization errors converge asymptotically to zero, and the estimated parameters converge asymptotically to some constants, i.e., the trajectory of the response system can synchronize asymptotically with the trajectory of the drive system.

7.3.3 Adaptive Synchronization Between Two Different Delayed Liao Chaotic Systems

The drive delayed Liao chaotic system [3] in this section is described by the following differential equation:

$$\dot{x}(t) = -\alpha x(t) + af(x(t) - bx(t-\tau) + c), \tag{7.30}$$

where $\tau > 0$ denotes the bounded delay, α, a, and b are the parameters of the system (7.30), and c is a constant. We assume that α is the unknown parameter of system (7.30), and $a > 0$ and $b > 0$ are the known parameters of the system. The initial conditions of system (7.30) are given by $x(t) = \varphi_x(t) \in C([-\tau, 0], \mathbb{R})$, where $C([-\tau, 0], \mathbb{R})$ denotes the set of all continuous functions from $[-\tau, 0]$ to $\mathbb{R}$. The function $f(x)$ is a one-dimensional and continuous nonlinear function.

The response delayed Liao chaotic system is described by the following equation:

$$\dot{y}(t) = -\beta y(t) + mg\left(y(t) - ny(t-\tau) + c\right) + u(t), \tag{7.31}$$

where $\tau > 0$ denotes the bounded delay and β, m, and n are the parameters of system (7.31). We assume that β is the unknown parameter of system (7.31), and $m > 0$ and $n > 0$ are the known parameters of the system. The initial conditions of system (7.31) are given by $y(t) = \varphi_y(t) \in C([-\rho,\, 0],\, \mathbb{R})$. The function $g(y)$ is a one-dimensional and continuous nonlinear function, and $u(t)$ denotes the designed controller that will realize the synchronization of system (7.30) and system (7.31), i.e., $\lim_{t\to\infty} \|e(t)\| = \lim_{t\to\infty} \|y(t) - x(t)\| = 0$.

Theorem 7.5 ([16]). By the following controller:

$$\begin{aligned} u(t) =& -\hat{\alpha}x(t) + af\left(x(t) - bx(t-\tau) + c\right) \\ &+ \hat{\beta}y(t) - mg\left(y(t) - ny(t-\tau) + c\right) - ke(t) \end{aligned} \tag{7.32}$$

and the parameter update law

$$\begin{cases} \dot{\hat{\alpha}} = e^{\mathrm{T}}(t)x(t), \\ \dot{\hat{\beta}} = -e^{\mathrm{T}}(t)y(t), \end{cases} \tag{7.33}$$

the drive system (7.30) and the response system (7.31) will be globally asymptotically synchronized. Here, $k > 0$ and $\hat{\alpha}$ and $\hat{\beta}$ are the estimated values of unknown parameters α and β, respectively.

Proof. From (7.30)–(7.32), we get the error system as follows:

$$\dot{e}(t) = -\left(\beta - \hat{\beta}\right)y(t) + (\alpha - \hat{\alpha})x(t) - ke(t). \tag{7.34}$$

Define

$$\varepsilon := (e^{\mathrm{T}}(t), \hat{\alpha}, \hat{\beta})^{\mathrm{T}}.$$

If a Lyapunov function candidate is chosen as

$$V(\varepsilon) = \frac{1}{2}\left(e^{\mathrm{T}}(t)e(t) + (\alpha - \hat{\alpha})^2 + \left(\beta - \hat{\beta}\right)^2\right),$$

then the time derivative of $V(\varepsilon)$ along the trajectory of the error system (7.34) is as follows:

$$\begin{aligned} \dot{V}(\varepsilon) &= e^{\mathrm{T}}(t)\dot{e}(t) + (\alpha - \hat{\alpha})\left(-\dot{\hat{\alpha}}\right) + \left(\beta - \hat{\beta}\right)\left(-\dot{\hat{\beta}}\right) \\ &= e^{\mathrm{T}}(t)\left(-\left(\beta - \hat{\beta}\right)y(t) + (\alpha - \hat{\alpha})x(t)\right) \\ &\quad - (\alpha - \hat{\alpha})\dot{\hat{\alpha}} - \left(\beta - \hat{\beta}\right)\dot{\hat{\beta}} - ke^{\mathrm{T}}(t)e(t). \end{aligned} \tag{7.35}$$

Substituting (7.33) into (7.35), we obtain

$$\dot{V}(\varepsilon) = -ke^{\mathrm{T}}(t)\,e(t) \leq 0, \tag{7.36}$$

which implies that $\dot{V}(\varepsilon) < 0$ for all $e(t) \neq 0$. It is clear that $e(t) \in L_\infty$, $\hat{\alpha} \in L_\infty$, and $\hat{\beta} \in L_\infty$. From the fact that

$$\int_0^t \|e(s)\|^2 \mathrm{d}s = \frac{[V(e(0)) - V(e(t))]}{k} \leq \frac{V(e(0))}{k},$$

we can easily show that $e(t) \in L_2$. From (7.34) we have $\dot{e}(t) \in L_\infty$. Thus, by Barbalat's lemma, we have $\lim_{t\to\infty} e(t) = 0$. This means that the proposed controller (7.32) can globally asymptotically synchronize the system (7.30) and the system (7.31). This completes the proof. □

Assume that the following delayed Liao chaotic system [8] is the drive system:

$$\dot{x} = -\alpha x(t) + af(x(t) - bx(t-\tau) + c), \tag{7.37}$$

where the activation function is

$$f(x) = \sum_{i=1}^{2} a_i \left[\tanh(x + k_i) - \tanh(x - k_i)\right].$$

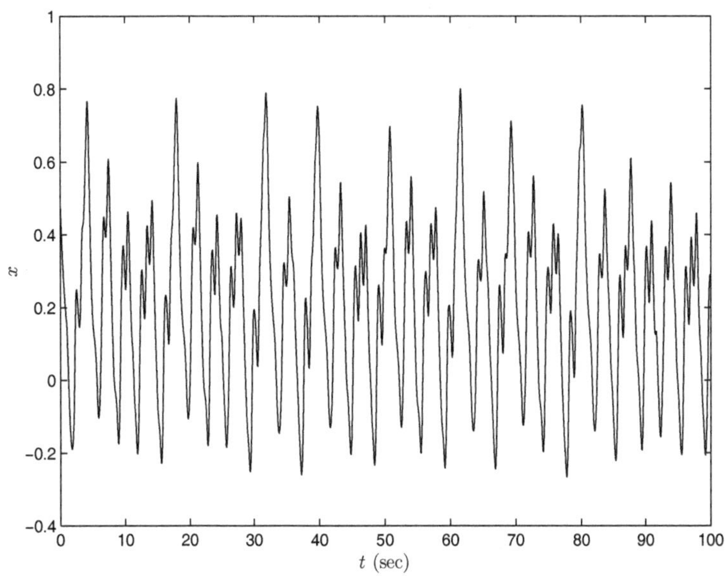

Fig. 7.27 The state curve of the drive chaotic system (7.37)

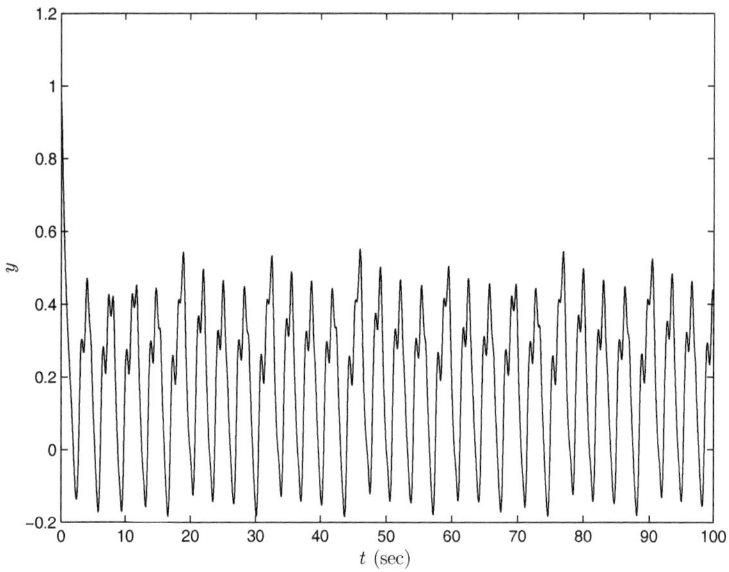

Fig. 7.28 The state curve of the response chaotic system (7.38)

The initial conditions of the drive system (7.37) are taken as $x(s) = 0.5$, $s \in [-1,0]$, $\alpha = 1$, $a = 3$, $b = 4.5$, $a_1 = 2$, $a_2 = -1.5$, $k_1 = 1$, $k_2 = 4/3$, $c = 0$, and $\tau = 1$. The state curve of the drive system (7.37) is shown in Fig. 7.27.

Take the following delayed Liao chaotic system [12] as the response system:

$$\dot{y}(t) = -\beta y(t) + mg\left(y(t) - ny(t-\tau) + c\right), \tag{7.38}$$

where the activation function

$$g(y) = \sum_{i=1}^{2} a_i \left[\arctan\left(y + k_i\right) - \arctan\left(y - k_i\right)\right].$$

The initial conditions of the response system (7.38) are taken as $y(s) = 1$, $s \in [-1,0]$, $\beta = 1$, $m = 3$, $n = 4.5$, $a_1 = 2$, $a_2 = -1.5$, $k_1 = 1$, $k_2 = 4/3$, $c = 0$, and $\tau = 1$. The state curve of the response system (7.38) is shown in Fig. 7.28.

According to Theorem 7.5, the initial values of the drive system and the response system are $x(0) = 0.7$ and $y(0) = 1.2$. Choose $\hat{\alpha}(0) = 0.7$, $\hat{\beta}(0) = 1.1$, and $k = 1$. The synchronization error curve of the drive system (7.37) and the response system (7.38) is shown in Fig. 7.29. The curves of the estimated parameters $\hat{\alpha}$ and $\hat{\beta}$ are shown in Figs. 7.30 and 7.31, respectively. From simulation results we can see that the synchronization errors converge asymptotically to zero, and the estimated parameters converge asymptotically to some constants, i.e., the trajectory of

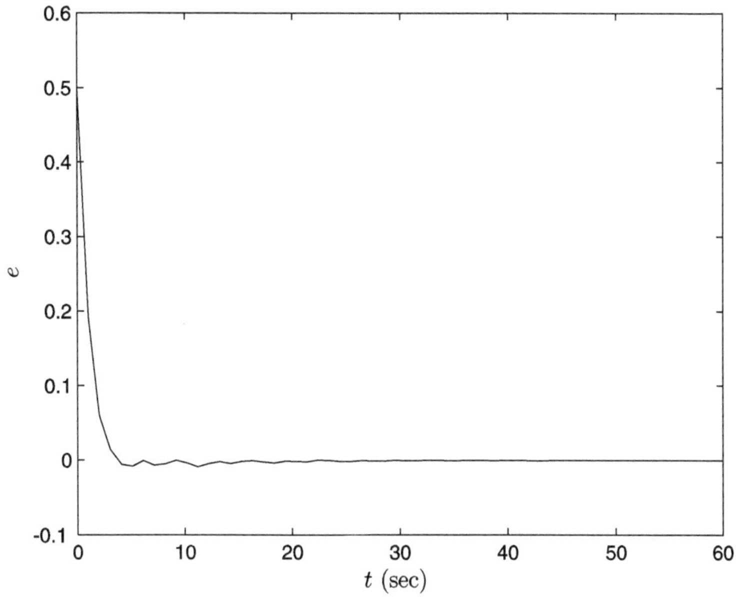

Fig. 7.29 The synchronization error curve of the drive system (7.37) and the response system (7.38)

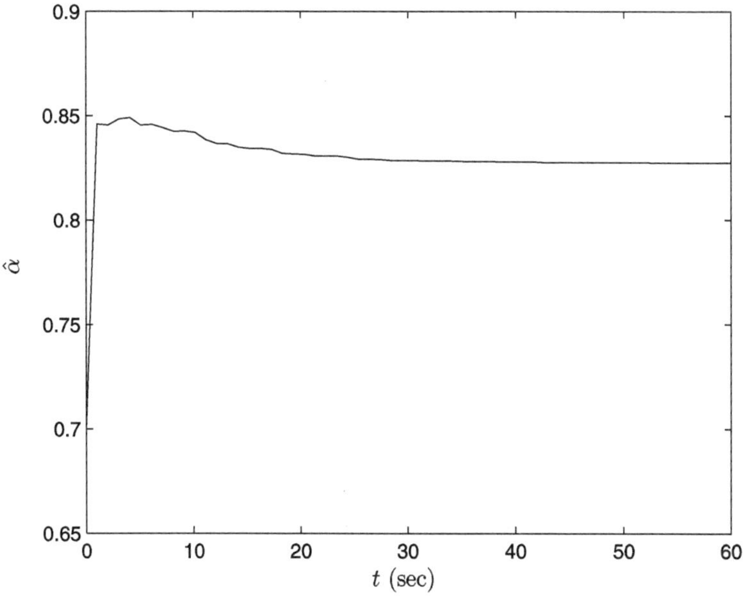

Fig. 7.30 The curve of the adaptive parameter $\hat{\alpha}(t)$

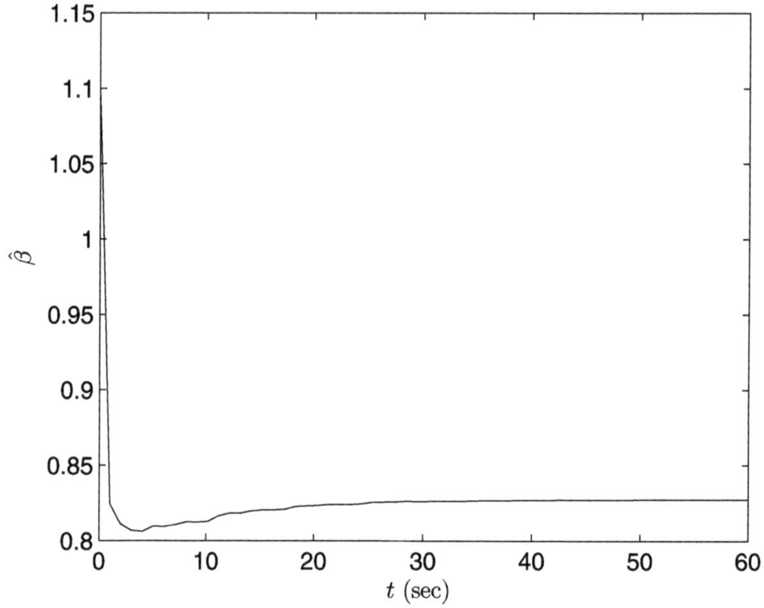

Fig. 7.31 The curve of the adaptive parameter $\hat{\beta}(t)$

the response system can synchronize asymptotically with the trajectory of the drive system.

7.4 Synchronization of Chaotic Delayed Neural Networks

7.4.1 Synchronization of a Class of Chaotic Delayed Neural Networks Based on Inverse Optimal Control Theory

Recently, there is increasing interest in the study of dynamical properties of delayed neural networks (DNNs). Most previous studies have predominantly concentrated on the stability analysis and periodic oscillations of this kind of network [1, 20, 21, 22]. It has been shown that such networks can exhibit some complicated dynamics and even chaotic behaviors. In particular, by appropriately choosing the network parameters and time delays, the dynamical behavior of these networks can be made very complicated [2, 11]. Synchronization of chaotic DNNs has potential applications in several areas such as secure communications, associative memory, combinatorial optimization, etc. [9, 10]. So, synchronization of this class of chaotic DNNs [15, 17] is essential and useful in real-world applications.

In this section, the problem of synchronizing two identical chaotic DNNs is studied. Under the framework of the inverse optimal control approach, synchronization controllers and adaptive updating laws of parameters with known and unknown parameters are designed, respectively [18].

7.4.1.1 Problem Description

A class of delayed chaotic neural networks in this section is described by the following differential equation:

$$\dot{x}_i(t) = -c_i x_i(t) + \sum_{j=1}^{n} a_{ij} g_j(x_j(t)) + \sum_{j=1}^{n} b_{ij} g_j(x_j(t-\tau_j)) + U_i, \tag{7.39}$$

where n denotes the number of neurons in the network, x_i denotes the state variable associated with the ith neuron, $\tau_j \geq 0$ denotes bounded delay and $\rho = \max\{\tau_j\}$, $c_i > 0$, a_{ij} indicates the strength of the neuron interconnections within the network, b_{ij} indicates the strength of the neuron interconnections within the network with constant delay parameter τ_j, $i, j = 1, \ldots, n$, and U_i is a constant input vector function. The activation function $g_j(x_j)$ satisfies

$$0 \leq \frac{g_j(\xi) - g_j(\zeta)}{\xi - \zeta} \leq \delta_j,$$

for all $\xi, \zeta \in \mathbb{R}$, and $\xi \neq \zeta$. The initial conditions of system (7.39) are given by $x_i(t) = \phi_i(t) \in C([-\rho, 0], \mathbb{R})$.

Chaotic dynamics is extremely sensitive to initial conditions. Even infinitesimal changes in the initial conditions will lead to an exponential divergence of orbits. In order to observe the synchronization behavior in this class of chaotic DNNs, we study two chaotic DNNs where the drive system and the response system have identical dynamical structures. The drive system's state variables are denoted by x_i and the response system's state variables are denoted by y_i.

7.4.1.2 Synchronization of a Class of Chaotic DNNs with Known Parameters

Suppose that the drive system has the form of (7.39). The response system is described by the following equation:

$$\dot{y}_i(t) = -c_i y_i(t) + \sum_{j=1}^{n} a_{ij} g_j(y_j(t)) + \sum_{j=1}^{n} b_{ij} g_j(y_j(t-\tau_j)) + U_i + u_i(t), \tag{7.40}$$

where the initial conditions of system (7.40) are given by $y_i(t) = \varphi_i(t) \in C([-\rho, 0],$ $\mathbb{R})$, and $u_i(t)$ denotes the designed controller that will realize the synchronization of system (7.39) and system (7.40).

Let us define the synchronization error signal as $e_i(t) = y_i(t) - x_i(t)$, where $x_i(t)$ and $y_i(t)$ are the ith state variables of the drive and response neural networks, respectively. The fact that $e_i(t) \to 0$ as $t \to \infty$ means that the drive neural network and the response neural network are synchronized. The goal of the controller design is to obtain $u_i(t)$ so that $\lim_{t\to\infty} \|e_i(t)\| = \lim_{t\to\infty} \|y_i(t) - x_i(t)\| = 0$, where $\|\cdot\|$ denotes the Euclidean vector norm.

Therefore, the error system between (7.39) and (7.40) can be expressed as

$$\dot{e}_i(t) = -c_i e_i(t) + \sum_{j=1}^{n} a_{ij} f_j(e_j(t)) + \sum_{j=1}^{n} b_{ij} f_j(e_j(t-\tau_j)) + u_i(t), \tag{7.41}$$

where $f_j(e_j(t)) = g_j(e_j(t) + x_j(t)) - g_j(x_j(t))$ and $f_j(0) = 0$, or by the following compact form:

$$\dot{e}(t) = -Ce(t) + Af(e(t)) + Bf(e(t-\tau)) + u(t), \tag{7.42}$$

where $C = \mathrm{diag}\{c_1, \ldots, c_n\}$, $A = (a_{ij})_{n\times n}$, $B = (b_{ij})_{n\times n}$, $\tau = (\tau_1, \ldots, \tau_n)^{\mathrm{T}}$, $u(t) = (u_1(t), \ldots, u_n(t))^{\mathrm{T}}$, $f(e(t)) = (f_1(e_1(t)), \ldots, f_n(e_n(t)))^{\mathrm{T}}$, and

$$f(e(t-\tau)) = (f_1(e_1(t-\tau_1)), \ldots, f_n(e_n(t-\tau_n)))^{\mathrm{T}}.$$

Before stating the main result, the following lemma is needed.

Lemma 7.1. For any matrices X, $Y \in R^{n\times k}$ and $Q \in R^{n\times n}$ with $Q = Q^{\mathrm{T}} > 0$, the following inequality holds:

$$X^{\mathrm{T}}Y + Y^{\mathrm{T}}X \le X^{\mathrm{T}}QX + Y^{\mathrm{T}}Q^{-1}Y. \qquad \square$$

About the synchronization of systems (7.39) and (7.40), we have the following theorem.

Theorem 7.6 ([18]). If the controller is designed as follows:

$$u(t) = -\left(AA^{\mathrm{T}} + BB^{\mathrm{T}} + 2\Sigma^2\right) e(t), \tag{7.43}$$

systems (7.39) and (7.40) will be globally asymptotically synchronized, where $\Sigma = \mathrm{diag}\{\delta_1, \ldots, \delta_n\}$.

Proof. Now construct a Lyapunov function of the form

$$V(e) = \frac{1}{2} e^{\mathrm{T}}(t) e(t) + \frac{1}{2} \sum_{j=1}^{n} \int_{t-\tau_j}^{t} \delta_j^2 e_j^2(s)\,\mathrm{d}s.$$

Then, the time derivative of $V(e)$ along the trajectory of the error system (7.42) is as follows:

$$\begin{aligned}\dot{V}(e) &= e^{\mathrm{T}}(t)\dot{e}(t) + \frac{1}{2}e^{\mathrm{T}}(t)\Sigma^2 e(t) - \frac{1}{2}e^{\mathrm{T}}(t-\tau)\Sigma^2 e(t-\tau)\\ &= -e^{\mathrm{T}}(t)Ce(t) + e^{\mathrm{T}}(t)(Af(e(t)) + Bf(e(t-\tau)))\\ &\quad + \frac{1}{2}e^{\mathrm{T}}(t)\Sigma^2 e(t) - \frac{1}{2}e^{\mathrm{T}}(t-\tau)\Sigma^2 e(t-\tau) + e^{\mathrm{T}}(t)u(t)\\ &:= L_{\bar{f}}V + (L_g V)u(t),\end{aligned} \tag{7.44}$$

where

$$\begin{aligned}L_{\bar{f}}V := &-e^{\mathrm{T}}(t)Ce(t) + e^{\mathrm{T}}(t)(Af(e(t)) + Bf(e(t-\tau)))\\ &+ \frac{1}{2}e^{\mathrm{T}}(t)\Sigma^2 e(t) - \frac{1}{2}e^{\mathrm{T}}(t-\tau)\Sigma^2 e(t-\tau),\end{aligned} \tag{7.45}$$

$$L_g V := e^{\mathrm{T}}(t). \tag{7.46}$$

Applying Lemma 7.1 with $Q = I$, we get

$$\begin{aligned}\dot{V}(e) \le &-e^{\mathrm{T}}(t)Ce(t) + \frac{1}{2}e^{\mathrm{T}}(t)AA^{\mathrm{T}}e(t) + \frac{1}{2}f^{\mathrm{T}}(e(t))f(e(t))\\ &+ \frac{1}{2}e^{\mathrm{T}}(t)BB^{\mathrm{T}}e(t) + \frac{1}{2}f^{\mathrm{T}}(e(t-\tau))f(e(t-\tau))\\ &+ \frac{1}{2}e^{\mathrm{T}}(t)\Sigma^2 e(t) - \frac{1}{2}e^{\mathrm{T}}(t-\tau)\Sigma^2 e(t-\tau) + e^{\mathrm{T}}(t)u(t)\\ \le &-e^{\mathrm{T}}(t)Ce(t) + \frac{1}{2}e^{\mathrm{T}}(t)AA^{\mathrm{T}}e(t) + \frac{1}{2}e^{\mathrm{T}}(t)\Sigma^2 e(t)\\ &+ \frac{1}{2}e^{\mathrm{T}}(t)BB^{\mathrm{T}}e(t) + \frac{1}{2}e^{\mathrm{T}}(t-\tau)\Sigma^2 e(t-\tau)\\ &+ \frac{1}{2}e^{\mathrm{T}}(t)\Sigma^2 e(t) - \frac{1}{2}e^{\mathrm{T}}(t-\tau)\Sigma^2 e(t-\tau) + e^{\mathrm{T}}(t)u(t)\\ \le &\, e^{\mathrm{T}}(t)\left(-C + \frac{1}{2}AA^{\mathrm{T}} + \frac{1}{2}BB^{\mathrm{T}} + \Sigma^2\right)e(t) + e^{\mathrm{T}}(t)u(t).\end{aligned} \tag{7.47}$$

We design a linear state feedback controller as follows:

$$u = -\beta R^{-1}(e)(L_g V)^{\mathrm{T}}, \tag{7.48}$$

where $\beta \ge 2$ is a constant. Denoting

$$R^{-1}(e) = \frac{1}{\beta}\left(AA^{\mathrm{T}} + BB^{\mathrm{T}} + 2\Sigma^2\right) \tag{7.49}$$

and substituting (7.46) and (7.49) into (7.48), we can derive

$$u = -\beta R^{-1}(e)(L_g V)^{\mathrm{T}} = -\left(AA^{\mathrm{T}} + BB^{\mathrm{T}} + 2\Sigma^2\right)e(t). \tag{7.50}$$

Substituting (7.50) into (7.47), we obtain

$$\begin{aligned}\dot{V}(e) &\leq -e^{\mathrm{T}}(t)\left(C+\frac{1}{2}AA^{\mathrm{T}}+\frac{1}{2}BB^{\mathrm{T}}+\Sigma^2\right)e(t)\\ &\leq -\lambda_{\min}\left(C+\frac{1}{2}AA^{\mathrm{T}}+\frac{1}{2}BB^{\mathrm{T}}+\Sigma^2\right)\|e\|^2\\ &\leq 0,\end{aligned}$$

which implies that $\dot{V}(e) < 0$ for all $e(t) \neq 0$. Since $V(\varepsilon)$ is positive definite and $\dot{V}(\varepsilon)$ is negative semidefinite, it follows that $e(t) \in L_\infty$. From the fact that

$$\begin{aligned}\int_0^t \|e(s)\|^2 \mathrm{d}s &= \frac{V(e(0)) - V(e(t))}{\lambda_{\min}\left(C+\frac{1}{2}AA^{\mathrm{T}}+\frac{1}{2}BB^{\mathrm{T}}+\Sigma^2\right)}\\ &\leq \frac{V(e(0))}{\lambda_{\min}\left(C+\frac{1}{2}AA^{\mathrm{T}}+\frac{1}{2}BB^{\mathrm{T}}+\Sigma^2\right)},\end{aligned}$$

we can easily show that $e(t) \in L_2$. From (7.41) we have $\dot{e}(t) \in L_\infty$. Thus, by Barbalat's lemma, we have $\lim_{t\to\infty} e(t) = 0$, i.e., $\lim_{t\to\infty} \|e(t)\| = \lim_{t\to\infty} \|y(t) - x(t)\| = 0$. This means that the proposed controller (7.50) can globally asymptotically synchronize the system (7.39) and the system (7.40). This completes the proof. □

Under the framework of the inverse optimal control approach, the controller (7.43) can be proved to optimize a cost functional of the following form:

$$J(u) = \lim_{t\to\infty}\left\{E(x(t)) + \int_0^t \left(l(x(\tau)) + u^{\mathrm{T}}(\tau)R(x(\tau))u(\tau)\right)\mathrm{d}\tau\right\}, \tag{7.51}$$

where $R = R^{\mathrm{T}} > 0$, $u(0) = 0$, and $l(x(\tau))$ and $E(x(t))$ are positive-definite and radially unbounded functions [5].

Theorem 7.7 ([18]). The following cost functional for system (7.42):

$$J(u) = \lim_{t\to\infty}\left\{2\beta V(e(t)) + \int_0^t \left(l(e(\tau)) + u^{\mathrm{T}}(\tau)R(e(\tau))u(\tau)\right)\mathrm{d}\tau\right\} \tag{7.52}$$

will be minimized by controller (7.43), where

$$l(e) = -2\beta L_{\bar{f}}V + \beta^2 (L_gV)R^{-1}(e)(L_gV)^{\mathrm{T}} \tag{7.53}$$

and

$$R(e) = \beta\left(AA^{\mathrm{T}} + BB^{\mathrm{T}} + 2\Sigma^2\right)^{-1}, \quad \beta \geq 2. \tag{7.54}$$

Proof. First, $E(e(t)) = 2\beta V(e(t))$ is a positive-definite and radially unbounded function. We will prove that $l(e)$ is a radially unbounded function. Using (7.45), (7.46), and (7.54), we obtain

$$\begin{aligned}
l(e) \geq & -2\beta e^{\mathrm{T}}(t)\left(-C+\frac{1}{2}AA^{\mathrm{T}}+\frac{1}{2}BB^{\mathrm{T}}+\Sigma^2\right)e(t) \\
& +\beta e^{\mathrm{T}}(t)\left(AA^{\mathrm{T}}+BB^{\mathrm{T}}+2\Sigma^2\right)e(t) \\
= & \, 2\beta e^{\mathrm{T}}(t)Ce(t).
\end{aligned}$$

This means that $l(e)$ is radially unbounded. Substituting (7.48) into (7.44), we get

$$\dot{V}(e) = L_{\bar{f}}V - \beta(L_gV)R^{-1}(e)(L_gV)^{\mathrm{T}}.$$

Multiplying both sides of the above equation by -2β, we obtain

$$-2\beta\dot{V}(e) = -2\beta L_{\bar{f}}V + 2\beta^2(L_gV)R^{-1}(e)(L_gV)^{\mathrm{T}}. \tag{7.55}$$

Considering (7.50) and (7.53), from (7.55) we get

$$l(e) + u^{\mathrm{T}}R(e)u = -2\beta\dot{V}(e). \tag{7.56}$$

Substituting (7.56) into (7.52), we have

$$\begin{aligned}
J(u) &= \lim_{t\to\infty}\left\{2\beta V(e(t)) + \int_0^t \left(-2\beta\dot{V}(e(\tau))\right)\mathrm{d}\tau\right\} \\
&= \lim_{t\to\infty}\left\{2\beta V(e(t)) + \left(-2\beta V(e(t)) + 2\beta V(e(0))\right)\right\} \\
&= 2\beta V(e(0)).
\end{aligned}$$

Thus, the minimum of the cost functional is $J_{\min}(u) = 2\beta V(e(0))$ for the optimal controller (7.43). The proof is completed. □

The following illustrative example is used to demonstrate the effectiveness of the above method.

We consider the delayed chaotic cellular neural network [2] as follows:

$$\begin{aligned}
\begin{pmatrix}\dot{x}_1(t)\\ \dot{x}_2(t)\end{pmatrix} = & -\begin{pmatrix}1 & 0\\ 0 & 1\end{pmatrix}\begin{pmatrix}x_1(t)\\ x_2(t)\end{pmatrix} + \begin{pmatrix}1+\frac{\pi}{4} & 20\\ 0.1 & 1+\frac{\pi}{4}\end{pmatrix}\begin{pmatrix}g_1(x_1(t))\\ g_2(x_2(t))\end{pmatrix} \\
& + \begin{pmatrix}-\frac{1.3\sqrt{2}\pi}{4} & 0.1\\ 0.1 & -\frac{1.3\sqrt{2}\pi}{4}\end{pmatrix}\begin{pmatrix}g_1(x_1(t-1))\\ g_2(x_2(t-1))\end{pmatrix},
\end{aligned} \tag{7.57}$$

where $g_i(x_i(t)) = 0.5(|x_i(t)+1| - |x_i(t)-1|)$, $i = 1, 2$.

The chaotic behavior of the drive system is shown in Fig. 7.32, which is plotted with the initial condition $(x_1(s),\ x_2(s))^{\mathrm{T}} = (0.01,\ 0.1)^{\mathrm{T}}$ for $-1 \leq s \leq 0$. The curves of the drive system's state variables x_1 and x_2 are shown in Fig. 7.33. The chaotic behavior of the response system is shown in Fig. 7.34, which is plotted with the initial condition $(y_1(s),\ y_2(s))^{\mathrm{T}} = (2,\ -0.2)^{\mathrm{T}}$ for $-1 \leq s \leq 0$.

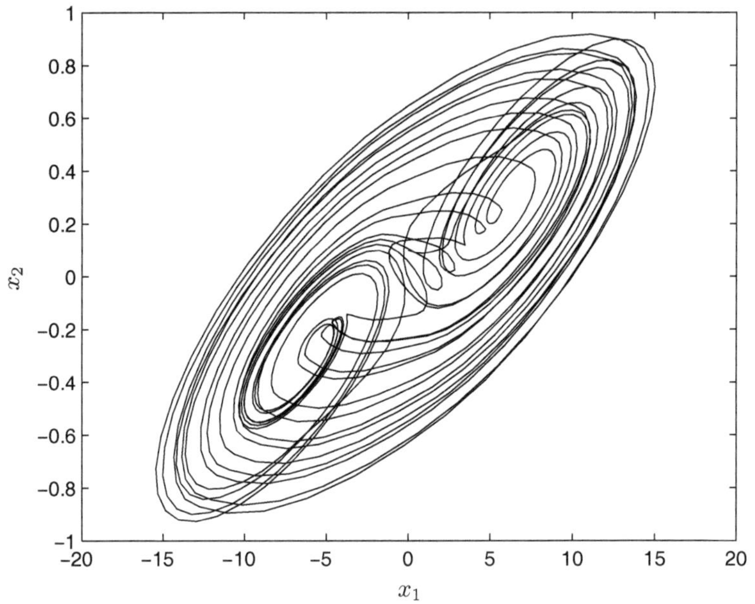

Fig. 7.32 Chaotic behavior of the drive delayed chaotic cellular neural network

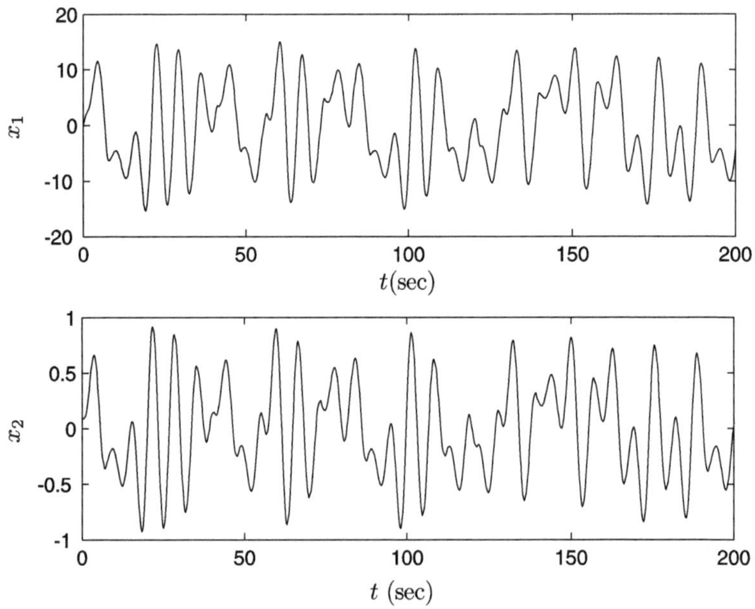

Fig. 7.33 The curves of the drive system's state variables x_1 and x_2

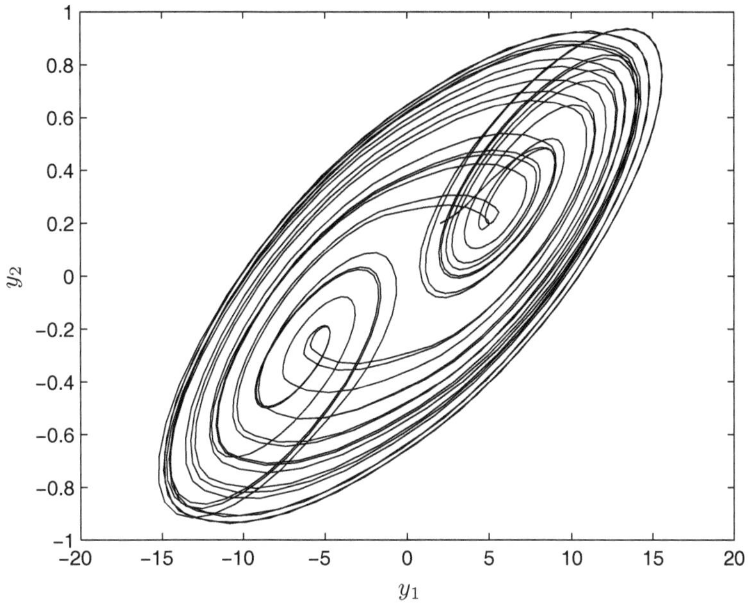

Fig. 7.34 Chaotic behavior of the response delayed chaotic cellular neural network

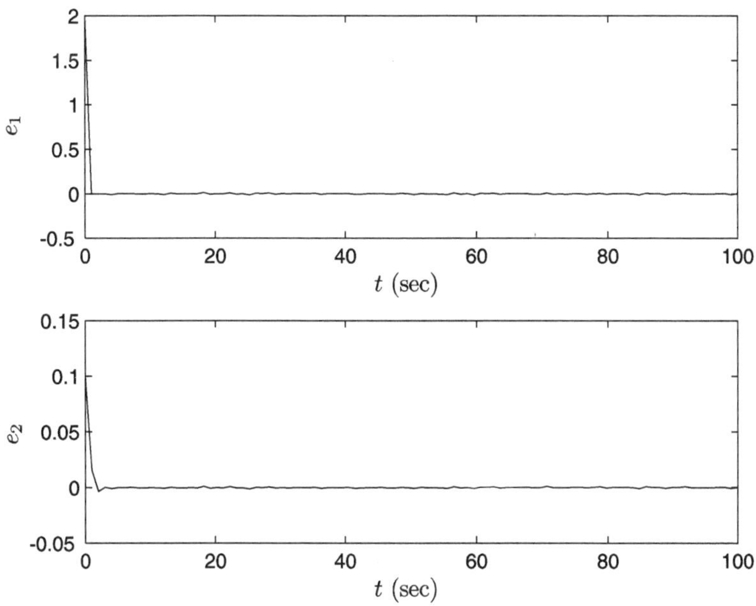

Fig. 7.35 The error curves of the delayed chaotic cellular neural network

The synchronization error curves of the response system and the drive system are shown in Fig. 7.35. From simulation results we can see that the synchronization errors converge asymptotically to zero, i.e., the trajectory of the response delayed chaotic cellular neural network can synchronize asymptotically with the trajectory of the drive delayed chaotic cellular neural network.

7.4.1.3 Synchronization of a Class of Nonidentical Chaotic DNNs

In the above subsection, the parameters of the two neural networks to be synchronized are known and are identical. Because the synchronized systems are inevitably perturbed by external factors and cannot be exactly identical, synchronizing two nonidentical delayed chaotic neural networks is more essential and useful in real-world applications. In this section, we consider the synchronization problem of a class of chaotic DNNs in which the parameters of the response system are tunable. Suppose that the drive system has the form of (7.39), and the response system is described by the following equation:

$$\dot{y}_i(t) = -\hat{c}_i y_i(t) + \sum_{j=1}^{n} \hat{a}_{ij} g_j(y_j(t)) + \sum_{j=1}^{n} \hat{b}_{ij} g_j(y_j(t-\tau_j)) + U_i + u_i(t), \tag{7.58}$$

where $\hat{c}_i$, $\hat{a}_{ij}$, and $\hat{b}_{ij}$ are tunable parameters of the response system whose updating laws need to be designed, and $u_i(t)$ denotes the designed controller that will achieve the synchronization of system (7.39) and system (7.58).

Let us define the synchronization error signal as $e_i(t) = y_i(t) - x_i(t)$. The fact that $e_i(t) \to 0$ as $t \to \infty$ means that the drive neural network and the response neural network are synchronized. Therefore, the error system between (7.39) and (7.58) can be expressed as

$$\begin{aligned}\dot{e}_i(t) =& -c_i e_i(t) + \sum_{j=1}^{n} a_{ij} f_j(e_j(t)) + \sum_{j=1}^{n} b_{ij} f_j(e_j(t-\tau_j)) \\ & + (c_i - \hat{c}_i) y_i(t) + \sum_{j=1}^{n} (\hat{a}_{ij} - a_{ij}) g_j(y_j(t)) \\ & + \sum_{j=1}^{n} (\hat{b}_{ij} - b_{ij}) g_j(y_j(t-\tau_j)) + u_i(t),\end{aligned} \tag{7.59}$$

where $f_j(e_j(t)) = g_j(e_j(t) + x_j(t)) - g_j(x_j(t))$, and $f_j(0) = 0$, or by the following compact form:

$$\begin{aligned}\dot{e}(t) =& -Ce(t) + Af(e(t)) + Bf(e(t-\tau)) + \tilde{C}y(t) + \tilde{A}g(y(t)) \\ & + \tilde{B}g(y(t-\tau)) + u(t),\end{aligned} \tag{7.60}$$

where $C = \mathrm{diag}\{c_1, \ldots, c_n\}$, $A = (a_{ij})_{n\times n}$, $B = (b_{ij})_{n\times n}$, $\tau = (\tau_1, \ldots, \tau_n)^{\mathrm{T}}$, $\tilde{C}(t) = \mathrm{diag}\{\tilde{c}_1(t), \tilde{c}_2(t), \ldots, \tilde{c}_n(t)\}$, $\tilde{A}(t) = (\tilde{a}_{ij}(t))_{n\times n}$, $\tilde{B}(t) = (\tilde{b}_{ij}(t))_{n\times n}$, $\tilde{c}_i = c_i - \hat{c}_i$, in which $\tilde{a}_{ij} = \hat{a}_{ij} - a_{ij}$ and $\tilde{b}_{ij} = \hat{b}_{ij} - b_{ij}$ are the estimated errors of unknown parameters c_i, a_{ij}, and b_{ij}, respectively, $f(e(t)) = (f_1(e_1(t)), \ldots, f_n(e_n(t)))^{\mathrm{T}}$, $f(e(t-\tau)) = (f_1(e_1(t-\tau_1)), \ldots, f_n(e_n(t-\tau_n)))^{\mathrm{T}}$, and $u(t) = (u_1(t), \ldots, u_n(t))^{\mathrm{T}}$.

Theorem 7.8 ([16]). By the following controller:

$$u(t) = -\left(AA^{\mathrm{T}} + BB^{\mathrm{T}} + 2\Sigma^2\right) e(t) \tag{7.61}$$

and the parameter update law

$$\begin{cases} \dot{\hat{c}}_i = e_i(t) y_i(t), \\ \dot{\hat{a}}_{ij} = -e_i(t) g_j(y_j(t)), \\ \dot{\hat{b}}_{ij} = -e_i(t) g_j(y_j(t-\tau_j)), \end{cases} \tag{7.62}$$

the systems (7.39) and (7.58) will be globally asymptotically synchronized. Here, $i, j = 1, \ldots, n$ and $\Sigma = \mathrm{diag}\{\delta_1, \ldots, \delta_n\}$.

Proof. Define

$$\begin{aligned} \varepsilon := \big(&e^{\mathrm{T}}(t), \tilde{c}_1, \ldots, \tilde{c}_n, \tilde{a}_{11}, \ldots, \tilde{a}_{1n}, \ldots, \tilde{a}_{n1}, \ldots, \tilde{a}_{nn}, \\ &\tilde{b}_{11}, \ldots, \tilde{b}_{1n}, \ldots, \tilde{b}_{n1}, \ldots, \tilde{b}_{nn}\big)^{\mathrm{T}}. \end{aligned}$$

We choose

$$\begin{aligned} V(\varepsilon) = {}& \frac{1}{2} e^{\mathrm{T}}(t) e(t) + \frac{1}{2} \sum_{j=1}^{n} \int_{t-\tau_j}^{t} \delta_j^2 e_j^2(s)\,\mathrm{d}s \\ & + \frac{1}{2} \sum_{i=1}^{n} \tilde{c}_i^2 + \frac{1}{2} \sum_{i,j=1}^{n} \tilde{a}_{ij}^2 + \frac{1}{2} \sum_{i,j=1}^{n} \tilde{b}_{ij}^2. \end{aligned}$$

Its time derivative can be derived as follows:

$$\begin{aligned} \dot{V}(\varepsilon) = {}& e^{\mathrm{T}}(t) \dot{e}(t) + \frac{1}{2} e^{\mathrm{T}}(t) \Sigma^2 e(t) - \frac{1}{2} e^{\mathrm{T}}(t-\tau) \Sigma^2 e(t-\tau) \\ & + \sum_{i=1}^{n} \tilde{c}_i \dot{\tilde{c}}_i + \sum_{i,j=1}^{n} \tilde{a}_{ij} \dot{\tilde{a}}_{ij} + \sum_{i,j=1}^{n} \tilde{b}_{ij} \dot{\tilde{b}}_{ij} \\ = {}& - e^{\mathrm{T}}(t) C e(t) + e^{\mathrm{T}}(t) \big(A f(e(t)) + B f(e(t-\tau))\big) + e^{\mathrm{T}}(t) u(t) \\ & + e^{\mathrm{T}}(t) \left(\tilde{C} y(t) + \tilde{A} g(y(t)) + \tilde{B} g(y(t-\tau))\right) + \frac{1}{2} e^{\mathrm{T}}(t) \Sigma^2 e(t) \\ & - \frac{1}{2} e^{\mathrm{T}}(t-\tau) \Sigma^2 e(t-\tau) - \sum_{i=1}^{n} \tilde{c}_i \dot{\hat{c}}_i \\ & + \sum_{i,j=1}^{n} \tilde{a}_{ij} \dot{\hat{a}}_{ij} + \sum_{i,j=1}^{n} \tilde{b}_{ij} \dot{\hat{b}}_{ij} \\ := {}& L_{\tilde{f}} V + (L_g V) u(t), \end{aligned}$$

where

$$\begin{aligned}
L_{\bar{f}}V := & -e^{\mathrm{T}}(t)Ce(t)+e^{\mathrm{T}}(t)\left(Af(e(t))+Bf(e(t-\tau))\right) \\
& +e^{\mathrm{T}}(t)\left(\tilde{C}y(t)+\tilde{A}g(y(t))+\tilde{B}g(y(t-\tau))\right)+\frac{1}{2}e^{\mathrm{T}}(t)\Sigma^2 e(t) \\
& -\frac{1}{2}e^{\mathrm{T}}(t-\tau)\Sigma^2 e(t-\tau)-\sum_{i=1}^{n}\tilde{c}_i\dot{\hat{c}}_i \\
& +\sum_{i,j=1}^{n}\tilde{a}_{ij}\dot{\hat{a}}_{ij}+\sum_{i,j=1}^{n}\tilde{b}_{ij}\dot{\hat{b}}_{ij}
\end{aligned}$$

and

$$L_gV := e^{\mathrm{T}}(t).$$

Applying Lemma 7.1 with $Q=I$, we get

$$\begin{aligned}
\dot{V}(\varepsilon) \leq & -e^{\mathrm{T}}(t)Ce(t)+\frac{1}{2}e^{\mathrm{T}}(t)AA^{\mathrm{T}}e(t)+\frac{1}{2}f^{\mathrm{T}}(e(t))f(e(t)) \\
& +\frac{1}{2}e^{\mathrm{T}}(t)BB^{\mathrm{T}}e(t)+\frac{1}{2}f^{\mathrm{T}}(e(t-\tau))f(e(t-\tau)) \\
& +\frac{1}{2}e^{\mathrm{T}}(t)\Sigma^2 e(t)-\frac{1}{2}e^{\mathrm{T}}(t-\tau)\Sigma^2 e(t-\tau)+e^{\mathrm{T}}(t)u(t) \\
& +e^{\mathrm{T}}(t)\left(\tilde{C}y(t)+\tilde{A}g(y(t))+\tilde{B}g(y(t-\tau))\right) \\
& -\sum_{i=1}^{n}\tilde{c}_i\dot{\hat{c}}_i+\sum_{i,j=1}^{n}\tilde{a}_{ij}\dot{\hat{a}}_{ij}+\sum_{i,j=1}^{n}\tilde{b}_{ij}\dot{\hat{b}}_{ij} \\
\leq & -e^{\mathrm{T}}(t)Ce(t)+\frac{1}{2}e^{\mathrm{T}}(t)AA^{\mathrm{T}}e(t)+\frac{1}{2}e^{\mathrm{T}}(t)\Sigma^2 e(t) \\
& +\frac{1}{2}e^{\mathrm{T}}(t)BB^{\mathrm{T}}e(t)+\frac{1}{2}e^{\mathrm{T}}(t-\tau)\Sigma^2 e(t-\tau) \\
& +\frac{1}{2}e^{\mathrm{T}}(t)\Sigma^2 e(t)-\frac{1}{2}e^{\mathrm{T}}(t-\tau)\Sigma^2 e(t-\tau)+e^{\mathrm{T}}(t)u(t) \\
& +\sum_{i=1}^{n}\tilde{c}_i\left(e_i(t)y_i(t)-\dot{\hat{c}}_i\right)+\sum_{i,j=1}^{n}\tilde{a}_{ij}\left(\dot{\hat{a}}_{ij}+e_i(t)g_j(y_j(t))\right) \\
& +\sum_{i,j=1}^{n}\tilde{b}_{ij}\left(\dot{\hat{b}}_{ij}+e_i(t)g_j(y_j(t-\tau_j))\right).
\end{aligned}$$

Substituting (7.62) into the above inequality, we obtain

$$\dot{V}(\varepsilon)\leq e^{\mathrm{T}}(t)\left(-C+\frac{1}{2}AA^{\mathrm{T}}+\frac{1}{2}BB^{\mathrm{T}}+\Sigma^2\right)e(t)+e^{\mathrm{T}}(t)u(t). \tag{7.63}$$

Substituting (7.61) into (7.63), we obtain

$$\begin{aligned}\dot{V}(\varepsilon) &\leq -e^{\mathrm{T}}(t)\left(C+\frac{1}{2}AA^{\mathrm{T}}+\frac{1}{2}BB^{\mathrm{T}}+\Sigma^2\right)e(t)\\ &\leq -\lambda_{\min}\left(C+\frac{1}{2}AA^{\mathrm{T}}+\frac{1}{2}BB^{\mathrm{T}}+\Sigma^2\right)\|e\|^2\\ &\leq 0,\end{aligned}$$

which implies that $\dot{V}(\varepsilon) < 0$ for all $e(t) \neq 0$. Since $V(\varepsilon)$ is positive definite and $\dot{V}(\varepsilon)$ is negative semidefinite, it follows that $e(t) \in L_\infty$, $\tilde{c}_i(t) \in L_\infty$, $\tilde{a}_{ij}(t) \in L_\infty$, and $\tilde{b}_i(t) \in L_\infty$. From the fact that

$$\begin{aligned}\int_0^t \|e(s)\|^2\,\mathrm{d}s &= \frac{V(e(0)) - V(e(t))}{\lambda_{\min}\left(C+\frac{1}{2}AA^{\mathrm{T}}+\frac{1}{2}BB^{\mathrm{T}}+\Sigma^2\right)}\\ &\leq \frac{V(e(0))}{\lambda_{\min}\left(C+\frac{1}{2}AA^{\mathrm{T}}+\frac{1}{2}BB^{\mathrm{T}}+\Sigma^2\right)},\end{aligned}$$

we can easily show that $e(t) \in L_2$. From (7.59) we have $\dot{e}(t) \in L_\infty$. Thus, by Barbalat's lemma, we have $\lim_{t\to\infty} e(t) = 0$, i.e., $\lim_{t\to\infty} \|e(t)\| = \lim_{t\to\infty} \|y(t) - x(t)\| = 0$. This means that the proposed controller (7.61) can globally asymptotically synchronize the system (7.39) and the system (7.58). This completes the proof. □

Remark 7.1. Theorem 7.8 can similarly be proved by showing that the cost functional (7.52) is minimized by the optimal control laws (7.61) and (7.62). □

The following illustrative example is used to demonstrate the effectiveness of the above method.

We consider that the drive delayed chaotic cellular neural network has the form of (7.39). When the parameters and the initial conditions are the same as those in (7.57), the chaotic behavior of the drive system is shown in Fig. 7.32.

The response delayed chaotic cellular neural network is chosen in the form of (7.58). In this simulation, the initial values of 'unknown' parameter vectors of the response system are selected as

$$\hat{C}(0) = \begin{pmatrix} 1.1 & 0 \\ 0 & 1.1 \end{pmatrix}, \quad \hat{A}(0) = \begin{pmatrix} 1.1+\pi/4 & 20.1 \\ 0.11 & 1.1+\pi/4 \end{pmatrix},$$

$$\hat{B}(0) = \begin{pmatrix} -1.29\sqrt{2}\pi/4 & 0.11 \\ 0.11 & -1.29\sqrt{2}\pi/4 \end{pmatrix}.$$

The chaotic behavior of the response system is shown in Fig. 7.34, which is plotted with the initial condition $(y_1(s),\ y_2(s))^{\mathrm{T}} = (2,\ -0.2)^{\mathrm{T}}$ for $-1 \leq s \leq 0$.

The synchronization error curves of the response system and the drive system are shown in Fig. 7.36. The curves of $\hat{C}$, $\hat{A}$, and $\hat{B}$ are shown in Figs. 7.37–7.39, respectively. Obviously, the synchronization errors converge asymptotically to zero and the adaptive parameters converge asymptotically to some constants.

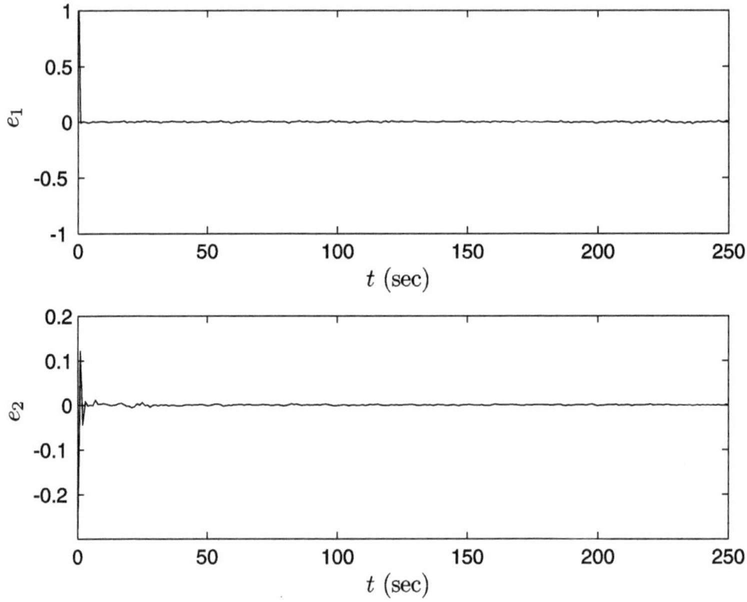

Fig. 7.36 The error curves of the delayed chaotic cellular neural network

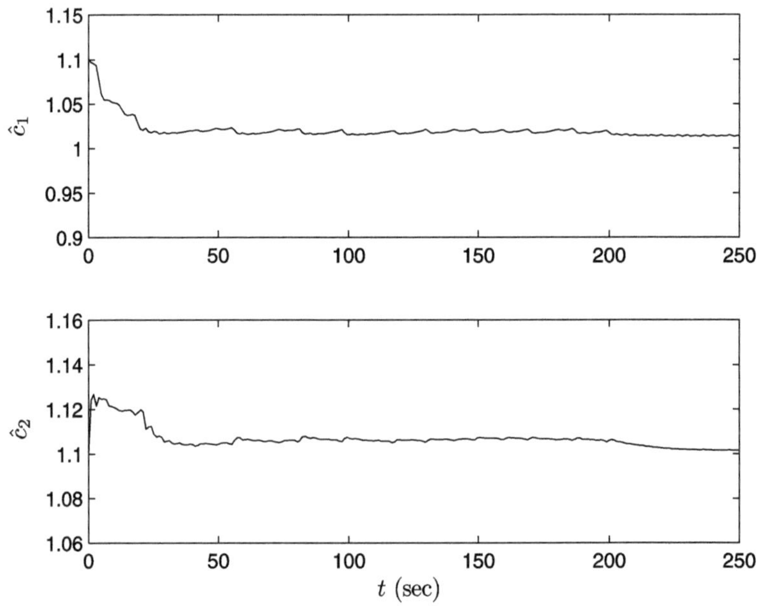

Fig. 7.37 The curves of parameter $\hat{C}$

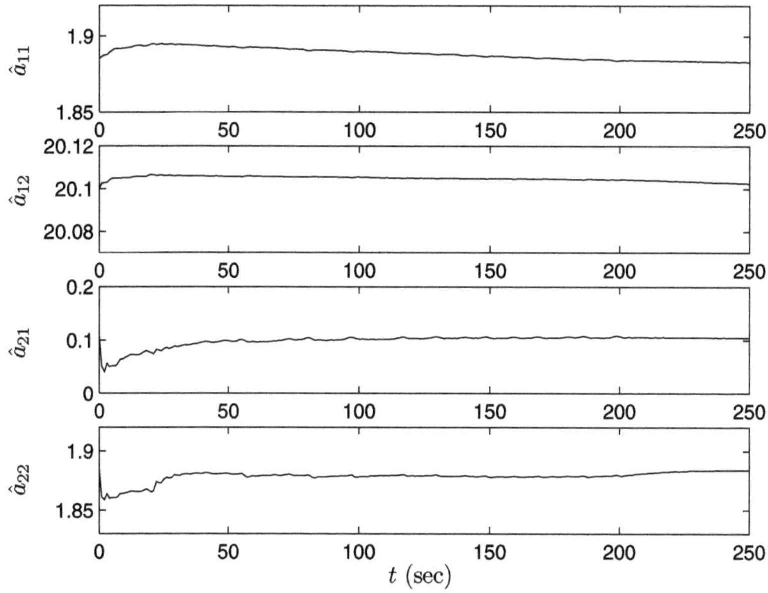

Fig. 7.38 The curves of parameter $\hat{A}$

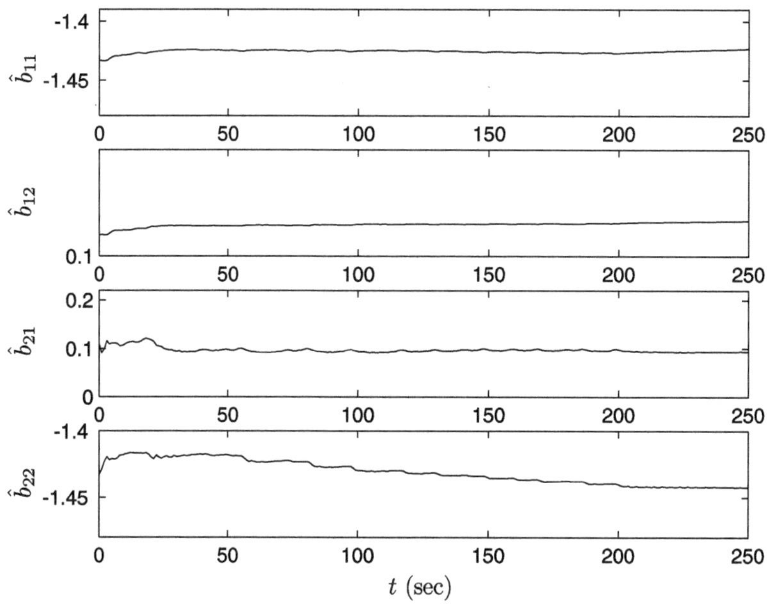

Fig. 7.39 The curves of parameter $\hat{B}$

7.4.2 Adaptive Synchronization Between Two Different Chaotic Delayed Neural Networks

In this section, the problem of synchronizing two different chaotic DNNs with unknown parameters is studied. By Lyapunov stability theory, the delay-independent and delay-dependent adaptive synchronization controllers and adaptive updating laws of parameters are designed, respectively [19].

7.4.2.1 Problem Description

A class of delayed chaotic neural networks is described by the following time-delay differential equation:

$$\dot{x}_i(t) = -c_i x_i(t) + \sum_{j=1}^{n} a_{ij} g_j(x_j(t)) + \sum_{j=1}^{n} b_{ij} g_j(x_j(t-\tau_j)) + H_i, \tag{7.64}$$

where $n \geq 2$ denotes the number of neurons in the network, $x_i(t)$ denotes the state variable associated with the ith neuron, $\tau_j \geq 0$ denotes the bounded delay, $\rho = \max\{\tau_j\}$, $c_i > 0$ denotes the self connection of neurons, a_{ij} indicates the strength of the neuron interconnections within the network, b_{ij} indicates the strength of the neuron interconnections within the network with constant delay τ_j, and H_i is an external constant input, $i,j = 1,\ldots,n$. The activation function $g_j\colon \mathbb{R} \to \mathbb{R}$, $j \in \{1,2,\ldots,n\}$, is bounded, and satisfies the condition $g_j(0) = 0$ and the Lipschitz condition with a Lipschitz constant $L_x > 0$, i.e., $\|g_j(\xi) - g_j(\varsigma)\| \leq L_x \|\xi - \varsigma\|$ for all $\xi, \varsigma \in \mathbb{R}$ and $\xi \neq \varsigma$. The initial conditions of system (7.64) are given by $x_i(t) = \phi_i(t) \in C([-\rho,0],\mathbb{R})$, where $C([-\rho,0],\mathbb{R})$ denotes the set of all continuous functions from $[-\rho,0]$ to $\mathbb{R}$.

Suppose that the drive system has the form of (7.64). The response system is described by the following equation:

$$\dot{y}_i(t) = -d_i y_i(t) + \sum_{j=1}^{n} p_{ij} f_j(y_j(t)) + \sum_{j=1}^{n} q_{ij} f_j(y_j(t-\tau_j)) + W_i + u_i(t), \tag{7.65}$$

where $n \geq 2$ denotes the number of neurons in the network, $y_i(t)$ denotes the state variable associated with the ith neuron, $d_i > 0$ denotes the self connection of neurons, p_{ij} and q_{ij} indicate the interconnection strengths among neurons without and with the bounded delay τ_j, W_i is an external constant input, and $u_i(t)$ denotes the external control input and will be appropriately designed to obtain a certain control objective, $i,j = 1,\ldots,n$. The activation function $f_j\colon \mathbb{R} \to \mathbb{R}$ ($j \in \{1,2,\ldots,n\}$) is bounded, and satisfies the condition $f_j(0) = 0$ and the Lipschitz condition with a Lipschitz constant $L_y > 0$, i.e., $\|f_j(\xi) - f_j(\varsigma)\| \leq L_y \|\xi - \varsigma\|$ for all $\xi, \varsigma \in \mathbb{R}$, and $\xi \neq \varsigma$. The initial conditions of system (7.65) are given by $y_i(t) = \varphi_i(t) \in C([-\rho,0],\mathbb{R})$.

Define the synchronization error signal $e_i(t) = y_i(t) - x_i(t)$. Our goal is to design a controller $u(t) = (u_1(t), u_2(t), \ldots, u_n(t))^{\mathrm{T}}$ such that the trajectory of the response delayed neural network (7.65) can synchronize asymptotically with the trajectory of the drive delayed neural network (7.64), i.e., $\lim_{t\to\infty} \|e(t)\| = 0$, where $e(t) = (e_1(t), e_2(t), \ldots, e_n(t))^{\mathrm{T}}$.

7.4.2.2 The Delay-independent Controller Design

In this section, we assume that c_i and a_{ij} are the unknown parameters of system (7.64), and b_{ij} is the known parameter of system (7.64). We assume that d_i and p_{ij} are the unknown parameters of the response system (7.65), and q_{ij} is the known parameter of the response system (7.65).

We will need the following assumption and definition.

Assumption 7.1. Both the states of the drive system and the response system are bounded, i.e., $\|x(t)\| \le B_d$ and $\|y(t)\| \le B_r$. □

Definition 7.1. A vector function $\Phi(x)$ is defined by

$$\Phi(x) = \begin{cases} \|x\|^{-1}x, & \text{if } x \neq 0, \\ 0, & \text{if } x = 0. \end{cases}$$

□

Clearly, we have $\Phi(0) = 0$ from Definition 7.1.

Theorem 7.9 ([19]). By the following controller:

$$\begin{aligned} u(t) = &-\hat{C}(t)x(t) + \hat{A}(t)g(x(t)) + \hat{D}(t)y(t) - \hat{P}(t)f(y(t)) \\ &-(m_1+m_2)\Phi(e) + H - W - Ke(t) \end{aligned} \tag{7.66}$$

and the parameter update law

$$\begin{cases} \dot{\hat{c}}_i(t) = e_i(t)x_i(t), \\ \dot{\hat{a}}_{ij}(t) = -e_i(t)g_j(x_j(t)), \\ \dot{\hat{d}}_i(t) = -e_i(t)y_i(t), \\ \dot{\hat{p}}_{ij}(t) = e_i(t)f_j(y_j(t)), \end{cases} \tag{7.67}$$

$i, j = 1, \ldots, n$, the response system (7.65) can synchronize asymptotically with the drive system (7.64), where the positive diagonal matrix $K = \mathrm{diag}\{k_1, k_2, \ldots, k_n\}$ is specified in advance, $m_1 = L_x B_d \|B\|$, $m_2 = L_y B_r \|Q\|$, $B = (b_{ij})_{n\times n}$, $Q = (q_{ij})_{n\times n}$, $H = (H_1, H_2, \ldots, H_n)^{\mathrm{T}}$, $W = (W_1, W_2, \ldots, W_n)^{\mathrm{T}}$, $\hat{C}(t) = \mathrm{diag}\{\hat{c}_1(t), \hat{c}_2(t), \ldots, \hat{c}_n(t)\}$, $\hat{A}(t) = (\hat{a}_{ij}(t))_{n\times n}$, $\hat{D}(t) = \mathrm{diag}\{\hat{d}_1(t), \hat{d}_2(t), \ldots, \hat{d}_n(t)\}$, $\hat{P}(t) = (\hat{p}_{ij}(t))_{n\times n}$, $\hat{c}_i(t)$, $\hat{a}_{ij}(t)$, $\hat{d}_i(t)$, and $\hat{p}_{ij}(t)$ are the estimated values of unknown parameters c_i, a_{ij}, d_i, and p_{ij}, respectively, $g(x(t)) = (g_1(x_1), \ldots, g_n(x_n))^{\mathrm{T}}$, $f(y(t)) = (f_1(y_1), \ldots, f_n(y_n))^{\mathrm{T}}$, and $\Phi(e)$ is the same vector function as defined in Definition 7.1.

Proof. From (7.64)–(7.66), we get the error system as follows:

$$
\begin{aligned}
\dot{e}_i(t) = & -(\hat{c}_i(t) - c_i)\, x_i(t) + \sum_{j=1}^{n} (\hat{a}_{ij}(t) - a_{ij})\, g_j(x_j(t)) \\
& - \sum_{j=1}^{n} b_{ij} g_j(x_j(t-\tau_j)) + (\hat{d}_i(t) - d_i)\, y_i(t) \\
& - \sum_{j=1}^{n} (\hat{p}_{ij}(t) - p_{ij})\, f_j(y_j(t)) + \sum_{j=1}^{n} q_{ij} f_j(y_j(t-\tau_j)) \\
& - (m_1 + m_2)\Phi(e) - k_i e_i(t).
\end{aligned}
\tag{7.68}
$$

Let $\tilde{c}_i(t) = \hat{c}_i(t) - c_i$, $\tilde{a}_{ij}(t) = \hat{a}_{ij}(t) - a_{ij}$, $\tilde{d}_i(t) = \hat{d}_i(t) - d_i$, and $\tilde{p}_{ij}(t) = \hat{p}_{ij}(t) - p_{ij}$ be the estimated errors of unknown parameters c_i, a_{ij}, d_i, and p_{ij}, respectively. We obtain the following compact form:

$$
\begin{aligned}
\dot{e}(t) = & -\tilde{C}(t)x(t) + \tilde{A}(t)g(x(t)) - Bg(x(t-\tau)) \\
& + \tilde{D}(t)y(t) - \tilde{P}(t)f(y(t)) + Qf(y(t-\tau)) \\
& - (m_1 + m_2)\Phi(e) - Ke(t),
\end{aligned}
\tag{7.69}
$$

where $\tau = (\tau_1, \ldots, \tau_n)^{\mathrm{T}}$, $\tilde{C}(t) = \mathrm{diag}\{\tilde{c}_1(t), \tilde{c}_2(t), \ldots, \tilde{c}_n(t)\}$, $\tilde{D}(t) = \mathrm{diag}\{\tilde{d}_1(t), \tilde{d}_2(t), \ldots, \tilde{d}_n(t)\}$, $\tilde{A}(t) = (\tilde{a}_{ij}(t))_{n\times n}$, $\tilde{P}(t) = (\tilde{p}_{ij}(t))_{n\times n}$, $g(x(t-\tau)) = (g_1(x_1(t-\tau_1)), \ldots, g_n(x_n(t-\tau_n)))^{\mathrm{T}}$, and $f(y(t-\tau)) = (f_1(y_1(t-\tau_1)), \ldots, f_n(y_n(t-\tau_n)))^{\mathrm{T}}$.

Define

$$
\begin{aligned}
\varepsilon := (& e^{\mathrm{T}}(t), \tilde{c}_1(t), \ldots, \tilde{c}_n(t), \tilde{a}_{11}(t), \ldots, \tilde{a}_{1n}(t), \ldots, \tilde{a}_{n1}(t), \ldots, \tilde{a}_{nn}(t), \\
& \tilde{d}_1(t), \ldots, \tilde{d}_n(t), \tilde{p}_{11}(t), \ldots, \tilde{p}_{1n}(t), \ldots, \tilde{p}_{n1}(t), \ldots, \tilde{p}_{nn}(t))^{\mathrm{T}}.
\end{aligned}
$$

If a Lyapunov function candidate is chosen as

$$
\begin{aligned}
V(\varepsilon) = & \frac{1}{2}\Bigg(e^{\mathrm{T}}(t)e(t) + \sum_{i=1}^{n} \tilde{c}_i^2(t) + \sum_{i,j=1}^{n} \tilde{a}_{ij}^2(t) \\
& + \sum_{i=1}^{n} \tilde{d}_i^2(t) + \sum_{i,j=1}^{n} \tilde{p}_{ij}^2(t) \Bigg),
\end{aligned}
\tag{7.70}
$$

then the time derivative of $V(\varepsilon)$ along the trajectory of the error system (7.69) is as follows:

$$
\begin{aligned}
\dot{V}(\varepsilon) = & e^{\mathrm{T}}(t)\dot{e}(t) + \sum_{i=1}^{n} \tilde{c}_i(t)\dot{\tilde{c}}_i(t) + \sum_{i,j=1}^{n} \tilde{a}_{ij}(t)\dot{\tilde{a}}_{ij}(t) \\
& + \sum_{i=1}^{n} \tilde{d}_i(t)\dot{\tilde{d}}_i(t) + \sum_{i,j=1}^{n} \tilde{p}_{ij}(t)\dot{\tilde{p}}_{ij}(t)
\end{aligned}
$$

$$
\begin{aligned}
&= e^{\mathrm{T}}(t)\left(-\tilde{C}(t)x(t) + \tilde{A}(t)g(x(t)) + \tilde{D}(t)y(t) - \tilde{P}(t)f(y(t))\right) \\
&\quad + e^{\mathrm{T}}(t)\left(-Bg(x(t-\tau)) + Qf(y(t-\tau)) - (m_1+m_2)\Phi(e)\right) \\
&\quad + \sum_{i=1}^{n} \tilde{c}_i(t)\dot{\hat{c}}_i(t) + \sum_{i,j=1}^{n} \tilde{a}_{ij}(t)\dot{\hat{a}}_{ij}(t) - e^{\mathrm{T}}(t)Ke(t) \\
&\quad + \sum_{i=1}^{n} \tilde{d}_i(t)\dot{\hat{d}}_i(t) + \sum_{i,j=1}^{n} \tilde{p}_{ij}(t)\dot{\hat{p}}_{ij}(t) \\
&= \sum_{i=1}^{n} \tilde{c}_i(t)\left(-e_i(t)x_i(t) + \dot{\hat{c}}_i(t)\right) + \sum_{i=1}^{n} \tilde{d}_i(t)\left(e_i(t)y_i(t) + \dot{\hat{d}}_i(t)\right) \\
&\quad + \sum_{i,j=1}^{n} \tilde{a}_{ij}(t)\left(e_i(t)g_j(x_j(t)) + \dot{\hat{a}}_{ij}(t)\right) \\
&\quad + \sum_{i,j=1}^{n} \tilde{p}_{ij}(t)\left(-e_i(t)f_j(y_j(t)) + \dot{\hat{p}}_{ij}(t)\right) \\
&\quad + e^{\mathrm{T}}(t)\left(-Bg(x(t-\tau)) + Qf(y(t-\tau))\right. \\
&\quad \left. - (m_1+m_2)\Phi(e)\right) - e^{\mathrm{T}}(t)Ke(t).
\end{aligned}
\tag{7.71}
$$

Substituting (7.67) into (7.71), we obtain

$$
\begin{aligned}
\dot{V}(\varepsilon) &\le \|e(t)\|\left(L_x\|-B\|\,\|x(t-\tau)\| + L_y\|Q\|\,\|y(t-\tau)\|\right) \\
&\quad - e^{\mathrm{T}}(t)m_1\Phi(e) - e^{\mathrm{T}}(t)m_2\Phi(e) - e^{\mathrm{T}}(t)Ke(t) \\
&\le \|e(t)\|\left(L_x\|B\|B_d + L_y\|Q\|B_r\right) - m_1\|e(t)\| \\
&\quad - m_2\|e(t)\| - e^{\mathrm{T}}(t)Ke(t) \\
&= -e^{\mathrm{T}}(t)Ke(t) \le 0,
\end{aligned}
$$

which implies that $\dot{V}(\varepsilon) < 0$ for all $e(t) \neq 0$. Since $V(\varepsilon)$ is positive definite and $\dot{V}(\varepsilon)$ is negative semidefinite, it follows that $e(t) \in L_\infty$, $\tilde{c}_i(t) \in L_\infty$, $\tilde{a}_{ij}(t) \in L_\infty$, $\tilde{d}_i(t) \in L_\infty$, and $\tilde{p}_{ij}(t) \in L_\infty$. From the fact that

$$
\int_0^t \|e(s)\|^2 \mathrm{d}s = \frac{V(e(0)) - V(e(t))}{\min\{k_i\}} \le \frac{V(e(0))}{\min\{k_i\}},
$$

we can easily show that $e(t) \in L_2$. From (7.69) we have $\dot{e}(t) \in L_\infty$. Thus, by Barbalat's lemma, we have $\lim_{t\to\infty} e(t) = 0$, i.e., $\lim_{t\to\infty}\|e(t)\| = \lim_{t\to\infty}\|y(t) - x(t)\| = 0$. This means that the proposed controller (7.66) can globally asymptotically synchronize the system (7.64) and the system (7.65). This completes the proof. □

The following illustrative example is used to demonstrate the effectiveness of the above method.

Assume that the following delayed chaotic cellular neural network [2] is the drive system:

$$\begin{pmatrix} \dot{x}_1(t) \\ \dot{x}_2(t) \end{pmatrix} = -C\begin{pmatrix} x_1(t) \\ x_2(t) \end{pmatrix} + A\begin{pmatrix} g_1(x_1(t)) \\ g_2(x_2(t)) \end{pmatrix} + B\begin{pmatrix} g_1(x_1(t-\tau_1)) \\ g_2(x_2(t-\tau_2)) \end{pmatrix}, \tag{7.72}$$

where $g_j(x_j(t)) = 0.5(|x_j(t)+1| - |x_j(t)-1|)$, $j = 1,2$, and the following delayed chaotic Hopfield neural network [11] is the response system:

$$\begin{pmatrix} \dot{y}_1(t) \\ \dot{y}_2(t) \end{pmatrix} = -D\begin{pmatrix} y_1(t) \\ y_2(t) \end{pmatrix} + P\begin{pmatrix} f_1(y_1(t)) \\ f_2(y_2(t)) \end{pmatrix} + Q\begin{pmatrix} f_1(y_1(t-\tau_3)) \\ f_2(y_2(t-\tau_4)) \end{pmatrix} + u(t), \tag{7.73}$$

where $f_j(y_j(t)) = \tanh(y_j(t))$, $j = 1,2$.

For numerical simulations, we use the fourth-order Runge–Kutta method to solve the systems of differential equations (7.66), (7.67), (7.72), and (7.73). The 'unknown' parameters of the drive system (7.72) and the response system (7.73) without controller $u(t)$ are selected as $C = \operatorname{diag}[1,1]$, $D = \operatorname{diag}\{1,1\}$,

$$A = \begin{pmatrix} 1+\pi/4 & 20 \\ 0.1 & 1+\pi/4 \end{pmatrix}, \quad B = \begin{pmatrix} -1.3\sqrt{2}\pi/4 & 0.1 \\ 0.1 & -1.3\sqrt{2}\pi/4 \end{pmatrix},$$

$$P = \begin{pmatrix} 2 & -0.1 \\ -5 & 4.5 \end{pmatrix}, \quad Q = \begin{pmatrix} -1.5 & -0.1 \\ -0.2 & -4 \end{pmatrix},$$

and $\tau_1 = \tau_2 = \tau_3 = \tau_4 = 1$. The initial conditions of the drive system (7.72) and the response system (7.73) are taken as $(x_1(s), x_2(s))^{\mathrm{T}} = (0.01,\ 0.1)^{\mathrm{T}}$ and $(y_1(s), y_2(s))^{\mathrm{T}} = (0.4,\ 0.6)^{\mathrm{T}}$ for $-1 \le s \le 0$, respectively. The chaotic behavior of the drive system (7.72) is shown in Fig. 7.32, and the response system (7.73) without the controller $u(t)$ is shown in Fig. 7.40.

In the example, we choose $B_d = 16.0312$ and $B_r = 6.0828$ from Fig. 7.32 and Fig. 7.40, respectively. According to Theorem 7.9, the initial values of the controller and the parameter update law are selected as follows: $\hat{C}(0) = \operatorname{diag}\{1.1, 1.1\}$, $\hat{D}(0) = \operatorname{diag}\{1.1, 1.1\}$, $K = \operatorname{diag}\{1,1\}$, $L_x = L_y = 1$,

$$\hat{A}(0) = \begin{pmatrix} 1.885 & 20.1 \\ 0.11 & 1.885 \end{pmatrix}, \quad \text{and} \quad \hat{P}(0) = \begin{pmatrix} 2.1 & -0.09 \\ -4.9 & 4.5 \end{pmatrix}.$$

The synchronization error curves of the drive system (7.72) and the response system (7.73) with the controller $u(t)$ are shown in Fig. 7.41. The curves of the estimated parameters $\hat{C}$, $\hat{A}$, $\hat{D}$, and $\hat{P}$ are shown in Figs. 7.42–7.45, respectively. From simulation results we can see that the synchronization errors converge asymptotically to zero, i.e., the trajectory of the response delayed chaotic Hopfield neural network can synchronize asymptotically with the trajectory of the drive delayed chaotic cellular neural network. The estimated parameters converge asymptotically to some constants, which shows the effectiveness of the adaptive synchronization scheme.

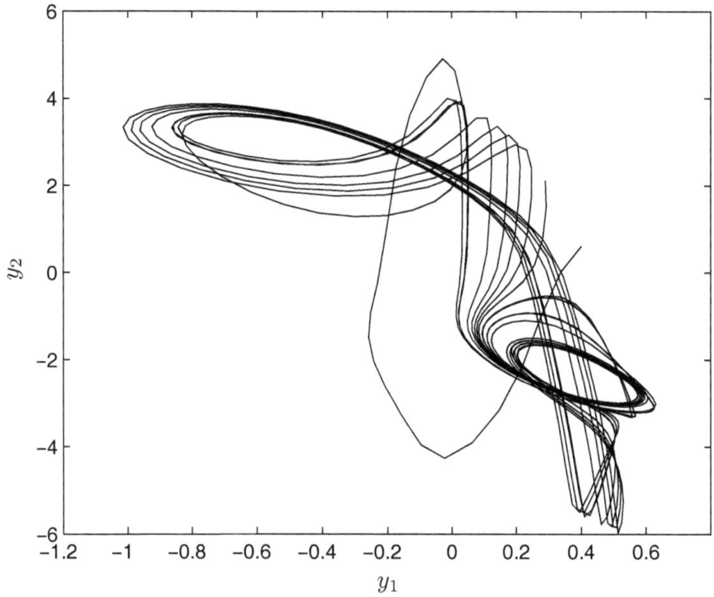

Fig. 7.40 Chaotic behavior of the response system (7.73)

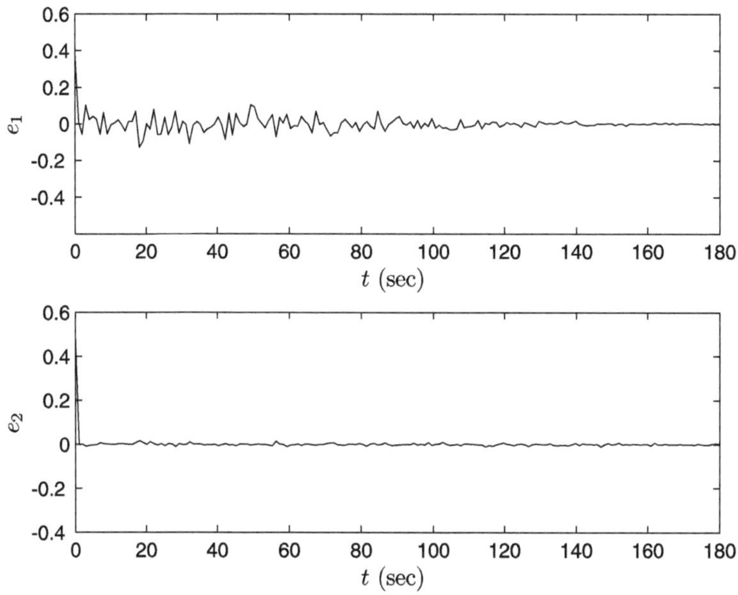

Fig. 7.41 The synchronization error curves of the drive system (7.72) and the response system (7.73)

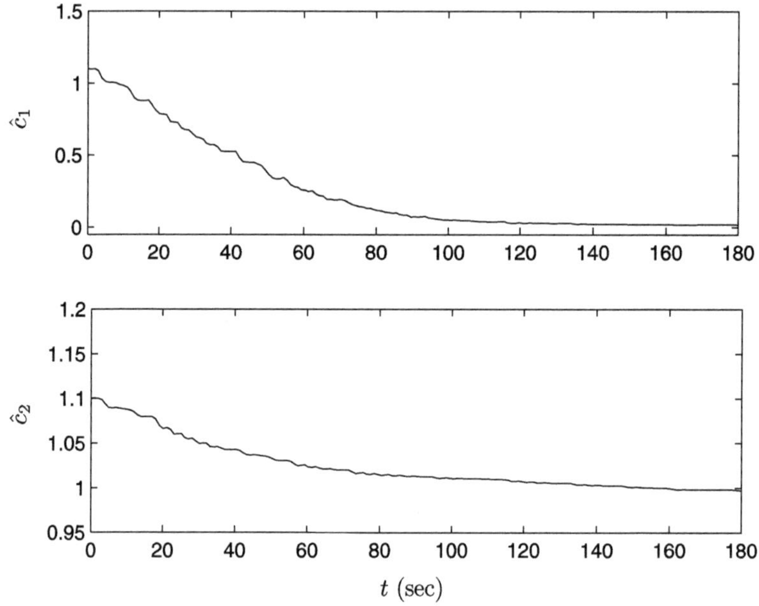

Fig. 7.42 The curves of parameter $\hat{C}(t)$

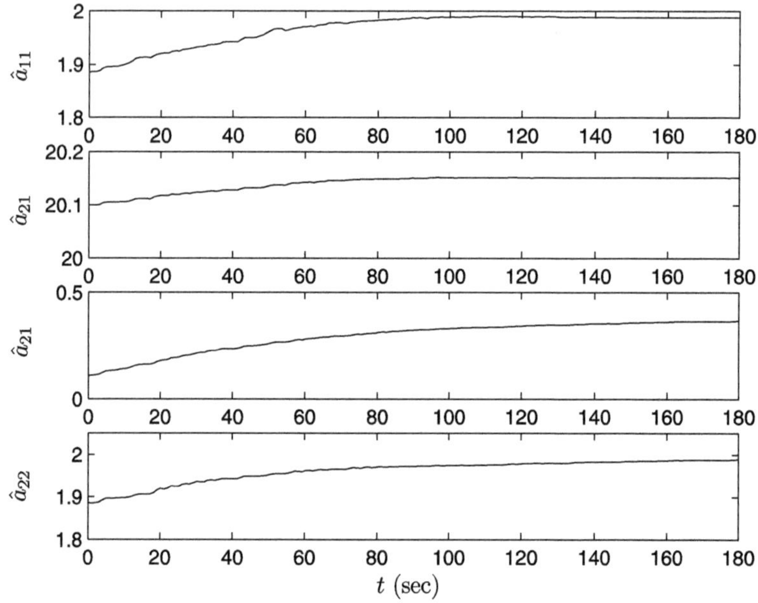

Fig. 7.43 The curves of parameter $\hat{A}(t)$

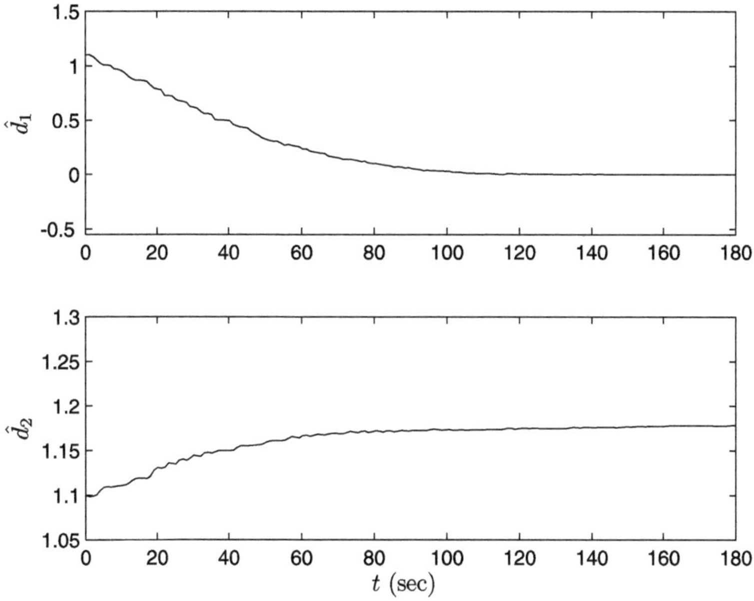

Fig. 7.44 The curves of parameter $\hat{D}(t)$

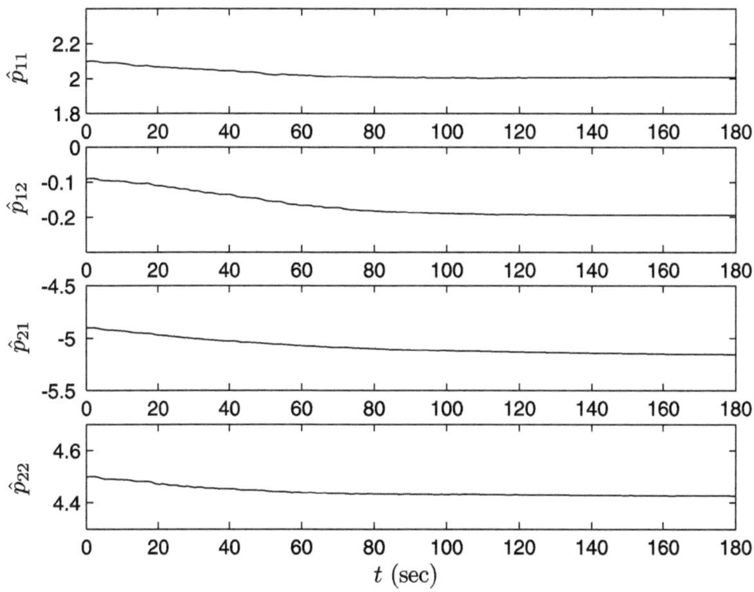

Fig. 7.45 The curves of parameter $\hat{P}(t)$

7.4.2.3 The Delay-dependent Controller Design

In this section, we assume that c_i, a_{ij}, and b_{ij} are the unknown parameters of system (7.64) and d_i, p_{ij}, and q_{ij} are the unknown parameters of the response system (7.65).

Theorem 7.10 ([16]). By the following controller:

$$\begin{aligned} u(t) = &-\hat{C}x(t) + \hat{A}g(x(t)) + \hat{B}g(x(t-\tau)) + \hat{D}y(t) \\ &- \hat{P}f(y(t)) - \hat{Q}f(y(t-\tau)) + H - W - Ke(t) \end{aligned} \tag{7.74}$$

and the parameter update law

$$\begin{cases} \dot{\hat{c}}_i = e_i(t)x_i(t), \\ \dot{\hat{a}}_{ij} = -e_i(t)g_j(x_j(t)), \\ \dot{\hat{b}}_{ij} = -e_i(t)g_j(x_j(t-\tau_j)), \\ \dot{\hat{d}}_i = -e_i(t)y_i(t), \\ \dot{\hat{p}}_{ij} = e_i(t)f_j(y_j(t)), \\ \dot{\hat{q}}_{ij} = e_i(t)f_j(y_j(t-\tau_j)), \end{cases} \tag{7.75}$$

$i, j = 1, \ldots, n$, the response system (7.65) can synchronize asymptotically with the drive system (7.64), where the positive diagonal matrix $K = \mathrm{diag}\{k_1, k_2, \ldots, k_n\}$ is specified in advance, $\tau = (\tau_1, \tau_2, \ldots, \tau_n)^{\mathrm{T}}$, $H = (H_1, H_2, \ldots, H_n)^{\mathrm{T}}$, $W = (W_1, W_2, \ldots, W_n)^{\mathrm{T}}$, $\hat{C}(t) = \mathrm{diag}\{\hat{c}_1(t), \hat{c}_2(t), \ldots, \hat{c}_n(t)\}$, $\hat{A}(t) = (\hat{a}_{ij}(t))_{n\times n}$, $\hat{B}(t) = (\hat{b}_{ij}(t))_{n\times n}$, $\hat{D}(t) = \mathrm{diag}\{\hat{d}_1(t), \hat{d}_2(t), \ldots, \hat{d}_n(t)\}$, $\hat{P}(t) = (\hat{p}_{ij}(t))_{n\times n}$, $\hat{Q}(t) = (\hat{q}_{ij}(t))_{n\times n}$, $\hat{c}_i(t)$, $\hat{a}_{ij}(t)$, $\hat{b}_{ij}(t)$, $\hat{d}_i(t)$, $\hat{p}_{ij}(t)$, and $\hat{q}_{ij}(t)$ are the estimated values of unknown parameters c_i, a_{ij}, b_{ij}, d_i, p_{ij}, and q_{ij}, respectively, $g(x(t)) = (g_1(x_1(t)), \ldots, g_n(x_n(t)))^{\mathrm{T}}$, $f(y(t)) = (f_1(y_1(t)), \ldots, f_n(y_n(t)))^{\mathrm{T}}$, $g(x(t-\tau)) = (g_1(x_1(t-\tau_1)), \ldots, g_n(x_n(t-\tau_n)))^{\mathrm{T}}$, and $f(y(t-\tau)) = (f_1(y_1(t-\tau_1)), \ldots, f_n(y_n(t-\tau_n)))^{\mathrm{T}}$.

Proof. From (7.64), (7.65), and (7.74), we get the error system as follows:

$$\begin{aligned} \dot{e}_i(t) = &-(\hat{c}_i - c_i)x_i(t) + \sum_{j=1}^{n}(\hat{a}_{ij} - a_{ij})g_j(x_j(t)) - k_i e_i(t) \\ &+ \sum_{j=1}^{n}(\hat{b}_{ij} - b_{ij})g_j(x_j(t-\tau_j)) + (\hat{d}_i - d_i)y_i(t) \\ &- \sum_{j=1}^{n}(\hat{p}_{ij} - p_{ij})f_j(y_j(t)) - \sum_{j=1}^{n}(\hat{q}_{ij} - q_{ij})f_j(y_j(t-\tau_j)). \end{aligned}$$

Let $\tilde{c}_i(t) = \hat{c}_i(t) - c_i$, $\tilde{a}_{ij}(t) = \hat{a}_{ij}(t) - a_{ij}$, $\tilde{b}_{ij}(t) = \hat{b}_{ij}(t) - b_{ij}$, $\tilde{d}_i(t) = \hat{d}_i(t) - d_i$, $\tilde{p}_{ij}(t) = \hat{p}_{ij}(t) - p_{ij}$, and $\tilde{q}_{ij}(t) = \hat{q}_{ij}(t) - q_{ij}$ be the estimated errors of unknown parameters c_i, a_{ij}, b_{ij}, d_i, p_{ij}, and q_{ij}, respectively.

We obtain the following compact form:

$$\begin{aligned} \dot{e}(t) = &-\tilde{C}(t)x(t) + \tilde{A}(t)g(x(t)) + \tilde{B}(t)g(x(t-\tau)) \\ &+ \tilde{D}(t)y(t) - \tilde{P}(t)f(y(t)) - \tilde{Q}(t)f(y(t-\tau)) - Ke(t), \end{aligned} \tag{7.76}$$

where $\tilde{C}(t) = \operatorname{diag}\{\tilde{c}_1(t), \tilde{c}_2(t), \ldots, \tilde{c}_n(t)\}$, $\tilde{D}(t) = \operatorname{diag}\{\tilde{d}_1(t), \tilde{d}_2(t), \ldots, \tilde{d}_n(t)\}$, $\tilde{A}(t) = (\tilde{a}_{ij}(t))_{n\times n}$, $\tilde{B}(t) = (\tilde{b}_{ij}(t))_{n\times n}$, $\tilde{P}(t) = (\tilde{p}_{ij}(t))_{n\times n}$, and $\tilde{Q}(t) = (\tilde{q}_{ij}(t))_{n\times n}$. Define

$$\begin{aligned}\varepsilon := [\, & e^{\mathrm{T}}(t), \tilde{c}_1(t), \ldots, \tilde{c}_n(t), \tilde{a}_{11}(t), \ldots, \tilde{a}_{1n}(t), \ldots, \tilde{a}_{n1}(t), \ldots, \tilde{a}_{nn}(t),\\ & \tilde{b}_{11}(t), \ldots, \tilde{b}_{1n}(t), \ldots, \tilde{b}_{n1}(t), \ldots, \tilde{b}_{nn}(t), \tilde{d}_1(t), \ldots, \tilde{d}_n(t),\\ & \tilde{p}_{11}(t), \ldots, \tilde{p}_{1n}(t), \ldots, \tilde{p}_{n1}(t), \ldots, \tilde{p}_{nn}(t),\\ & \tilde{q}_{11}(t), \ldots, \tilde{q}_{1n}(t), \ldots, \tilde{q}_{n1}(t), \ldots, \tilde{q}_{nn}(t)]^{\mathrm{T}}.\end{aligned}$$

If a Lyapunov function candidate is chosen as

$$\begin{aligned}V(\varepsilon) = & \frac{1}{2}\left(e^{\mathrm{T}}(t)e(t) + \sum_{i=1}^{n} \tilde{c}_i^2(t) + \sum_{i,j=1}^{n} \tilde{a}_{ij}^2(t) + \sum_{i,j=1}^{n} \tilde{b}_{ij}^2(t) + \sum_{i=1}^{n} \tilde{d}_i^2(t) \right.\\ & \left. + \sum_{i,j=1}^{n} \tilde{q}_{ij}^2(t) + \sum_{i,j=1}^{n} \tilde{p}_{ij}^2(t) \right),\end{aligned}$$

then the time derivative of $V(\varepsilon)$ along the trajectory of the error system (7.76) is as follows:

$$\begin{aligned}\dot{V}(\varepsilon) = & \, e^{\mathrm{T}}(t)\dot{e}(t) + \sum_{i=1}^{n} \tilde{c}_i(t)\dot{\tilde{c}}_i(t) + \sum_{i,j=1}^{n} \tilde{a}_{ij}(t)\dot{\tilde{a}}_{ij}(t)\\ & + \sum_{i,j=1}^{n} \tilde{b}_{ij}(t)\dot{\tilde{b}}_{ij}(t) + \sum_{i=1}^{n} \tilde{d}_i(t)\dot{\tilde{d}}_i(t)\\ & + \sum_{i,j=1}^{n} \tilde{p}_{ij}(t)\dot{\tilde{p}}_{ij}(t) + \sum_{i,j=1}^{n} \tilde{q}_{ij}(t)\dot{\tilde{q}}_{ij}(t)\\ = & \, e^{\mathrm{T}}(t)\left(-\tilde{C}(t)x(t) + \tilde{A}(t)g(x(t)) + \tilde{D}(t)y(t) - \tilde{P}(t)f(y(t))\right.\\ & \left. + \tilde{B}g(x(t-\tau)) - \tilde{Q}f(y(t-\tau))\right)\\ & + \sum_{i=1}^{n} \tilde{c}_i(t)\dot{\hat{c}}_i(t) + \sum_{i,j=1}^{n} \tilde{a}_{ij}(t)\dot{\hat{a}}_{ij}(t) - e^{\mathrm{T}}(t)Ke(t)\\ & + \sum_{i,j=1}^{n} \tilde{b}_{ij}(t)\dot{\hat{b}}_{ij}(t) + \sum_{i=1}^{n} \tilde{d}_i(t)\dot{\hat{d}}_i(t)\\ & + \sum_{i,j=1}^{n} \tilde{p}_{ij}(t)\dot{\hat{p}}_{ij}(t) + \sum_{i,j=1}^{n} \tilde{q}_{ij}(t)\dot{\hat{q}}_{ij}(t)\\ = & \sum_{i=1}^{n} \tilde{c}_i(t)\left(-e_i(t)x_i(t) + \dot{\hat{c}}_i(t)\right)\\ & + \sum_{i=1}^{n} \tilde{d}_i(t)\left(e_i(t)y_i(t) + \dot{\hat{d}}_i(t)\right)\end{aligned}$$

$$
\begin{aligned}
&+\sum_{i,j=1}^{n}\tilde{a}_{ij}(t)\left(e_i(t)g_j(x_j(t))+\dot{\hat{a}}_{ij}(t)\right)\\
&+\sum_{i,j=1}^{n}\tilde{b}_{ij}(t)\left(e_i(t)g_j(x_j(t-\tau_j))+\dot{\hat{b}}_{ij}(t)\right)\\
&+\sum_{i,j=1}^{n}\tilde{p}_{ij}(t)\left(-e_i(t)f_j(y_j(t))+\dot{\hat{p}}_{ij}(t)\right)\\
&+\sum_{i,j=1}^{n}\tilde{q}_{ij}(t)\left(-e_i(t)f_j(y_j(t-\tau_j))+\dot{\hat{q}}_{ij}(t)\right)-e^{\mathrm{T}}(t)Ke(t). \qquad (7.77)
\end{aligned}
$$

Substituting (7.75) into (7.77), we obtain

$$\dot{V}(\varepsilon)=-e^{\mathrm{T}}(t)Ke(t)\le 0,$$

which implies that $\dot{V}(\varepsilon)<0$ for all $e(t)\neq 0$. Since $V(\varepsilon)$ is positive definite and $\dot{V}(\varepsilon)$ is negative semidefinite, it follows that $e(t)\in L_\infty$, $\tilde{c}_i(t)\in L_\infty$, $\tilde{a}_{ij}(t)\in L_\infty$, $\tilde{b}_{ij}(t)\in L_\infty$, $\tilde{d}_i(t)\in L_\infty$, $\tilde{p}_{ij}(t)\in L_\infty$, and $\tilde{q}_{ij}(t)\in L_\infty$. From the fact that

$$\int_0^t \| e(s)\|^2\,\mathrm{d}s=\frac{V(e(0))-V(e(t))}{\min\{k_i\}}\le\frac{V(e(0))}{\min\{k_i\}},$$

we can easily show that $e(t)\in L_2$. From (7.76) we have $\dot{e}(t)\in L_\infty$. Thus, by Barbalat's lemma, we have $\lim\limits_{t\to\infty}e(t)=0$, i.e., $\lim\limits_{t\to\infty}\|e(t)\|=\lim\limits_{t\to\infty}\|y(t)-x(t)\|=0$. This means that the proposed controller (7.74) and (7.75) can globally asymptotically synchronize the system (7.64) and the system (7.65). This completes the proof. □

The following illustrative example is used to demonstrate the effectiveness of the above method.

Suppose that the drive delayed chaotic cellular neural network has the form of (7.72). We consider that the response delayed chaotic Hopfield neural network has the form of (7.73).

According to Theorem 7.10, the initial values of the controller and the parameter update law are selected as follows: $\hat{C}(0)=\mathrm{diag}\{1.1,1.1\}$, $\hat{D}(0)=\mathrm{diag}\{1.1,1.1\}$, $K=\mathrm{diag}\{1,1\}$,

$$\hat{A}(0)=\begin{pmatrix}1.885 & 20.1\\ 0.11 & 1.885\end{pmatrix},\quad \tilde{B}(0)=\begin{pmatrix}-1.432 & 0.11\\ 0.11 & -1.432\end{pmatrix},$$

$$\hat{P}(0)=\begin{pmatrix}2.1 & -0.09\\ -4.9 & 4.5\end{pmatrix},\quad\text{and}\quad \tilde{Q}(0)=\begin{pmatrix}-1.56 & -0.09\\ -0.19 & -4.1\end{pmatrix}.$$

The synchronization error curves of the drive system (7.72) and the response system (7.73) with the controller $u(t)$ are shown in Fig. 7.46. The curves of the estimated parameters $\hat{C}$, $\hat{A}$, $\hat{B}$, $\hat{D}$, $\hat{P}$, and $\hat{Q}$ are shown in Figs. 7.47–7.52, respectively. From simulation results we can see that the synchronization errors converge asymptotically to zero, i.e., the trajectory of the response delayed chaotic Hopfield neural

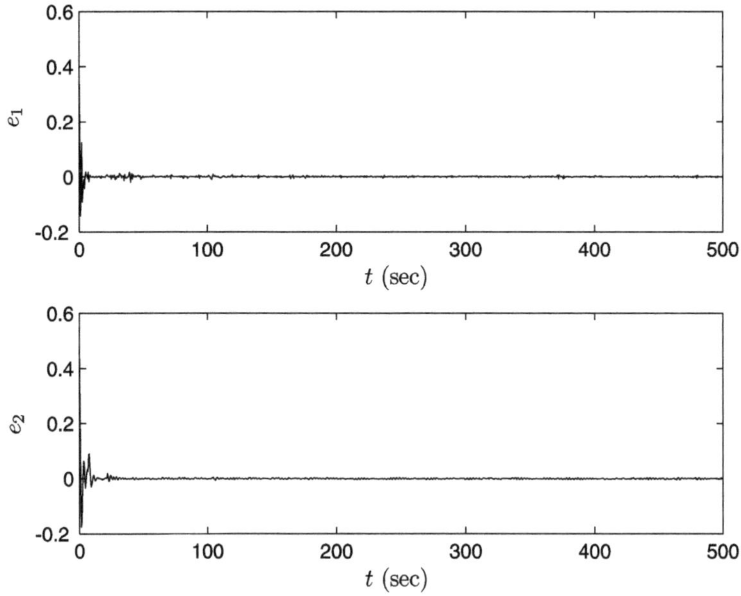

Fig. 7.46 The synchronization error curves of the drive system (7.72) and the response system (7.73)

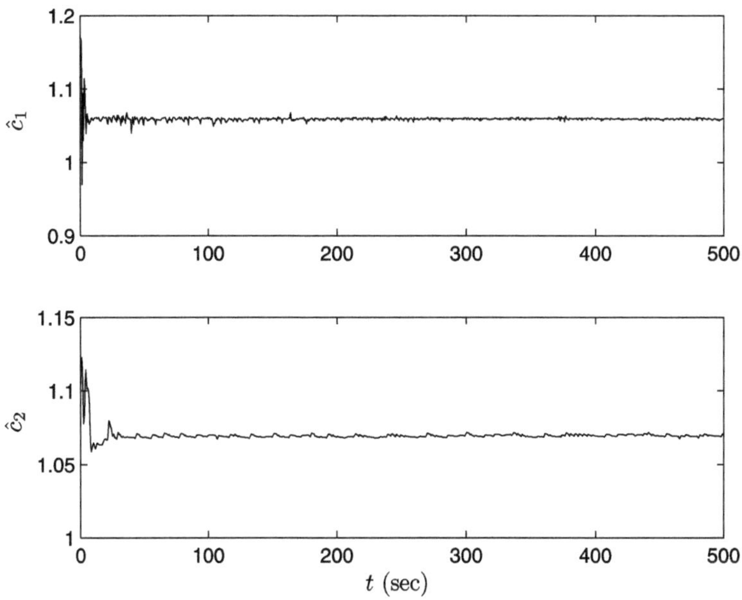

Fig. 7.47 The curves of parameter $\hat{C}(t)$

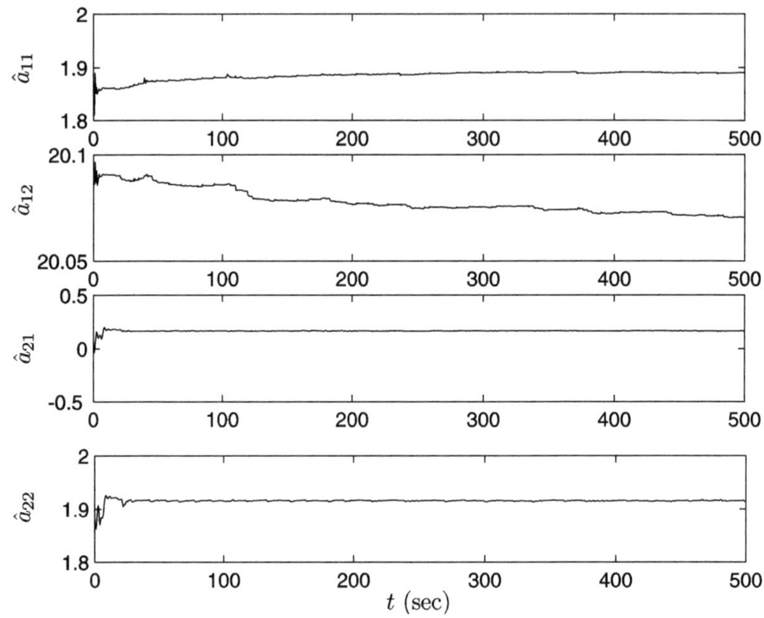

Fig. 7.48 The curves of parameter $\hat{A}(t)$

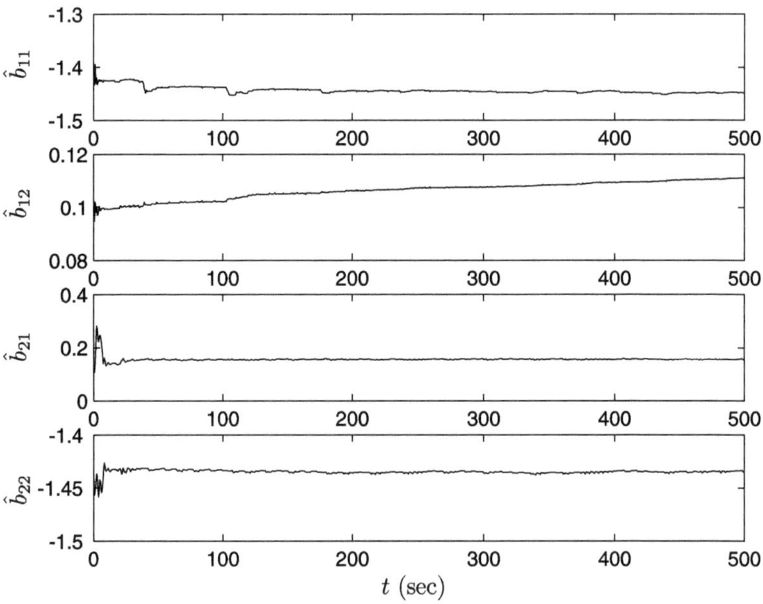

Fig. 7.49 The curves of parameter $\hat{B}(t)$

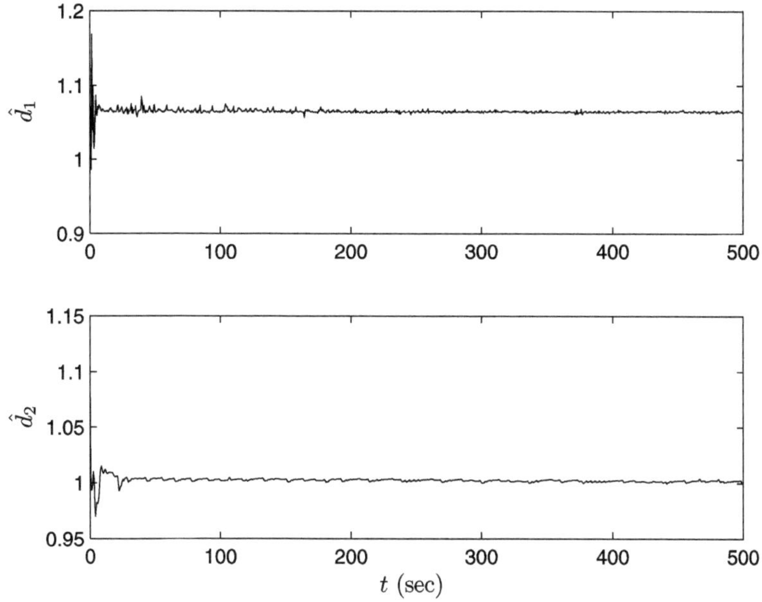

Fig. 7.50 The curves of parameter $\hat{D}(t)$

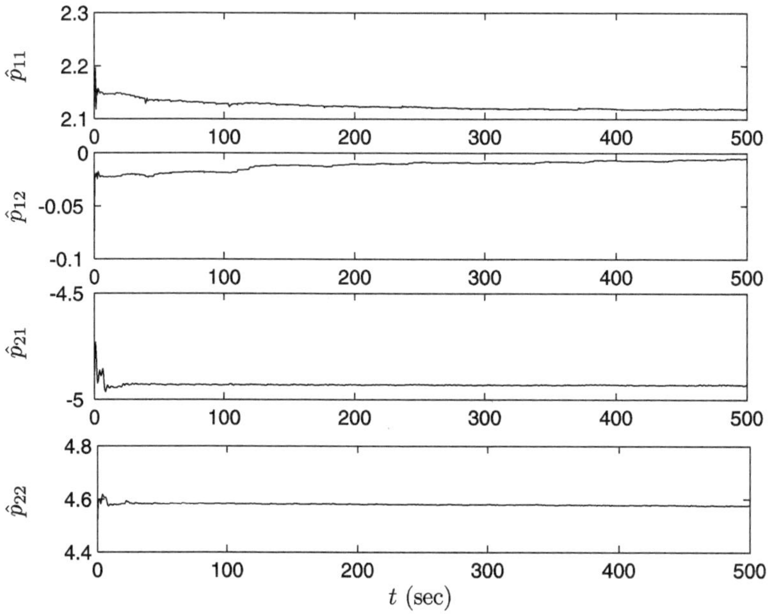

Fig. 7.51 The curves of parameter $\hat{P}(t)$

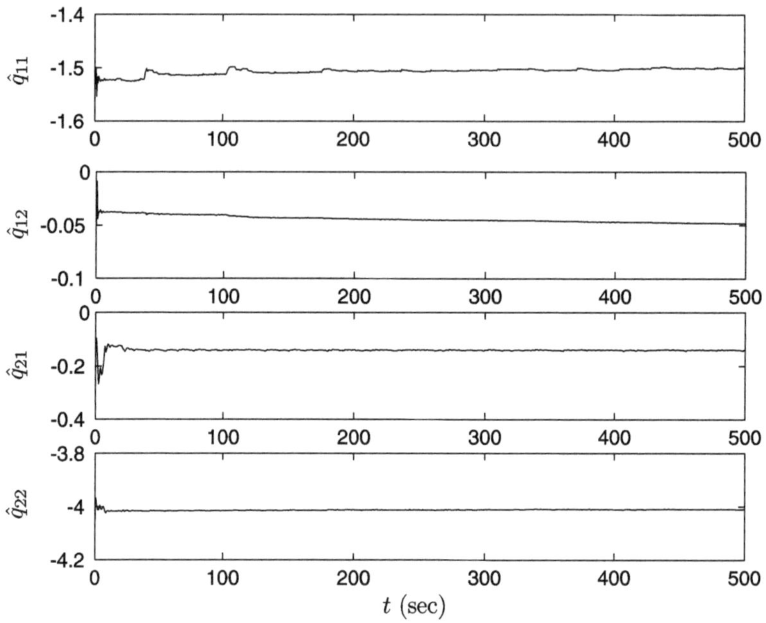

Fig. 7.52 The curves of parameter $\hat{Q}(t)$

network can synchronize asymptotically with the trajectory of the drive delayed chaotic cellular neural network. The estimated parameters converge asymptotically to some constants, which shows the effectiveness of the adaptive synchronization scheme.

7.5 Summary

In this chapter, the problem of adaptive synchronization of delayed chaotic systems is studied. Many examples are presented to show how to design the synchronization controllers when the systems to be synchronized have different structures and different parameters. The method proposed in Sect. 7.2 can adaptively synchronize two identical delay chaotic systems with unknown parameters. The methods developed in Sects. 7.3 and 7.4.2 can adaptively synchronize different chaotic systems. In Sect. 7.4.1, a controller based on inverse optimal control theory is designed, which not only can synchronize two delayed chaotic systems, but also can optimize a meaningful performance index. The numerical simulations have shown that the proposed methods are effective and can be used for a quite wide class of delayed chaotic systems.

References

1. Cao J (2000) Periodic solution and exponential stability of delayed CNNs. Phys Lett A 270:157–163
2. Gilli M (1993) Strange attractors in delayed cellular neural networks. IEEE Trans Circuits Syst I 40:849–853
3. Gopalsmay K, Issic LKC (1994) Convergence under dynamical thresholds with delays. IEEE Trans Neural Networks 8:341–348
4. Ikeda K (1979) Multiple-valued stationary state and its instability of the transmitted light by a ring cavity system. Opt Commun 30:257–261
5. Krstic M, Li ZH (1998) Inverse optimal design of input-to-state stabilizing nonlinear controllers. IEEE Trans Autom Control 43:336–351
6. Li CD, Liao XF, Wong KW (2004) Chaotic lag synchronization of coupled time-delayed systems and its applications in secure communication. Physica D 194:187–202
7. Li CD, Liao XF, Zhang R (2005) A unified approach for impulsive lag synchronization of chaotic systems with time delay. Chaos Solitons Fractals 23:1177–1184
8. Liao XF, Wong KW, Leung CS, Wu Z (2001) Hopf bifurcation and chaos in a single delayed neuron equation with non-monotonic activation function. Chaos Solitons Fractals 12:1535–1547
9. Liu D, Zhang Y, Zhang HG (2005) Self-learning call admission control for CDMA cellular networks. IEEE Trans Neural Networks 16:1219–1228
10. Liu D, Xiong X, DasGupta B, Zhang HG (2006) Motif discoveries in unaligned molecular sequences using self-organizing neural networks. IEEE Trans Neural Networks 17:919–928
11. Lu HT (2002) Chaotic attractors in delayed neural networks. Phys Lett A 298:109–116
12. Peng J, Liao XF (2003) Synchronization of a coupled time-delay chaotic system and its application to secure communications. Comput Res Dev 40:263–268
13. Ren HP, Liu D, Han CZ (2006) Anticontrol of chaos via direct time delay feedback. Acta Phys Sinica 55:2694–2708
14. Song Y, Wei J (2004) Bifurcation analysis for Chen's system with delayed feedback and its application to control of chaos. Chaos Solitons Fractals 22:75–91
15. Wang ZS, Zhang HG, Wang ZL (2006) Global synchronization of a class of chaotic neural networks. Acta Phys Sinica 55:2687–2693
16. Xie YH (2007) Researches on control and synchronization of several kinds of nonlinear chaotic systems with time delay. Ph.D. dissertation, Northeastern University, Shenyang
17. Zhang HG, Guan HX, Wang ZS (2007) Adaptive synchronization of neural networks with different attractors. Prog Nat Sci 17:687–695
18. Zhang HG, Xie Y, Liu D (2006) Synchronization of a class of delayed chaotic neural networks with fully unknown parameters. Dyn Contin Discrete Impuls Syst B 13:297–308
19. Zhang HG, Xie Y, Wang ZL, Zheng CD (2007) Adaptive synchronization between two different chaotic neural networks with time-delay. IEEE Trans Neural Networks 18:1841–1845
20. Zhang HG, Wang ZS, Liu D (2007) Robust exponential stability of cellular neural networks with multiple time varying delays. IEEE Trans Circuits Syst II 54:730–734
21. Zhang HG, Wang ZS (2007) Global asymptotic stability of delayed cellular neural networks. IEEE Trans Neural Networks 18:947–950
22. Zhang Y (2002) Absolute periodicity and absolute stability of delayed neural networks. IEEE Trans Circuits Syst I 49:256–261

Chapter 8
Synchronizing Chaotic Systems Based on Fuzzy Models

Abstract A motivation for using fuzzy systems and fuzzy control stems in part from the fact that they are particularly suitable for industrial processes when the physical systems or qualitative criteria are too complex to model and they have provided an efficient and effective way in the control of complex uncertain nonlinear or ill-defined systems. In recent years, fuzzy logic systems have received much attention from control theorists as a powerful tool for nonlinear control. In this chapter, we first introduce fuzzy modeling methods for some classical chaotic systems via the Takagi–Sugeno (T–S) fuzzy model. Next, we model some hyperchaotic systems using the T–S fuzzy model and then, based on these fuzzy models, we develop an H_∞ synchronization method for two different hyperchaotic systems. Finally, the problem of synchronizing a class of time-delayed chaotic systems based on the T–S fuzzy model is considered.

8.1 Introduction

In recent years, fuzzy logic systems have received much attention from the control community as a powerful tool for nonlinear control. A motivation for using fuzzy systems and fuzzy control stems in part from the fact that they are particularly suitable for industrial processes when the physical systems or qualitative criteria are too complex to model and they have provided an efficient and effective way in the control of complex uncertain nonlinear or ill-defined systems.

Among various kinds of fuzzy control or system methods, the T–S fuzzy system is widely accepted as a powerful tool for the design of fuzzy control [12, 19, 20, 21, 24]. The T–S fuzzy model is frequently used for mathematical simplicity of analysis. In this type of fuzzy model, local dynamics in different state–space regions is represented by linear models [13]. The overall model of the system is achieved by fuzzy blending of these linear models.

It is well known that, since the pioneering work of Pecora and Carroll [11], synchronization of two chaotic dynamical systems is one of the most important applications of chaos and has been paid much attention. Various control methods have been applied to synchronize two chaotic or hyperchaotic systems, such as linear feedback control, nonlinear feedback control, and impulsive control [1, 22, 23]. Among these synchronization methods, some must use high gains in designing parameters, others need Lipschitz conditions of nonlinear terms to be satisfied. Moreover, many results aforementioned dealt with one or two kinds of specified chaotic systems. There are few unified methods suitable for synchronization of a variety of chaotic systems.

In particular, for chaotic systems that evolve within a bounded region of the state space, the T–S fuzzy model can represent the nonlinear dynamics by a small set of linear subsystems coupled with linguistic variables. The idea of controlling nonlinear dynamical systems based on the T–S fuzzy model can always be divided into two steps: (i) represent the nonlinear system using the T–S fuzzy system and (ii) design a controller for the fuzzy system. Based on the parallel distributed compensation (PDC) scheme, a lot of research on the synchronization of fuzzy systems has been carried out.

In this chapter, we first introduce fuzzy modeling methods for some classical chaotic systems via the T–S fuzzy model. Next, we model some hyperchaotic systems with the T–S fuzzy model and then, based on the fuzzy model, we develop an H_∞ synchronization method for two different hyperchaotic systems. Finally, the problem of synchronizing a class of time-delayed chaotic systems based on the T–S fuzzy model is considered.

8.2 Modeling Chaotic Systems via T–S Fuzzy Models

The T–S fuzzy dynamical model, which originated from Takagi and Sugeno, is described by fuzzy IF–THEN rules in which the consequent parts represent local linear models. To realize a fuzzy model-based design, chaotic systems should first be represented exactly by T–S fuzzy models. From the investigation of many well-known continuous-time and discrete-time chaotic systems, we find that nonlinear terms have a common variable or depend only on one variable. If we take it as the premise variable of fuzzy rules, a simple fuzzy dynamical model can be obtained and will represent exactly chaotic systems. We observe that all the well-known chaotic systems can be applied to synchronization and secure communications either by a fuzzy driving signal or by a crisp driving signal. In the following, we will introduce how to represent many well-known chaotic systems by the T–S fuzzy model [9].

Consider a general chaotic dynamical system as follows:

$$sx(t) = f(x(t)) + g(x(t))u(t) + C, \tag{8.1}$$

where $sx(t)$ is $\dot{x}(t)$ in continuous-time or $x(t+1)$ in discrete-time systems, respectively; $x(t)$ and $u(t)$ are the state and control input vectors, respectively; C is a constant vector; and $f(\cdot)$ and $g(\cdot)$ are nonlinear functions with appropriate dimensions. Then, the fuzzy model is composed of the following rules:

$$\begin{aligned} R^i: &\ \text{IF } z(t) \text{ is } M_i, \\ &\ \text{THEN } sx(t) = A_i x(t) + B_i u(t) + b_i, \end{aligned} \tag{8.2}$$

where $x(t) \in \mathbb{R}^n$ and $u(t) \in \mathbb{R}^m$ denote the state vector and the input vector, respectively, $A_i \in \mathbb{R}^{n\times n}$ and $B_i \in \mathbb{R}^{n\times m}$ are the known system matrix and input matrix with appropriate dimensions, respectively, $z(t)$ is a premise variable, M_i is a fuzzy set, b_i is a constant vector, $i = 1,2,\ldots,r$, and r is the number of fuzzy rules.

By taking a standard fuzzy inference strategy, i.e., using a singleton fuzzifier, product fuzzy inference, and center average defuzzifier, the final continuous-time fuzzy T–S system is inferred as follows:

$$sx(t) = \frac{\sum_{i=1}^{r} \omega_i(z(t))[A_i x(t) + B_i u(t) + b_i]}{\sum_{i=1}^{r} \omega_i(z(t))}, \tag{8.3}$$

where $\omega_i(z(t)) = M_i(z(t))$, in which $M_i(z(t))$ is the degree of membership of $z(t)$ in M_i, with

$$\begin{cases} \sum\limits_{i=1}^{r} \omega_i(z(t)) > 0, \\ \omega_i(z(t)) \geq 0, \quad i = 1,2,\ldots,r. \end{cases}$$

By using $\mu_i(z(t)) = \omega_i(z(t))/\sum_{i=1}^{r}\omega_i(z(t))$ instead of $\omega_i(z(t))$, system (8.3) is rewritten as

$$sx(t) = \sum_{i=1}^{r} \mu_i(z(t))[A_i x(t) + B_i u(t) + b_i].$$

The premise variables are assumed to be independent of the input vector $u(t)$ in this chapter. However, it should be noted that all the stability conditions derived in this chapter can be applied even to the case that the premise variables depend on the input vector. This assumption is employed to avoid a complicated defuzzification process of fuzzy controllers.

Assumption 8.1. Considering the boundedness of chaotic systems, the fuzzy set is chosen as the region of the system trajectory in the set

$$\Omega := \{x(t) \in \mathbb{R}^n : \|x(t)\| \leqslant \varepsilon\}.$$

For chaotic systems, the existence of the parameter ε is natural. Therefore, nonlinear systems can be represented exactly by the T–S fuzzy model. □

In the process of constructing a T–S fuzzy model (8.2) which represents exactly the nonlinear system (8.1), we will focus on nonlinear terms of the nonlinear system. The consistency of nonlinear terms in the system and its associated fuzzy representation is emphasized here. Without loss of generality, fuzzy modeling methods are proposed for three kinds of nonlinear terms [9]; that is (i) only one variable in a nonlinear term; (ii) multi-variables in a nonlinear term; and (iii) multiple nonlinear terms in a system.

Case 1: The Hénon Map (Only One Variable in a Nonlinear Term) is given by

$$\begin{cases} x_1(t+1) = 1.4 - x_1^2(t) + 0.3x_2(t), \\ x_2(t+1) = x_1(t). \end{cases}$$

Consider a single scalar nonlinear function $f(x(t)) = x_1^2(t)$ which depends only on one state variable x_k in the Hénon map ($x_k = x_1$). Let the nonlinear term $f(x_k)$ take the form $\phi(x_k)\tilde{x}_k$, where

$$\tilde{x}_k = \begin{cases} x_k, \text{ if } \lim\limits_{\substack{x\in\Omega \\ x_k\to 0}} \dfrac{f(x_k)}{x_k} \in L_\infty, \\ 1, \text{ otherwise.} \end{cases}$$

Take x_k, which forms the function $\phi(x_k)$, as the premise variable. Then, the fuzzy representation is composed of the following fuzzy rules:

$$\begin{aligned} &\text{IF } x_k \text{ is } F_i, \\ &\text{THEN } \hat{f} = d_i\tilde{x}_k, \end{aligned}$$

for $i = 1, 2, \ldots, r$, where F_i is the fuzzy set, $\hat{f}(x_k)$ is a fuzzy representation of $f(x_k)$, d_i is a constant coefficient to be determined, and r is the number of fuzzy rules. The fuzzy inferred output is written as

$$\hat{f}(x_k) = \frac{\sum\limits_{i=1}^{r} \mu_i(x_k)d_i}{\sum\limits_{i=1}^{r} \mu_i(x_k)}\tilde{x}_k.$$

Without loss of generality, it is required that $\sum\limits_{i=1}^{r} \mu_i(x_k) = 1$, $\mu_i(x_k) \in [0, 1]$, which further yields $\phi(x_k) = \sum\limits_{i=1}^{r} \mu_i(x_k)d_i$. We let $r = 2$ and specify the membership functions. From $\mu_1(x_k) + \mu_2(x_k) = 1$ and $\mu_1(x_k)d_1 + \mu_2(x_k)d_2 = \phi(x_k)$, we have

$$\begin{aligned} \mu_1 &= \frac{-d_2}{d_1 - d_2} + \frac{1}{d_1 - d_2}\phi(x_k), \\ \mu_2 &= 1 - \mu_1. \end{aligned}$$

For instance, let $d = d_1 = -d_2$ in which d is the upper bound of $\phi(x_k)$, i.e., $d = \sup_{x\in\Omega}|\phi(x_k)|$. The fuzzy sets can be chosen as

$$\begin{aligned} F_1 &= \frac{1}{2}\left(1 + \frac{\phi(x_k)}{d}\right), \\ F_2 &= \frac{1}{2}\left(1 - \frac{\phi(x_k)}{d}\right). \end{aligned}$$

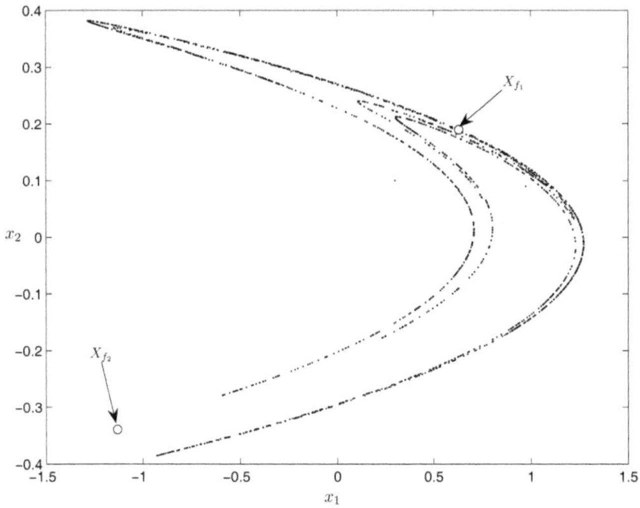

Fig. 8.1 The chaotic attractor of the fuzzy Hénon map

Let x_1 be the premise variable. Then, the equivalent fuzzy model of the Hénon map can be constructed as

$$\begin{aligned}&\text{IF } x_1 \text{ is } F_i,\\&\text{THEN } \dot{x}(t) = A_i x(t) + b_i,\end{aligned}$$

$i = 1,2$, where $d = 2$, $\phi(x_1) = x_1$, $A_1 = \begin{pmatrix} -2 & 0.3 \\ 1 & 0 \end{pmatrix}$, $A_2 = \begin{pmatrix} 2 & 0.3 \\ 1 & 0 \end{pmatrix}$, $b_1 = b_2 = \begin{pmatrix} 1.4 \\ 0 \end{pmatrix}$, and the fuzzy sets are

$$\begin{aligned}F_1 &= \frac{1}{2}\left(1 + \frac{x_1}{2}\right),\\F_2 &= \frac{1}{2}\left(1 - \frac{x_1}{2}\right).\end{aligned}$$

The chaotic attractor of the fuzzy Hénon map is shown in Fig. 8.1.

Case 2: The Rössler System (Multi-variables in a Nonlinear Term) is given by

$$\begin{cases}\dot{x}_1 = -x_2 - x_3,\\\dot{x}_2 = x_1 + 0.2x_2,\\\dot{x}_3 = 0.2 + x_3(x_1 - 5).\end{cases}$$

Consider a single scalar nonlinear function $f(x)$ in which $x=(x_1,x_2,\dots,x_n)^{\mathrm{T}}$, where many state variables appear in it. Assume that the nonlinear term $f(x)$ can be expressed as $f(x)=\phi_1(x_1)\phi_2(x_2)\cdots\phi_n(x_n)\tilde{x}_k$, where

$$\tilde{x}_k=\begin{cases} x_k, \text{ if } \lim\limits_{\substack{x\in\Omega\\ x_k\to 0}} \dfrac{f(x_k)}{x_k}\in L_\infty \text{ for some } k=1,2,\dots,n,\\ 1, \text{ otherwise.}\end{cases}$$

Let the variables x_k in $\phi_k(\cdot)$, for $k=1,2,\dots,n$, be premise variables. Then, the ith rule of the fuzzy system will be of the following form:

$$\begin{aligned}&\text{IF } x_1 \text{ is } F_{1i} \text{ and},\dots,\text{and } x_n \text{ is } F_{ni},\\ &\text{THEN } \hat{f}=d_i\tilde{x}_k,\end{aligned}$$

$i=1,2,\dots,r$, where F_{ki} $(k=1,2,\dots,n)$ is the fuzzy set, $\hat{f}$ is a fuzzy representation of $f(x)$, and d_i is to be determined later. The final output of the fuzzy system is inferred as follows:

$$\hat{f}(x)=\frac{\sum\limits_{i=1}^{r}\mu_i(x)d_i}{\sum\limits_{i=1}^{r}\mu_i(x)}\tilde{x}_k,$$

where $\mu_i(x)=\prod_{k=1}^{n}F_{ki}(x)/\sum_{i=1}^{r}\prod_{k=1}^{n}F_{ki}(x)$ and $F_{ki}(x_k)$ is the degree of membership of x_k in $f(x)$. The membership functions and coefficients can be chosen such that $F_{ki}(x_k)\in[0,1]$, $\sum_{i=1}^{r}\mu_i(x)=1$, and $\phi_1(x_1)\phi_2(x_2)\cdots\phi_n(x_n)=\sum_{i=1}^{r}\mu_i(x)d_i$.

The nonlinear term x_1x_3 can have the extracted variable as x_1 or x_3. Here, we choose $\phi(x_1)=x_1$ and let x_1 be the premise variable of fuzzy rules, which satisfies $x_1(t)\in[-d,d]$. The fuzzy model which represents exactly the Rössler system is

$$\begin{aligned}&\text{IF } x_1 \text{ is } F_i,\\ &\text{THEN } \dot{x}(t)=A_ix(t)+b_i,\end{aligned}$$

for $i=1,2$, where $x(t)=(x_1(t),x_2(t),x_3(t))^{\mathrm{T}}$, $d=10.5$,

$$A_1=\begin{pmatrix}0&-1&-1\\1&0.2&0\\0&0&5.5\end{pmatrix},\quad A_2=\begin{pmatrix}0&-1&-1\\1&0.2&0\\0&0&-15.5\end{pmatrix},$$

$$b_1=b_2=\begin{pmatrix}0\\0\\0.2\end{pmatrix},$$

$\phi(x_1)=x_1$, and the fuzzy sets are chosen as

$$F_1=\frac{1}{2}\left(1+\frac{x_1}{10.5}\right),\quad F_2=\frac{1}{2}\left(1-\frac{x_1}{10.5}\right).$$

The chaotic attractor of the fuzzy Rössler system is shown in Fig. 8.2.

Case 3: The Transformed Rössler System (Multiple Nonlinear Terms in a System) is given by

$$\begin{cases} \dot{x}_1 = -x_2 - \exp(x_3), \\ \dot{x}_2 = x_1 + a x_2, \\ \dot{x}_3 = x_1 + c\exp(-x_3) - b, \end{cases}$$

where $a = 0.2$, $b = 5$, and $c = 0.2$. More than one nonlinear term would be simultaneously considered in the system. When a nonlinear $m \times 1$ vector $f(x) = (f_1(x), f_2(x), \ldots, f_m(x))^{\mathrm{T}}$ is considered, each element of $f(x)$ is assumed to satisfy $f_j(x) = \phi_{1j}(x_1)\phi_{2j}(x_2)\cdots\phi_{nj}(x_n)\tilde{x}_j$, $j = 1,2,\ldots,m$. According to Cases 1 and 2, the fuzzy system representing the nonlinear terms is described as follows:

IF x_1 is F_{1i} and,...,and x_n is F_{ni},
THEN $\hat{f} = (d_{k1}\tilde{x}_{k1},\ d_{k2}\tilde{x}_{k2}, \ldots, d_{km}\tilde{x}_{km})^{\mathrm{T}}$ $(k = 1,2,\ldots,n)$,

$i = 1,2,\ldots,r$. The final output of the fuzzy system is inferred as follows:

$$\hat{f}(x) = \left(\frac{\sum_{i=1}^{r} \mu_i(x) d_{k1}}{\sum_{i=1}^{r} \mu_i(x)} \tilde{x}_{k1},\ \frac{\sum_{i=1}^{r} \mu_i(x) d_{k2}}{\sum_{i=1}^{r} \mu_i(x)} \tilde{x}_{k2},\ \ldots,\ \frac{\sum_{i=1}^{r} \mu_i(x) d_{km}}{\sum_{i=1}^{r} \mu_i(x)} \tilde{x}_{km} \right)^{\mathrm{T}},$$

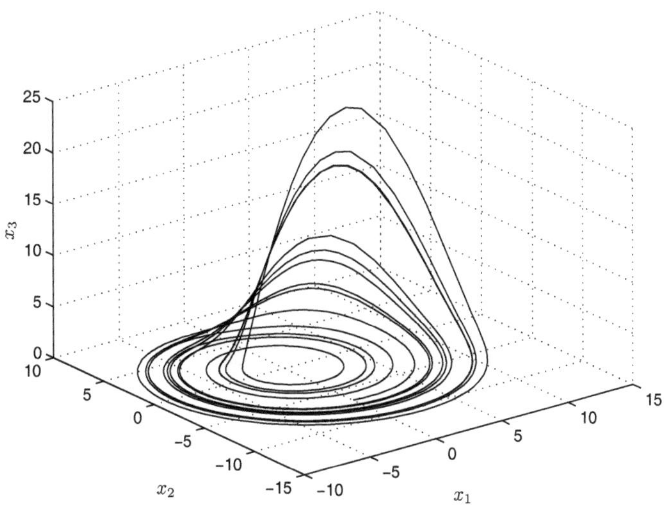

Fig. 8.2 The attractor of the fuzzy Rössler system

with $\mu_i(x) = \prod_{k=1}^{n} F_{ki}(x_k) / \sum_{i=1}^{r} \prod_{k=1}^{n} F_{ki}(x_k)$. The remaining procedure is the same as that for Case 2. It is noted that if the nonlinear terms $f_j(x)$, for $j = 1, 2, \ldots, m$, have a common factor, the number of fuzzy rules may be reduced. Since the Rössler system depends on the common variable $x_3(t)$, the premise variable is set as $x_3(t)$. However, the variable $x_3(t)$ of two nonlinear terms cannot be extracted due to $\lim_{\substack{x \in \Omega \\ x_3 \to 0}} \frac{\exp(x_3)}{x_3} \notin L_\infty$ and $\lim_{\substack{x \in \Omega \\ x_3 \to 0}} \frac{\exp(-x_3)}{x_3} \notin L_\infty$, and thus bias terms will appear in the fuzzy model.

The fuzzy model which represents exactly the transformed Rössler system is

$$
\begin{aligned}
&\text{IF } x_3 \text{ is } F_i, \\
&\text{THEN } \dot{x}(t) = A_i x(t) + b_i,
\end{aligned}
$$

$i = 1, 2, 3, 4$, where $x(t) = (x_1(t), x_2(t), x_3(t))^{\mathrm{T}}$. The transformed fuzzy Rössler system is represented exactly by the fuzzy model with $r = 4$ and

$$
A_i = \begin{pmatrix} 0 & -1 & 0 \\ 1 & 0.2 & 0 \\ 1 & 0 & 0 \end{pmatrix}, \quad i = 1, 2, 3, 4,
$$

$$
b_1 = (-75, 0, 10)^{\mathrm{T}}, \; b_2 = (-75, 0, -20)^{\mathrm{T}},
$$

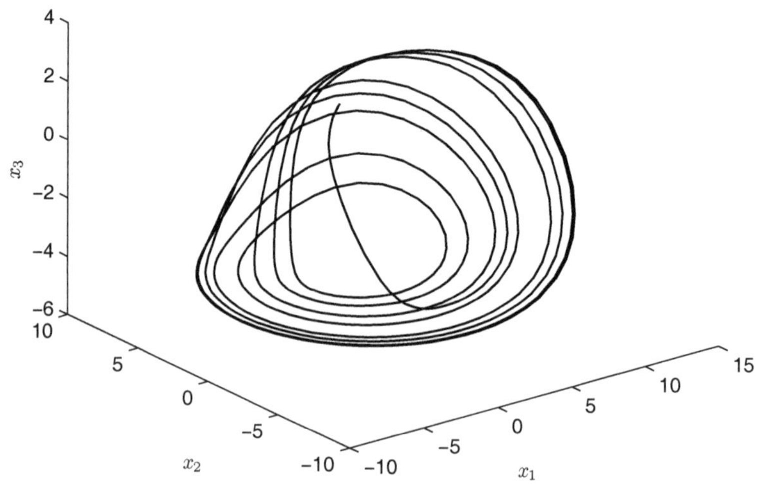

Fig. 8.3 The attractor of the fuzzy transformed Rössler's system

$$F_1 = \frac{1}{2^2}\left(1+\frac{\exp(x_3)}{75}\right)\left(1+\frac{\exp(-x_3)}{75}\right),$$
$$F_2 = \frac{1}{2^2}\left(1+\frac{\exp(x_3)}{75}\right)\left(1-\frac{\exp(-x_3)}{75}\right),$$
$$F_3 = \frac{1}{2^2}\left(1-\frac{\exp(x_3)}{75}\right)\left(1+\frac{\exp(-x_3)}{75}\right),$$
$$F_4 = \frac{1}{2^2}\left(1-\frac{\exp(x_3)}{75}\right)\left(1-\frac{\exp(-x_3)}{75}\right).$$

The chaotic attractor of the fuzzy Rössler system is shown in Fig. 8.3.

8.3 Synchronization of Hyperchaotic Systems via T–S Fuzzy Models

8.3.1 Fuzzy Modeling of Hyperchaotic Systems

In this subsection, we investigate the T–S fuzzy modeling of many typical hyperchaotic systems. In the following hyperchaotic systems, x_1, x_2, x_3, and x_4 are state variables, α, β, θ, σ, e, l, k, h, β_1, and β_2 are system parameters. It has been proven that all the following systems are hyperchaotic if their parameters are chosen appropriately. Assume that $x_1 \in [\tilde{c}_1 - \tilde{d}_1, \tilde{c}_1 + \tilde{d}_1]$, $x_2 \in [\tilde{c}_2 - \tilde{d}_2, \tilde{c}_2 + \tilde{d}_2]$, $\tilde{d}_1 > 0$, and $\tilde{d}_2 > 0$. According to the property of boundedness of hyperchaotic systems, this assumption is reasonable. The exact values of $\tilde{c}_1$, $\tilde{d}_1$, $\tilde{c}_2$, and $\tilde{d}_2$ can be obtained by simulation. Since there is no input in the autonomous hyperchaotic systems, the term $B_i u(t)$ in the T–S fuzzy model is omitted in the process of fuzzy modeling of hyperchaotic systems.

System 1: The Chen Hyperchaotic System [3]

$$\begin{cases} \dot{x}_1 = \alpha(x_2 - x_1) + x_4, \\ \dot{x}_2 = \sigma x_1 - x_1 x_3 + \theta x_2, \\ \dot{x}_3 = x_1 x_2 - \beta x_3, \\ \dot{x}_4 = x_2 x_3 + l x_4. \end{cases}$$

For the above hyperchaotic system, we derive the following T–S fuzzy model to represent it exactly:

$$R^1: \text{IF } x_1(t) \text{ is } F_{11} \text{ and } x_2(t) \text{ is } F_{12}, \text{ THEN } \dot{x}(t) = A_1 x(t) + b_1,$$
$$R^2: \text{IF } x_1(t) \text{ is } F_{21} \text{ and } x_2(t) \text{ is } F_{22}, \text{ THEN } \dot{x}(t) = A_2 x(t) + b_2,$$
$$R^3: \text{IF } x_1(t) \text{ is } F_{31} \text{ and } x_2(t) \text{ is } F_{32}, \text{ THEN } \dot{x}(t) = A_3 x(t) + b_3,$$
$$R^4: \text{IF } x_1(t) \text{ is } F_{41} \text{ and } x_2(t) \text{ is } F_{42}, \text{ THEN } \dot{x}(t) = A_4 x(t) + b_4,$$

where $x(t) = (x_1(t), x_2(t), x_3(t), x_4(t))^{\mathrm{T}}$,

$$A_1 = \begin{pmatrix} -\alpha & \alpha & 0 & 1 \\ \sigma & \theta & 2(-\tilde{c}_1 - \tilde{d}_1) & 0 \\ 0 & 2(\tilde{c}_1 + \tilde{d}_1) & -\beta & 0 \\ 0 & 0 & 0 & l \end{pmatrix},$$

$$A_2 = \begin{pmatrix} -\alpha & \alpha & 0 & 1 \\ \sigma & \theta & 2(-\tilde{c}_1 + \tilde{d}_1) & 0 \\ 0 & 2(\tilde{c}_1 - \tilde{d}_1) & -\beta & 0 \\ 0 & 0 & 0 & l \end{pmatrix},$$

$$A_3 = \begin{pmatrix} -\alpha & \alpha & 0 & 1 \\ \sigma & \theta & 0 & 0 \\ 0 & 0 & -\beta & 0 \\ 0 & 0 & 2(\tilde{c}_2 + \tilde{d}_2) & l \end{pmatrix}, \quad A_4 = \begin{pmatrix} -\alpha & \alpha & 0 & 1 \\ \sigma & \theta & 0 & 0 \\ 0 & 0 & -\beta & 0 \\ 0 & 0 & 2(\tilde{c}_2 - \tilde{d}_2) & l \end{pmatrix},$$

and

$$b_1 = b_2 = b_3 = b_4 = (0\ 0\ 0\ 0)^{\mathrm{T}}.$$

The membership functions are chosen as

$$F_{11}(x_1(t)) = \frac{1}{2}\left(1 - \frac{\tilde{c}_1 - x_1(t)}{\tilde{d}_1}\right), \quad F_{21}(x_1(t)) = \frac{1}{2}\left(1 + \frac{\tilde{c}_1 - x_1(t)}{\tilde{d}_1}\right),$$

$$F_{32}(x_2(t)) = \frac{1}{2}\left(1 - \frac{\tilde{c}_2 - x_2(t)}{\tilde{d}_2}\right), \quad F_{42}(x_2(t)) = \frac{1}{2}\left(1 + \frac{\tilde{c}_2 - x_2(t)}{\tilde{d}_2}\right),$$

$$F_{12}(x_2(t)) = F_{22}(x_2(t)) = F_{31}(x_1(t)) = F_{41}(x_1(t)) = 1.$$

System 2: The Liu Hyperchaotic System [7, 14]

$$\begin{cases} \dot{x}_1 = \alpha(x_2 - x_1), \\ \dot{x}_2 = \beta x_1 - k x_1 x_3 + x_4, \\ \dot{x}_3 = -\theta x_3 + h x_1^2, \\ \dot{x}_4 = -\sigma x_1. \end{cases}$$

For the above hyperchaotic system, we derive the following T–S fuzzy model to represent it exactly:

$$R^1: \text{ IF } x_1(t) \text{ is } F_1, \text{ THEN } \dot{x}(t) = A_1 x(t) + b_1,$$
$$R^2: \text{ IF } x_1(t) \text{ is } F_2, \text{ THEN } \dot{x}(t) = A_2 x(t) + b_2,$$

where $x(t) = (x_1(t), x_2(t), x_3(t), x_4(t))^{\mathrm{T}}$,

$$A_1 = \begin{pmatrix} -\alpha & \alpha & 0 & 0 \\ \beta & 0 & k(-\tilde{c}_1 - \tilde{d}_1) & 1 \\ h(\tilde{c}_1 + \tilde{d}_1) & 0 & -\theta & 0 \\ -\sigma & 0 & 0 & 0 \end{pmatrix},$$

$$A_2 = \begin{pmatrix} -\alpha & \alpha & 0 & 0 \\ \beta & 0 & k(-\tilde{c}_1 + \tilde{d}_1) & 1 \\ h(\tilde{c}_1 - \tilde{d}_1) & 0 & -\theta & 0 \\ -\sigma & 0 & 0 & 0 \end{pmatrix},$$

and

$$b_1 = b_2 = (0\ \ 0\ \ 0\ \ 0)^{\mathrm{T}}.$$

The membership functions are chosen as

$$F_1(x_1(t)) = \frac{1}{2}\left(1 - \frac{\tilde{c}_1 - x_1(t)}{\tilde{d}_1}\right),\ F_2(x_1(t)) = \frac{1}{2}\left(1 + \frac{\tilde{c}_1 - x_1(t)}{\tilde{d}_1}\right).$$

System 3: The Lü Hyperchaotic System (i) [18]

$$\begin{cases} \dot{x}_1 = \alpha(x_2 - x_1) + x_4, \\ \dot{x}_2 = -x_1x_3 + \theta x_2, \\ \dot{x}_3 = x_1x_2 - \beta x_3, \\ \dot{x}_4 = x_1x_3 + lx_4. \end{cases}$$

For the above hyperchaotic system, we derive the following T–S fuzzy model to represent it exactly:

$$R^1:\ \text{IF } x_1(t) \text{ is } F_1,\ \text{THEN } \dot{x}(t) = A_1x(t) + b_1,$$
$$R^2:\ \text{IF } x_1(t) \text{ is } F_2,\ \text{THEN } \dot{x}(t) = A_2x(t) + b_2,$$

where $x(t) = (x_1(t), x_2(t), x_3(t), x_4(t))^{\mathrm{T}}$,

$$A_1 = \begin{pmatrix} -\alpha & \alpha & 0 & 1 \\ 0 & \theta & -\tilde{c}_1 - \tilde{d}_1 & 0 \\ 0 & \tilde{c}_1 + \tilde{d}_1 & -\beta & 0 \\ 0 & 0 & \tilde{c}_1 + \tilde{d}_1 & l \end{pmatrix},\ A_2 = \begin{pmatrix} -\alpha & \alpha & 0 & 1 \\ 0 & \theta & -\tilde{c}_1 + \tilde{d}_1 & 0 \\ 0 & \tilde{c}_1 - \tilde{d}_1 & -\beta & 0 \\ 0 & 0 & \tilde{c}_1 - \tilde{d}_1 & l \end{pmatrix},$$

and

$$b_1 = b_2 = (0\ \ 0\ \ 0\ \ 0)^{\mathrm{T}}.$$

The membership functions are chosen as

$$F_1(x_1(t)) = \frac{1}{2}\left(1 - \frac{\tilde{c}_1 - x_1(t)}{\tilde{d}_1}\right),\ F_2(x_1(t)) = \frac{1}{2}\left(1 + \frac{\tilde{c}_1 - x_1(t)}{\tilde{d}_1}\right).$$

System 4: The Lü Hyperchaotic System (ii) [4]

$$\begin{cases} \dot{x}_1 = \alpha(x_2 - x_1), \\ \dot{x}_2 = -x_1x_3 + \theta x_2 + x_4, \\ \dot{x}_3 = x_1x_2 - \beta x_3, \\ \dot{x}_4 = x_3 - \sigma x_4. \end{cases}$$

For the above hyperchaotic system, we derive the following T–S fuzzy model to represent it exactly:

$$\begin{aligned} R^1 &: \text{ IF } x_1(t) \text{ is } F_1, \text{ THEN } \dot{x}(t) = A_1x(t) + b_1, \\ R^2 &: \text{ IF } x_1(t) \text{ is } F_2, \text{ THEN } \dot{x}(t) = A_2x(t) + b_2, \end{aligned}$$

where $x(t) = (x_1(t), x_2(t), x_3(t), x_4(t))^{\mathrm{T}}$,

$$A_1 = \begin{pmatrix} -\alpha & \alpha & 0 & 0 \\ 0 & \theta & -\tilde{c}_1 - \tilde{d}_1 & 1 \\ 0 & \tilde{c}_1 + \tilde{d}_1 & -\beta & 0 \\ 0 & 0 & 1 & -\sigma \end{pmatrix}, A_2 = \begin{pmatrix} -\alpha & \alpha & 0 & 0 \\ 0 & \theta & -\tilde{c}_1 + \tilde{d}_1 & 1 \\ 0 & \tilde{c}_1 - \tilde{d}_1 & -\beta & 0 \\ 0 & 0 & 1 & -\sigma \end{pmatrix},$$

and

$$b_1 = b_2 = (0\ 0\ 0\ 0)^{\mathrm{T}}.$$

The membership functions are chosen as

$$F_1(x_1(t)) = \frac{1}{2}\left(1 - \frac{\tilde{c}_1 - x_1(t)}{\tilde{d}_1}\right), \ F_2(x_1(t)) = \frac{1}{2}\left(1 + \frac{\tilde{c}_1 - x_1(t)}{\tilde{d}_1}\right).$$

System 5: The Modified Lü Hyperchaotic System [15]

$$\begin{cases} \dot{x}_1 = \alpha(x_2 - x_1 + x_2x_3), \\ \dot{x}_2 = -x_1x_3 + \beta x_2 + x_4, \\ \dot{x}_3 = x_1x_2 - \theta x_3, \\ \dot{x}_4 = -kx_1. \end{cases}$$

For the above hyperchaotic system, we derive the following T–S fuzzy model to represent it exactly:

$$\begin{aligned} R^1 &: \text{ IF } x_1(t) \text{ is } F_{11} \text{ and } x_2(t) \text{ is } F_{12}, \text{ THEN } \dot{x}(t) = A_1x(t) + b_1, \\ R^2 &: \text{ IF } x_1(t) \text{ is } F_{21} \text{ and } x_2(t) \text{ is } F_{22}, \text{ THEN } \dot{x}(t) = A_2x(t) + b_2, \\ R^3 &: \text{ IF } x_1(t) \text{ is } F_{31} \text{ and } x_2(t) \text{ is } F_{32}, \text{ THEN } \dot{x}(t) = A_3x(t) + b_3, \\ R^4 &: \text{ IF } x_1(t) \text{ is } F_{41} \text{ and } x_2(t) \text{ is } F_{42}, \text{ THEN } \dot{x}(t) = A_4x(t) + b_4, \end{aligned} \tag{8.4}$$

where $x(t) = (x_1(t), x_2(t), x_3(t), x_4(t))^{\mathrm{T}}$,

$$A_1 = \begin{pmatrix} -\alpha & \alpha & 0 & 0 \\ 0 & \beta & 2(-\tilde{c}_1 - \tilde{d}_1) & 1 \\ 0 & 2(\tilde{c}_1 + \tilde{d}_1) & -\theta & 0 \\ -k & 0 & 0 & 0 \end{pmatrix},$$

$$A_2 = \begin{pmatrix} -\alpha & \alpha & 0 & 0 \\ 0 & \beta & 2(-\tilde{c}_1 + \tilde{d}_1) & 1 \\ 0 & 2(\tilde{c}_1 - \tilde{d}_1) & -\theta & 0 \\ -k & 0 & 0 & 0 \end{pmatrix},$$

$$A_3 = \begin{pmatrix} -\alpha & \alpha & 2\alpha(\tilde{c}_2 + \tilde{d}_2) & 0 \\ 0 & \beta & 0 & 1 \\ 0 & 0 & -\theta & 0 \\ -k & 0 & 0 & 0 \end{pmatrix}, \quad A_4 = \begin{pmatrix} -\alpha & \alpha & 2\alpha(\tilde{c}_2 - \tilde{d}_2) & 0 \\ 0 & \beta & 0 & 1 \\ 0 & 0 & -\theta & 0 \\ -k & 0 & 0 & 0 \end{pmatrix},$$

and

$$b_1 = b_2 = b_3 = b_4 = (0\ 0\ 0\ 0)^{\mathrm{T}}.$$

The membership functions are chosen as

$$F_{11}(x_1(t)) = \frac{1}{2}\left(1 - \frac{\tilde{c}_1 - x_1(t)}{\tilde{d}_1}\right), \quad F_{21}(x_1(t)) = \frac{1}{2}\left(1 + \frac{\tilde{c}_1 - x_1(t)}{\tilde{d}_1}\right),$$

$$F_{32}(x_2(t)) = \frac{1}{2}\left(1 - \frac{\tilde{c}_2 - x_2(t)}{\tilde{d}_2}\right), \quad F_{42}(x_2(t)) = \frac{1}{2}\left(1 + \frac{\tilde{c}_2 - x_2(t)}{\tilde{d}_2}\right),$$

$$F_{12}(x_2(t)) = F_{22}(x_2(t)) = F_{31}(x_1(t)) = F_{41}(x_1(t)) = 1.$$

System 6: The Lorenz Hyperchaotic System [6, 7]

$$\begin{cases} \dot{x}_1 = -\alpha(x_1 - x_2) + x_4, \\ \dot{x}_2 = l x_1 - x_2 - x_1 x_3, \\ \dot{x}_3 = x_1 x_2 - \beta x_3, \\ \dot{x}_4 = -x_1 x_3 + \sigma x_4. \end{cases}$$

For the above hyperchaotic system, we derive the following T–S fuzzy model to represent it exactly:

$$R^1: \text{ IF } x_1(t) \text{ is } F_1, \text{ THEN } \dot{x}(t) = A_1 x(t) + b_1,$$
$$R^2: \text{ IF } x_1(t) \text{ is } F_2, \text{ THEN } \dot{x}(t) = A_2 x(t) + b_2,$$

where $x(t) = (x_1(t), x_2(t), x_3(t), x_4(t))^{\mathrm{T}}$,

$$A_1 = \begin{pmatrix} -\alpha & \alpha & 0 & 1 \\ l & -1 & -\tilde{c}_1 - \tilde{d}_1 & 0 \\ 0 & \tilde{c}_1 + \tilde{d}_1 & -\beta & 0 \\ 0 & 0 & -\tilde{c}_1 - \tilde{d}_1 & \sigma \end{pmatrix}, \quad A_2 = \begin{pmatrix} -\alpha & \alpha & 0 & 1 \\ l & -1 & -\tilde{c}_1 + \tilde{d}_1 & 0 \\ 0 & \tilde{c}_1 - \tilde{d}_1 & -\beta & 0 \\ 0 & 0 & -\tilde{c}_1 + \tilde{d}_1 & \sigma \end{pmatrix},$$

and

$$b_1 = b_2 = (0\ 0\ 0\ 0)^{\mathrm{T}}.$$

The membership functions are chosen as

$$F_1(x_1(t)) = \frac{1}{2}\left(1 - \frac{\tilde{c}_1 - x_1(t)}{\tilde{d}_1}\right), \quad F_2(x_1(t)) = \frac{1}{2}\left(1 + \frac{\tilde{c}_1 - x_1(t)}{\tilde{d}_1}\right).$$

System 7: A New Hyperchaotic System [2]

$$\begin{cases} \dot{x}_1 = \alpha(x_2 - x_1) + e x_2 x_3, \\ \dot{x}_2 = \theta x_1 + x_2 - \sigma x_1 x_3 + x_4, \\ \dot{x}_3 = x_1 x_2 - \beta x_3, \\ \dot{x}_4 = -k x_2. \end{cases}$$

For the above hyperchaotic system, we derive the following T–S fuzzy model to represent it exactly:

$$\begin{aligned} &R^1: \text{ IF } x_1(t) \text{ is } F_{11} \text{ and } x_2(t) \text{ is } F_{12}, \text{ THEN } \dot{x}(t) = A_1 x(t) + b_1, \\ &R^2: \text{ IF } x_1(t) \text{ is } F_{21} \text{ and } x_2(t) \text{ is } F_{22}, \text{ THEN } \dot{x}(t) = A_2 x(t) + b_2, \\ &R^3: \text{ IF } x_1(t) \text{ is } F_{31} \text{ and } x_2(t) \text{ is } F_{32}, \text{ THEN } \dot{x}(t) = A_3 x(t) + b_3, \\ &R^4: \text{ IF } x_1(t) \text{ is } F_{41} \text{ and } x_2(t) \text{ is } F_{42}, \text{ THEN } \dot{x}(t) = A_4 x(t) + b_4, \end{aligned} \tag{8.5}$$

where $x(t) = (x_1(t), x_2(t), x_3(t), x_4(t))^{\mathrm{T}}$,

$$A_1 = \begin{pmatrix} -\alpha & \alpha & 0 & 0 \\ \theta & 1 & 2\sigma(-\tilde{c}_1 - \tilde{d}_1) & 1 \\ 0 & 2(\tilde{c}_1 + \tilde{d}_1) & -\beta & 0 \\ 0 & -k & 0 & 0 \end{pmatrix},$$

$$A_2 = \begin{pmatrix} -\alpha & \alpha & 0 & 0 \\ \theta & 1 & 2\sigma(-\tilde{c}_1 + \tilde{d}_1) & 1 \\ 0 & 2(\tilde{c}_1 - \tilde{d}_1) & -\beta & 0 \\ 0 & -k & 0 & 0 \end{pmatrix},$$

$$A_3 = \begin{pmatrix} -\alpha & \alpha & 2e(\tilde{c}_2 + \tilde{d}_2) & 0 \\ \theta & 1 & 0 & 1 \\ 0 & 0 & -\beta & 0 \\ 0 & -k & 0 & 0 \end{pmatrix},$$

$$A_4 = \begin{pmatrix} -\alpha & \alpha & 2e(\tilde{c}_2 - \tilde{d}_2) & 0 \\ \theta & 1 & 0 & 1 \\ 0 & 0 & -\beta & 0 \\ 0 & -k & 0 & 0 \end{pmatrix},$$

and

$$b_1 = b_2 = b_3 = b_4 = (0\ 0\ 0\ 0)^{\mathrm{T}}.$$

The membership functions are chosen as

$$F_{11}(x_1(t)) = \frac{1}{2}\left(1 - \frac{\tilde{c}_1 - x_1(t)}{\tilde{d}_1}\right), \quad F_{21}(x_1(t)) = \frac{1}{2}\left(1 + \frac{\tilde{c}_1 - x_1(t)}{\tilde{d}_1}\right),$$

$$F_{32}(x_2(t)) = \frac{1}{2}\left(1 - \frac{\tilde{c}_2 - x_2(t)}{\tilde{d}_2}\right), \quad F_{42}(x_2(t)) = \frac{1}{2}\left(1 + \frac{\tilde{c}_2 - x_2(t)}{\tilde{d}_2}\right),$$

$$F_{12}(x_2(t)) = F_{22}(x_2(t)) = F_{31}(x_1(t)) = F_{41}(x_1(t)) = 1.$$

System 8: Another New Hyperchaotic System [5]

$$\begin{cases} \dot{x}_1 = \alpha(x_2 - x_1), \\ \dot{x}_2 = \sigma x_1 + \theta x_2 - x_1 x_3 - x_4, \\ \dot{x}_3 = x_1 x_2 - \beta x_3, \\ \dot{x}_4 = x_1 + k. \end{cases}$$

For the above hyperchaotic system, we derive the following T–S fuzzy model to represent it exactly:

$$R^1: \text{ IF } x_1(t) \text{ is } F_1, \text{ THEN } \dot{x}(t) = A_1 x(t) + b_1,$$
$$R^2: \text{ IF } x_1(t) \text{ is } F_2, \text{ THEN } \dot{x}(t) = A_2 x(t) + b_2,$$

where $x(t) = (x_1(t), x_2(t), x_3(t), x_4(t))^{\mathrm{T}}$,

$$A_1 = \begin{pmatrix} -\alpha & \alpha & 0 & 0 \\ \sigma & \theta & -\tilde{c}_1 - \tilde{d}_1 & -1 \\ 0 & \tilde{c}_1 + \tilde{d}_1 & -\beta & 0 \\ 1 & 0 & 0 & 0 \end{pmatrix},$$

$$A_2 = \begin{pmatrix} -\alpha & \alpha & 0 & 0 \\ \sigma & \theta & -\tilde{c}_1 + \tilde{d}_1 & -1 \\ 0 & \tilde{c}_1 - \tilde{d}_1 & -\beta & 0 \\ 1 & 0 & 0 & 0 \end{pmatrix},$$

and

$$b_1 = b_2 = (0 \ \ 0 \ \ 0 \ \ k)^{\mathrm{T}}.$$

The membership functions are chosen as:

$$F_1(x_1(t)) = \frac{1}{2}\left(1 - \frac{\tilde{c}_1 - x_1(t)}{\tilde{d}_1}\right), \quad F_2(x_1(t)) = \frac{1}{2}\left(1 + \frac{\tilde{c}_1 - x_1(t)}{\tilde{d}_1}\right).$$

System 9: The Generalized Hénon–Heiles System [16]

$$\begin{cases} \dot{x}_1 = x_3, \\ \dot{x}_2 = x_4, \\ \dot{x}_3 = -x_1 - 2x_1x_2, \\ \dot{x}_4 = -x_1^2 - x_2 + x_2^2. \end{cases}$$

For the above hyperchaotic system, we derive the following T–S fuzzy model to represent it exactly:

$$\begin{aligned} R^1 &: \text{IF } x_1(t) \text{ is } F_{11} \text{ and } x_2(t) \text{ is } F_{12}, \text{ THEN } \dot{x}(t) = A_1x(t) + b_1, \\ R^2 &: \text{IF } x_1(t) \text{ is } F_{21} \text{ and } x_2(t) \text{ is } F_{22}, \text{ THEN } \dot{x}(t) = A_2x(t) + b_2, \\ R^3 &: \text{IF } x_1(t) \text{ is } F_{31} \text{ and } x_2(t) \text{ is } F_{32}, \text{ THEN } \dot{x}(t) = A_3x(t) + b_3, \\ R^4 &: \text{IF } x_1(t) \text{ is } F_{41} \text{ and } x_2(t) \text{ is } F_{42}, \text{ THEN } \dot{x}(t) = A_4x(t) + b_4, \end{aligned}$$

where $x(t) = (x_1(t), x_2(t), x_3(t), x_4(t))^{\mathrm{T}}$,

$$A_1 = \begin{pmatrix} 0 & 0 & 1 & 0 \\ 0 & 0 & 0 & 1 \\ -1 & -4(\tilde{c}_1 + \tilde{d}_1) & 0 & 0 \\ -2(\tilde{c}_1 + \tilde{d}_1) & -1 & 0 & 0 \end{pmatrix},$$

$$A_2 = \begin{pmatrix} 0 & 0 & 1 & 0 \\ 0 & 0 & 0 & 1 \\ -1 & -4(\tilde{c}_1 - \tilde{d}_1) & 0 & 0 \\ -2(\tilde{c}_1 - \tilde{d}_1) & -1 & 0 & 0 \end{pmatrix},$$

$$A_3 = \begin{pmatrix} 0 & 0 & 1 & 0 \\ 0 & 0 & 0 & 1 \\ -1 & 0 & 0 & 0 \\ 0 & -1 + 2(\tilde{c}_2 + \tilde{d}_2) & 0 & 0 \end{pmatrix},$$

$$A_4 = \begin{pmatrix} 0 & 0 & 1 & 0 \\ 0 & 0 & 0 & 1 \\ -1 & 0 & 0 & 0 \\ 0 & -1 + 2(\tilde{c}_2 - \tilde{d}_2) & 0 & 0 \end{pmatrix},$$

and

$$b_1 = b_2 = b_3 = b_4 = (0\ 0\ 0\ 0)^{\mathrm{T}}.$$

The membership functions are chosen as

$$F_{11}(x_1(t)) = \frac{1}{2}\left(1 - \frac{\tilde{c}_1 - x_1(t)}{\tilde{d}_1}\right), \quad F_{21}(x_1(t)) = \frac{1}{2}\left(1 + \frac{\tilde{c}_1 - x_1(t)}{\tilde{d}_1}\right),$$

$$F_{32}(x_2(t)) = \frac{1}{2}\left(1 - \frac{\tilde{c}_2 - x_2(t)}{\tilde{d}_2}\right), \quad F_{42}(x_2(t)) = \frac{1}{2}\left(1 + \frac{\tilde{c}_2 - x_2(t)}{\tilde{d}_2}\right),$$

$$F_{12}(x_2(t)) = F_{22}(x_2(t)) = F_{31}(x_1(t)) = F_{41}(x_1(t)) = 1.$$

System 10: The Modified Rössler Hyperchaotic System (i) [10]

$$\begin{cases} \dot{x}_1 = -x_2 - x_3, \\ \dot{x}_2 = x_1 + \alpha x_2 + x_4, \\ \dot{x}_3 = \beta + \beta_1 x_1 + x_1 x_3, \\ \dot{x}_4 = -\theta x_3 + \sigma x_4. \end{cases} \tag{8.6}$$

For the above hyperchaotic system, we derive the following T–S fuzzy model to represent it exactly:

$$R^1: \text{ IF } x_1(t) \text{ is } F_1, \text{ THEN } \dot{x}(t) = A_1 x(t) + b_1,$$
$$R^2: \text{ IF } x_1(t) \text{ is } F_2, \text{ THEN } \dot{x}(t) = A_2 x(t) + b_2,$$

where $x(t) = (x_1(t), x_2(t), x_3(t), x_4(t))^{\mathrm{T}}$,

$$A_1 = \begin{pmatrix} 0 & -1 & -1 & 0 \\ 1 & \alpha & 0 & 1 \\ \beta_1 & 0 & \tilde{c}_1 + \tilde{d}_1 & 0 \\ 0 & 0 & -\theta & \sigma \end{pmatrix},$$

$$A_2 = \begin{pmatrix} 0 & -1 & -1 & 0 \\ 1 & \alpha & 0 & 1 \\ \beta_1 & 0 & \tilde{c}_1 - \tilde{d}_1 & 0 \\ 0 & 0 & -\theta & \sigma \end{pmatrix},$$

and

$$b_1 = b_2 = (0 \;\; 0 \;\; \beta \;\; 0)^{\mathrm{T}}.$$

The membership functions are chosen as

$$F_1(x_1(t)) = \frac{1}{2}\left(1 - \frac{\tilde{c}_1 - x_1(t)}{\tilde{d}_1}\right), \quad F_2(x_1(t)) = \frac{1}{2}\left(1 + \frac{\tilde{c}_1 - x_1(t)}{\tilde{d}_1}\right).$$

System 11: The Modified Rössler Hyperchaotic System (ii) [10]

$$\begin{cases} \dot{x}_1 = -x_2 - x_3, \\ \dot{x}_2 = x_1 + \alpha x_2 + x_4, \\ \dot{x}_3 = \beta + \beta_2 x_3 + x_1 x_3, \\ \dot{x}_4 = -\theta x_3 + \sigma x_4. \end{cases}$$

For the above hyperchaotic system, we derive the following T–S fuzzy model to represent it exactly:

$$R^1: \text{ IF } x_1(t) \text{ is } F_1, \text{ THEN } \dot{x}(t) = A_1 x(t) + b_1,$$
$$R^2: \text{ IF } x_1(t) \text{ is } F_2, \text{ THEN } \dot{x}(t) = A_2 x(t) + b_2,$$

where $x(t) = (x_1(t), x_2(t), x_3(t), x_4(t))^{\mathrm{T}}$,

$$A_1 = \begin{pmatrix} 0 & -1 & -1 & 0 \\ 1 & \alpha & 0 & 1 \\ 0 & 0 & \beta_2 + \tilde{c}_1 + \tilde{d}_1 & 0 \\ 0 & 0 & -\theta & \sigma \end{pmatrix}, \ A_2 = \begin{pmatrix} 0 & -1 & -1 & 0 \\ 1 & \alpha & 0 & 1 \\ 0 & 0 & \beta_2 + \tilde{c}_1 - \tilde{d}_1 & 0 \\ 0 & 0 & -\theta & \sigma \end{pmatrix},$$

and

$$b_1 = b_2 = (0 \ 0 \ \beta \ 0)^{\mathrm{T}}.$$

The membership functions are chosen as

$$F_1(x_1(t)) = \frac{1}{2}\left(1 - \frac{\tilde{c}_1 - x_1(t)}{\tilde{d}_1}\right), \ F_2(x_1(t)) = \frac{1}{2}\left(1 + \frac{\tilde{c}_1 - x_1(t)}{\tilde{d}_1}\right).$$

To summarize, all the aforementioned hyperchaotic systems can be represented exactly by the following T–S fuzzy models:

$$R^i: \text{ IF } z(t) \text{ is } F_i, \text{ THEN } \dot{x}(t) = A_i x(t) + b_i,$$

$i = 1, 2, \ldots, r$, where the premise variable $z(t)$ is a proper state variable, b_i is a constant vector, and r is the number of fuzzy IF–THEN rules. Then, applying the product inference rule, a singleton fuzzifier, and the center of gravity defuzzifier to the above fuzzy rule base, the overall fuzzy hyperchaotic systems can be inferred as

$$\dot{x}(t) = \sum_{i=1}^{r} \mu_i(z(t))[A_i x(t) + b_i], \tag{8.7}$$

where

$$\mu_i(z(t)) = \frac{F_i(z(t))}{\sum_{i=1}^{r} F_i(z(t))}.$$

Next, we give a systematic modeling method to represent exactly the above hyperchaotic systems by T–S fuzzy models. Note that all the aforementioned hyperchaotic systems can be expressed as follows:

$$\dot{x}(t) = Ax(t) + f(x) + b, \tag{8.8}$$

where $Ax(t)$ and $f(x)$ denote the linear part and the nonlinear part of hyperchaotic systems, respectively, and b is the constant vector of the hyperchaotic systems. Next, fuzzy modeling methods are summarized for several cases of the nonlinear part $f(x)$, as follows.

Case 1: Only one nonlinear term in $f(x)$. Systems 10 and 11 belong to this case. For this case, we can choose any state variable in the nonlinear term as premise variable. For example, in system 10, the unique nonlinear term is $x_1 x_3$. Then, we can choose either x_1 or x_3 as premise variable. If we choose x_1 as premise variable, then the membership functions are defined as

$$F_1(x_1(t)) = \frac{1}{2}\left(1 - \frac{\tilde{c}_1 - x_1(t)}{\tilde{d}_1}\right)$$

and

$$F_2(x_1(t)) = \frac{1}{2}\left(1 + \frac{\tilde{c}_1 - x_1(t)}{\tilde{d}_1}\right),$$

and the fuzzy IF–THEN rules are defined as:

$$R^1: \text{ IF } x_1(t) \text{ is } F_1, \text{ THEN } \dot{x}(t) = A_1 x(t) + b_1,$$
$$R^2: \text{ IF } x_1(t) \text{ is } F_2, \text{ THEN } \dot{x}(t) = A_2 x(t) + b_2,$$

where $x(t) = (x_1(t), x_2(t), x_3(t), x_4(t))^{\mathrm{T}}$. For the above fuzzy rule base, by taking a standard fuzzy inference strategy, i.e., using a singleton fuzzifier, product fuzzy inference, and center average defuzzifier, we get the following T–S fuzzy system:

$$\dot{x}(t) = \frac{\sum_{i=1}^{2} F_i(x_1(t))(A_i x(t) + b_i)}{\sum_{i=1}^{2} F_i(x_1(t))}. \tag{8.9}$$

The rest of the task is to choose appropriate A_1, A_2, b_1, and b_2 such that (8.9) can be equivalent to (8.8). For system 10, each part of (8.8) is presented as follows:

$$A = \begin{pmatrix} 0 & -1 & -1 & 0 \\ 1 & \alpha & 0 & 1 \\ \beta_1 & 0 & 0 & 0 \\ 0 & 0 & -\theta & \sigma \end{pmatrix}, \quad f(x) = \begin{pmatrix} 0 \\ 0 \\ x_1 x_3 \\ 0 \end{pmatrix}, \quad \text{and } b = \begin{pmatrix} 0 \\ 0 \\ \beta \\ 0 \end{pmatrix}.$$

Denote $A_1 = A_1' + A_1''$, $A_2 = A_2' + A_2''$, and let $A_1' = A_2' = A$ and $b_1 = b_2 = \beta$. Substituting them into (8.9), and making (8.9) equal to (8.8), we get the values of A_1'' and A_2'' as follows:

$$A_1'' = \begin{pmatrix} 0 & 0 & 0 & 0 \\ 0 & 0 & 0 & 0 \\ 0 & 0 & \tilde{c}_1 + \tilde{d}_1 & 0 \\ 0 & 0 & 0 & 0 \end{pmatrix}, \quad A_2'' = \begin{pmatrix} 0 & 0 & 0 & 0 \\ 0 & 0 & 0 & 0 \\ 0 & 0 & \tilde{c}_1 - \tilde{d}_1 & 0 \\ 0 & 0 & 0 & 0 \end{pmatrix}.$$

Consequently, we derive the exact T–S fuzzy models for this class of hyperchaotic systems.

Case 2: More than one nonlinear term in $f(x)$, but there is a common factor in these nonlinear terms. Systems 2, 3, 4, 6, and 8 belong to this case. For this case, we can choose the common factor in the nonlinear terms as premise variable. The remaining process of T–S fuzzy modeling of this class of hyperchaotic systems is similar to Case 1, and therefore it is omitted.

Case 3: More than one nonlinear term in $f(x)$, and there is no common factor in these nonlinear terms. Systems 1, 5, 7, and 9 belong to this case. For this case, we have to choose two state variables occurring in $f(x)$ as premise variables. For example, in system 1, we can choose x_1 and x_2 as premise variables and the membership functions are defined as

$$F_{11}(x_1(t)) = \frac{1}{2}\left(1 - \frac{\tilde{c}_1 - x_1(t)}{\tilde{d}_1}\right), \quad F_{21}(x_1(t)) = \frac{1}{2}\left(1 + \frac{\tilde{c}_1 - x_1(t)}{\tilde{d}_1}\right),$$

$$F_{32}(x_2(t)) = \frac{1}{2}\left(1 - \frac{\tilde{c}_2 - x_2(t)}{\tilde{d}_2}\right), \quad F_{42}(x_2(t)) = \frac{1}{2}\left(1 + \frac{\tilde{c}_2 - x_2(t)}{\tilde{d}_2}\right),$$

$$F_{12}(x_2(t)) = F_{22}(x_2(t)) = F_{31}(x_1(t)) = F_{41}(x_1(t)) = 1.$$

Furthermore, the following fuzzy IF–THEN rules are presented:

$$R^1: \text{IF } x_1(t) \text{ is } F_{11} \text{ and } x_2(t) \text{ is } F_{12}, \text{ THEN } \dot{x}(t) = A_1 x(t) + b_1,$$
$$R^2: \text{IF } x_1(t) \text{ is } F_{21} \text{ and } x_2(t) \text{ is } F_{22}, \text{ THEN } \dot{x}(t) = A_2 x(t) + b_2,$$
$$R^3: \text{IF } x_1(t) \text{ is } F_{31} \text{ and } x_2(t) \text{ is } F_{32}, \text{ THEN } \dot{x}(t) = A_3 x(t) + b_3,$$
$$R^4: \text{IF } x_1(t) \text{ is } F_{41} \text{ and } x_2(t) \text{ is } F_{42}, \text{ THEN } \dot{x}(t) = A_4 x(t) + b_4,$$

where $x(t) = (x_1(t), x_2(t), x_3(t), x_4(t))^{\mathrm{T}}$.

For the above fuzzy rule base, by taking a standard fuzzy inference strategy, i.e., using a singleton fuzzifier, product fuzzy inference, and center average defuzzifier, we get the following T–S fuzzy system:

$$\dot{x}(t) = \frac{\sum_{i=1}^{4} \prod_{j=1}^{2} F_{ij}(x_j(t)) (A_i x(t) + b)}{\sum_{i=1}^{4} \prod_{j=1}^{2} F_{ij}(x_j(t))}. \tag{8.10}$$

The remaining task is to choose appropriate A_i and b_i $(i = 1, 2, 3, 4)$ such that (8.10) can be equivalent to (8.8). For system 1, each part of (8.8) is represented as follows:

$$A = \begin{pmatrix} -\alpha & \alpha & 0 & 1 \\ \sigma & \theta & 0 & 0 \\ 0 & 0 & -\beta & 0 \\ 0 & 0 & 0 & l \end{pmatrix}, \quad f(x) = \begin{pmatrix} 0 \\ -x_1 x_3 \\ x_1 x_2 \\ x_2 x_3 \end{pmatrix}, \quad \text{and } b = \begin{pmatrix} 0 \\ 0 \\ 0 \\ 0 \end{pmatrix}.$$

Denote $A_i = A'_i + A''_i$, and let $A'_i = A$, $b_i = b$ $(i = 1, 2, 3, 4)$. Substituting them into (8.10) and making (8.10) equal to (8.8), we get the values of A''_i as follows:

$$A''_1 = \begin{pmatrix} 0 & 0 & 0 & 0 \\ 0 & 0 & 2(-\tilde{c}_1 - \tilde{d}_1) & 0 \\ 0 & 2(\tilde{c}_1 + \tilde{d}_1) & 0 & 0 \\ 0 & 0 & 0 & 0 \end{pmatrix},$$

$$A''_2 = \begin{pmatrix} 0 & 0 & 0 & 0 \\ 0 & 0 & 2(-\tilde{c}_1 + \tilde{d}_1) & 0 \\ 0 & 2(\tilde{c}_1 - \tilde{d}_1) & 0 & 0 \\ 0 & 0 & 0 & 0 \end{pmatrix},$$

$$A_3'' = \begin{pmatrix} 0 & 0 & 0 & 0 \\ 0 & 0 & 0 & 0 \\ 0 & 0 & 0 & 0 \\ 0 & 0 & 2(\tilde{c}_2 + \tilde{d}_2) & 0 \end{pmatrix}, \text{ and } A_4'' = \begin{pmatrix} 0 & 0 & 0 & 0 \\ 0 & 0 & 0 & 0 \\ 0 & 0 & 0 & 0 \\ 0 & 0 & 2(\tilde{c}_2 - \tilde{d}_2) & 0 \end{pmatrix}.$$

Consequently, we derive the exact T–S fuzzy model of this class of hyperchaotic systems.

8.3.2 Synchronization of Two Different Hyperchaotic Systems

In this section, we will investigate the synchronization of two different hyperchaotic systems based on the T–S fuzzy models [17]. First, an assumption is given as follows.

Consider the fuzzy hyperchaotic system (8.7) as the drive system, and the following fuzzy hyperchaotic system as the response system:

$$\dot{\hat{x}}(t) = \sum_{i=1}^{r} h_i(\hat{z}(t))[\hat{A}_i\hat{x}(t) + \hat{b}_i] + u(t). \tag{8.11}$$

Suppose in (8.11) $\hat{b}_i = \hat{b}$ and in (8.7) $b_i = b$. Based on the PDC technique, a fuzzy controller can be designed to realize the synchronization as follows:

Subcontroller $u_1(t)$:

$$R^i : \text{IF } z(t) \text{ is } F_i \text{ ,THEN } u_1(t) = K_i x(t),\ i = 1,2,\ldots,r,$$

Subcontroller $u_2(t)$:

$$R^i : \text{IF } \hat{z}(t) \text{ is } F_i \text{ ,THEN } u_2(t) = -H_i\hat{x}(t),\ i = 1,2,\ldots,r.$$

The overall fuzzy controller is constructed by a parallel connection, i.e.,

$$u(t) = u_1(t) + u_2(t) = \sum_{i=1}^{r} \mu_i(z(t))K_i x(t) - \sum_{i=1}^{r} h_i(\hat{z}(t))H_i\hat{x}(t). \tag{8.12}$$

Denote the error signal as $e(t) = x(t) - \hat{x}(t)$. Substituting (8.12) into (8.11), we can derive the closed-loop synchronization error system as follows:

$$\dot{e}(t) = \sum_{i=1}^{r} \mu_i(z(t))(A_i - K_i)e(t) + \sum_{i=1}^{r} h_i(\hat{z}(t))(\hat{A}_i - H_i)e(t) + \varpi(t), \tag{8.13}$$

where

$$\varpi(t) = \sum_{i=1}^{r} \mu_i(z(t))(A_i - K_i)\hat{x}(t) - \sum_{i=1}^{r} h_i(\hat{z}(t))(\hat{A}_i - H_i)x(t) + (b - \hat{b}).$$

For the synchronization error system (8.13), consider the following H_∞ performance index:

$$\int_0^{t_f} e^{\mathrm{T}}(t)e(t)\mathrm{d}t < \frac{1}{2}e^{\mathrm{T}}(0)Pe(0) + \frac{\gamma_1^2+\gamma_2^2}{2}\int_0^{t_f} \varpi^{\mathrm{T}}(t)\varpi(t)\mathrm{d}t, \tag{8.14}$$

where $\gamma_1 > 0$ and $\gamma_2 > 0$ are prescribed attenuation levels, and t_f is the terminal time.

Theorem 8.1. Consider the fuzzy error system (8.13). For the given constants $\gamma_1 > 0$ and $\gamma_2 > 0$, if there exist $P = P^{\mathrm{T}} > 0$, X_i, and Y_i with appropriate dimensions, such that the following linear matrix inequalities (LMIs) hold:

$$\begin{pmatrix} PA_i + A_i^{\mathrm{T}}P - X_i^{\mathrm{T}} - X_i + I & \frac{1}{2}P \\ \frac{1}{2}P & -\gamma_1^2 I \end{pmatrix} < 0 \tag{8.15}$$

and

$$\begin{pmatrix} P\hat{A}_i + \hat{A}_i^{\mathrm{T}}P - Y_i^{\mathrm{T}} - Y_i + I & \frac{1}{2}P \\ \frac{1}{2}P & -\gamma_2^2 I \end{pmatrix} < 0, \tag{8.16}$$

then we can choose $K_i = P^{-1}X_i$ and $H_i = P^{-1}Y_i$ in (8.12), such that the robust performance (8.14) is guaranteed for $i = 1, 2, \ldots, r$.

Proof. Now, we select the Lyapunov function as

$$V(t) = e^{\mathrm{T}}(t)Pe(t),$$

where P is a positive-definite matrix. The time derivative of $V(t)$ along the trajectory of (8.13) is given by

$$\begin{aligned}
\dot{V}(t) &= \dot{e}^{\mathrm{T}}(t)Pe(t) + e^{\mathrm{T}}(t)P\dot{e}(t) \\
&= e^{\mathrm{T}}(t)(\Xi + \Pi)^{\mathrm{T}}Pe(t) + e^{\mathrm{T}}(t)P(\Xi + \Pi)e(t) \\
&\quad + \varpi^{\mathrm{T}}(t)Pe(t) + e^{\mathrm{T}}(t)P\varpi(t) \\
&= \begin{pmatrix} e(t) \\ \varpi(t) \end{pmatrix}^{\mathrm{T}} \begin{pmatrix} (\Xi+\Pi)^{\mathrm{T}}P + P(\Xi+\Pi) & P \\ P & 0 \end{pmatrix} \begin{pmatrix} e(t) \\ \varpi(t) \end{pmatrix} \\
&= \begin{pmatrix} e(t) \\ \varpi(t) \end{pmatrix}^{\mathrm{T}} \begin{pmatrix} \Xi^{\mathrm{T}}P + P\Xi + \Pi^{\mathrm{T}}P + P\Pi & P \\ P & 0 \end{pmatrix} \begin{pmatrix} e(t) \\ \varpi(t) \end{pmatrix},
\end{aligned}$$

where

$$\Xi = \sum_{i=1}^{r} \mu_i(z(t))(A_i - K_i)$$

and

$$\Pi = \sum_{i=1}^{r} h_i(\hat{z}(t))(\hat{A}_i - H_i).$$

Based on (8.14), define

$$J = \int_0^{t_f} [2e^{\mathrm{T}}(t)e(t) - (\gamma_1^2 + \gamma_2^2)\varpi^{\mathrm{T}}(t)\varpi(t) + \dot{V}(t)]\mathrm{d}t - \int_0^{t_f} \dot{V}(t)\mathrm{d}t.$$

We have

$$\begin{aligned} J &= \int_0^{t_f} [2e^{\mathrm{T}}(t)e(t) - (\gamma_1^2 + \gamma_2^2)\varpi^{\mathrm{T}}(t)\varpi(t) + \dot{V}(t)]\mathrm{d}t - V(t_f) + V(0) \\ &\leq \int_0^{t_f} \left\{ \sum_{i=1}^{r} \mu_i(z(t)) \begin{pmatrix} e(t) \\ \varpi(t) \end{pmatrix}^{\mathrm{T}} \begin{pmatrix} \Omega & \frac{1}{2}P \\ \frac{1}{2}P & -\gamma_1^2 I \end{pmatrix} \begin{pmatrix} e(t) \\ \varpi(t) \end{pmatrix} \right. \\ &\quad \left. + \sum_{i=1}^{r} h_i(\hat{z}(t)) \begin{pmatrix} e(t) \\ \varpi(t) \end{pmatrix}^{\mathrm{T}} \begin{pmatrix} \Theta & \frac{1}{2}P \\ \frac{1}{2}P & -\gamma_2^2 I \end{pmatrix} \begin{pmatrix} e(t) \\ \varpi(t) \end{pmatrix} \right\} \mathrm{d}t + V(0), \end{aligned} \tag{8.17}$$

where

$$\Omega = (A_i - K_i)^{\mathrm{T}} P + P(A_i - K_i) + I$$

and

$$\Theta = (\hat{A}_i - H_i)^{\mathrm{T}} P + P(\hat{A}_i - H_i) + I.$$

It is easy to see that if the following inequalities hold:

$$\begin{pmatrix} \Omega & \frac{1}{2}P \\ \frac{1}{2}P & -\gamma_1^2 I \end{pmatrix} < 0 \tag{8.18}$$

and

$$\begin{pmatrix} \Theta & \frac{1}{2}P \\ \frac{1}{2}P & -\gamma_2^2 I \end{pmatrix} < 0, \tag{8.19}$$

then we have

$$J = \int_0^{t_f} \left(2e^{\mathrm{T}}(t)e(t) - (\gamma_1^2 + \gamma_2^2)\varpi^{\mathrm{T}}(t)\varpi(t)\right) \mathrm{d}t \leq V(0). \tag{8.20}$$

From (8.20), it is obvious that the H_∞ performance index (8.14) is guaranteed. Denote $X_i = PK_i$ and $Y_i = PH_i$. Then, (8.18) and (8.19) are equivalent to the LMIs (8.15) and (8.16), respectively. This completes the proof. □

8.3.3 Simulations

To illustrate the effectiveness of the theoretical analysis and design method, some examples are included for illustration. First, we investigate the T–S fuzzy modeling of system 2 (the Liu hyperchaotic system) and system 6 (the Lorenz hyperchaotic system). When the parameters are chosen as $\alpha = 10$, $\beta = 40$, $\theta = 2.5$, $\sigma = 10.6$, $k = 1$, and $h = 4$, system 2 has two positive Lyapunov exponents $\lambda_1 = 1.1491$ and $\lambda_2 = 0.12688$. Thus, system 2 is hyperchaotic in this case, whose attractors are

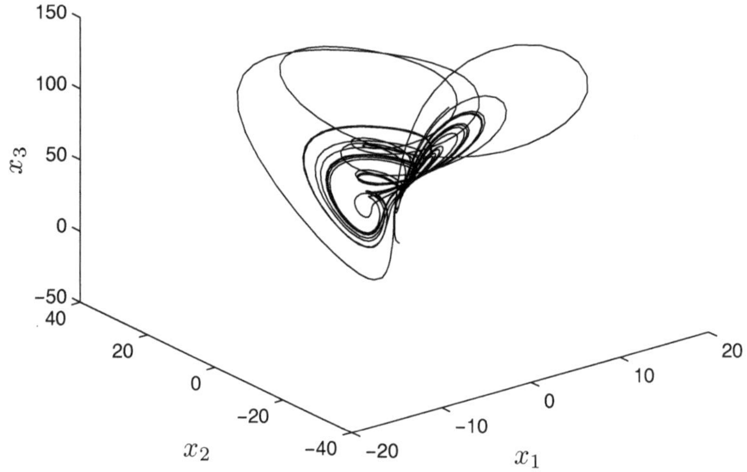

Fig. 8.4 Projection of the Liu hyperchaotic attractor on $x_1 - x_2 - x_3$

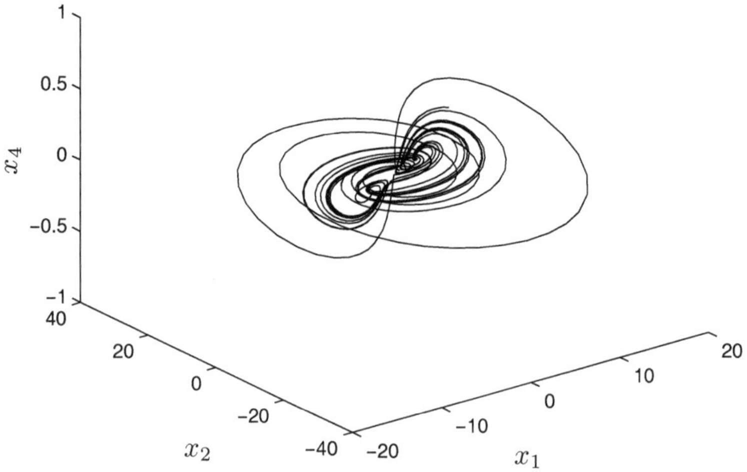

Fig. 8.5 Projection of the Liu hyperchaotic attractor on $x_1 - x_2 - x_4$

shown in Figs. 8.4 and 8.5. The initial state vector is chosen as $x(0) = (1,1,-1,0)^{\mathrm{T}}$. From the simulation we get $x_1 \in [\tilde{c}_1 - \tilde{d}_1, \tilde{c}_1 + \tilde{d}_1]$, where $\tilde{c}_1 = 1.2347$ and $\tilde{d}_1 = 18.6833$. Then, we can derive the T–S fuzzy models of system 2 as follows:

$$R^1: \text{ IF } x_1(t) \text{ is } F_1, \text{THEN } \dot{x}(t) = A_1 x(t) + b_1,$$
$$R^2: \text{ IF } x_1(t) \text{ is } F_2, \text{THEN } \dot{x}(t) = A_2 x(t) + b_2,$$

where $x(t) = (x_1(t), x_2(t), x_3(t), x_4(t))^{\mathrm{T}}$,

$$A_1 = \begin{pmatrix} -10 & 10 & 0 & 0 \\ 40 & 0 & -19.9180 & 1 \\ 79.6720 & 0 & -2.5 & 0 \\ -10.6 & 0 & 0 & 0 \end{pmatrix}, \quad A_2 = \begin{pmatrix} -10 & 10 & 0 & 0 \\ 40 & 0 & 17.4486 & 1 \\ -69.7944 & 0 & -2.5 & 0 \\ -10.6 & 0 & 0 & 0 \end{pmatrix},$$

and

$$b_1 = b_2 = (0\ 0\ 0\ 0)^{\mathrm{T}}.$$

The membership functions are chosen as

$$F_1(x_1(t)) = \frac{1}{2}\left(1 - \frac{1.2347 - x_1(t)}{18.6833}\right), \quad F_2(x_1(t)) = \frac{1}{2}\left(1 + \frac{1.2347 - x_1(t)}{18.6833}\right).$$

When the parameters are chosen as $\alpha = 10$, $l = 28$, $\beta = 8/3$, and $\sigma = 1.3$, system 6 has two positive Lyapunov exponents $\lambda_1 = 0.39854$ and $\lambda_2 = 0.24805$. Thus, system 6 is hyperchaotic in this case, which is shown in Figs. 8.6 and 8.7. The initial state vector is chosen as $x(0) = (1,2,1,2)^{\mathrm{T}}$. From the simulation we get $x_1 \in [\tilde{c}_1 - \tilde{d}_1, \tilde{c}_1 + \tilde{d}_1]$, where $\tilde{c}_1 = -1.4366$ and $\tilde{d}_1 = 28.4241$. Then, we can derive the T–S fuzzy models of systems 6 as follows:

$$R^1: \text{ IF } x_1(t) \text{ is } F_1, \text{THEN } \dot{x}(t) = A_1 x(t) + b_1,$$
$$R^2: \text{ IF } x_1(t) \text{ is } F_2, \text{THEN } \dot{x}(t) = A_2 x(t) + b_2,$$

where $x(t) = (x_1(t), x_2(t), x_3(t), x_4(t))^{\mathrm{T}}$,

$$A_1 = \begin{pmatrix} -10 & 10 & 0 & 1 \\ 28 & -1 & -26.9875 & 0 \\ 0 & 26.9875 & -8/3 & 0 \\ 0 & 0 & -26.9875 & 1.3 \end{pmatrix},$$

$$A_2 = \begin{pmatrix} -10 & 10 & 0 & 1 \\ 28 & -1 & 29.8607 & 0 \\ 0 & -29.8607 & -8/3 & 0 \\ 0 & 0 & 29.8607 & 1.3 \end{pmatrix},$$

and

$$b_1 = b_2 = (0\ 0\ 0\ 0)^{\mathrm{T}}.$$

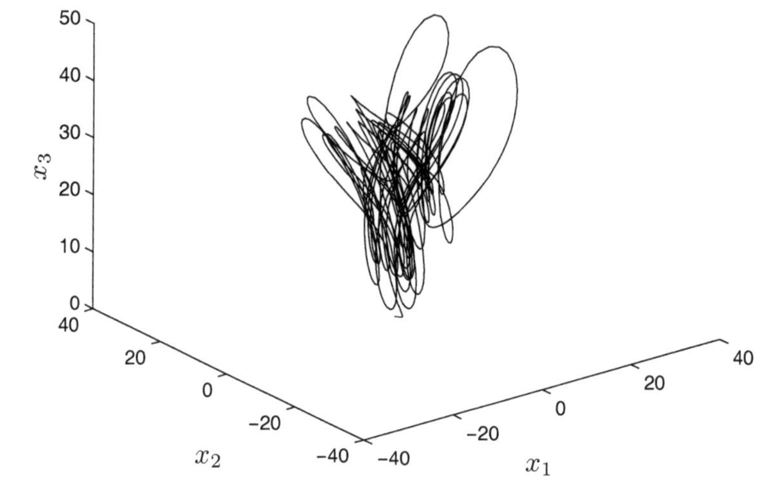

Fig. 8.6 Projection of the Lorenz hyperchaotic attractor on $x_1 - x_2 - x_3$

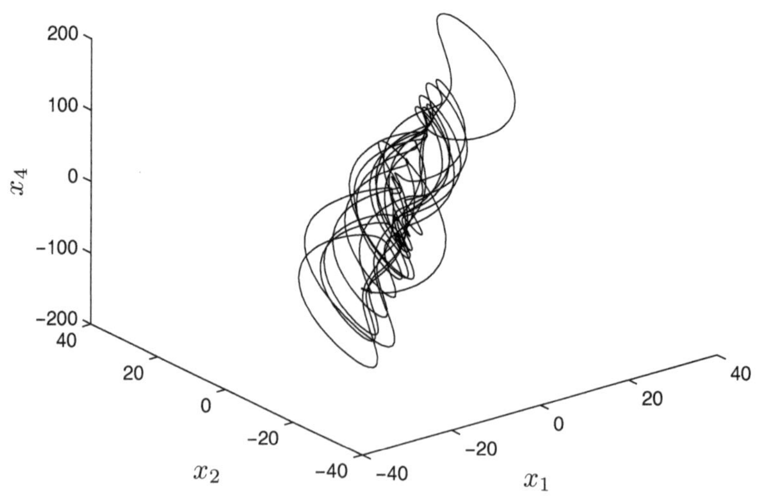

Fig. 8.7 Projection of the Lorenz hyperchaotic attractor on $x_1 - x_2 - x_4$

The membership functions are chosen as

$$F_1(x_1(t)) = \frac{1}{2}\left(1 - \frac{-1.4366 - x_1(t)}{28.4241}\right),$$

$$F_2(x_1(t)) = \frac{1}{2}\left(1 + \frac{-1.4366 - x_1(t)}{28.4241}\right).$$

In the following, based on the above T–S fuzzy hyperchaotic models, we investigate the synchronization between the Liu hyperchaotic system and the Lorenz hyperchaotic system. We consider the Liu hyperchaotic system as the drive system and the Lorenz hyperchaotic system as the response system. The initial conditions of the drive system and the response system are $x(0) = (15, 25, 100, 0)^T$ and $\hat{x}(0) = (-25, -20, 30, 100)^T$, respectively. According to Theorem 8.1, choosing $\gamma_1 = \gamma_2 = 0.1$, we get the positive-definite matrix P and feedback gains K_i and H_i $(i = 1, 2)$ computed by LMI Control Toolbox in Matlab® 7.0 as follows:

$$P = \begin{pmatrix} 1.6093 & -0.0410 & -0.0022 & -0.3284 \\ -0.0410 & 0.3789 & -0.0044 & 0.0092 \\ -0.0022 & -0.0044 & 0.1059 & -0.0003 \\ -0.3284 & 0.0092 & -0.0003 & 0.2155 \end{pmatrix},$$

$$K_1 = \begin{pmatrix} 39.3794 & 13.5798 & -0.2381 & 76.9472 \\ 40.6736 & 142.2773 & -1.6448 & -6.2164 \\ 29.4006 & -22.1059 & 516.3327 & -25.0879 \\ 68.0431 & 5.3914 & -13.0605 & 371.6015 \end{pmatrix},$$

$$K_2 = \begin{pmatrix} 38.9603 & 14.2497 & 2.9000 & 77.1718 \\ 40.6733 & 143.8961 & 14.6389 & -5.0789 \\ -27.7512 & 37.8918 & 514.4715 & 31.3370 \\ 66.1391 & 7.3065 & 18.7104 & 372.9590 \end{pmatrix},$$

$$H_1 = \begin{pmatrix} 35.7096 & 13.3922 & 2.1372 & 72.7527 \\ 39.6357 & 134.9720 & -3.9529 & -5.4449 \\ 51.7616 & -29.9640 & 487.2763 & -22.7680 \\ 62.5319 & 6.6327 & -8.1879 & 350.9440 \end{pmatrix},$$

and

$$H_2 = \begin{pmatrix} 35.3859 & 13.8039 & 0.4756 & 72.9124 \\ 38.7566 & 136.4886 & 16.1391 & -4.6764 \\ -50.9872 & 44.6941 & 485.6173 & 29.9071 \\ 61.7018 & 7.5811 & 14.2738 & 351.4143 \end{pmatrix}.$$

The synchronization error curves are presented in Figs. 8.8–8.11, which show that H_∞ synchronization between two different hyperchaotic systems is achieved.

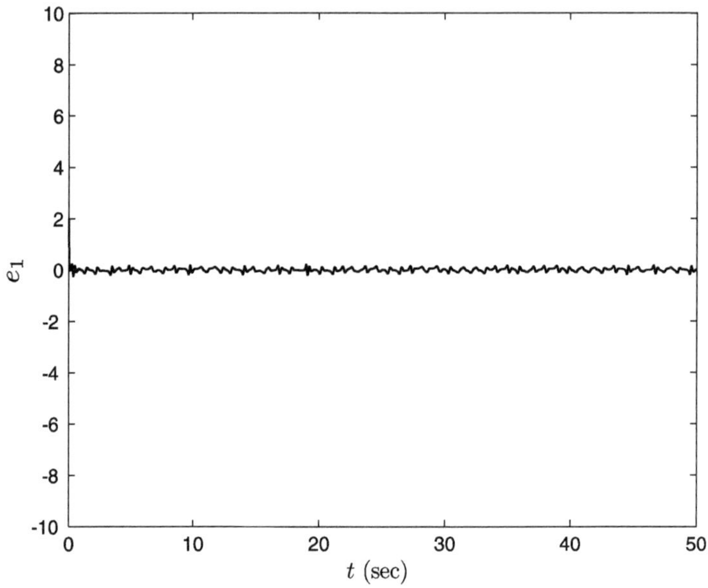

Fig. 8.8 Synchronization error $e_1(t)$ between two different hyperchaotic systems

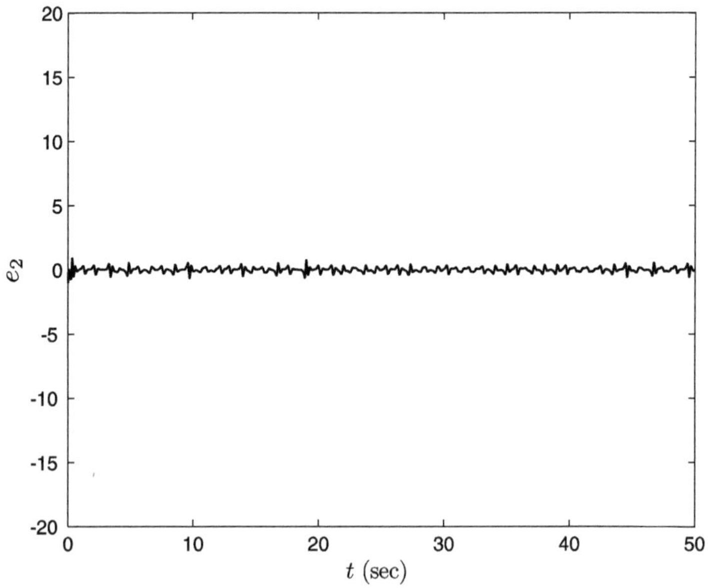

Fig. 8.9 Synchronization error $e_2(t)$ between two different hyperchaotic systems

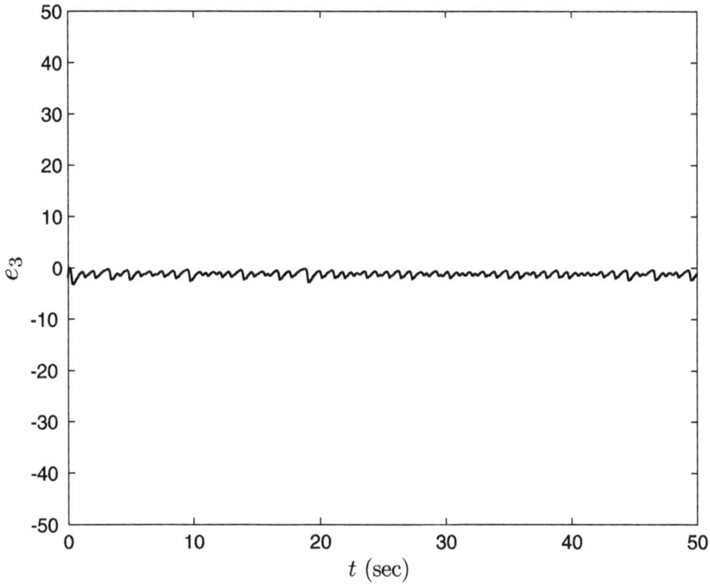

Fig. 8.10 Synchronization error $e_3(t)$ between two different hyperchaotic systems

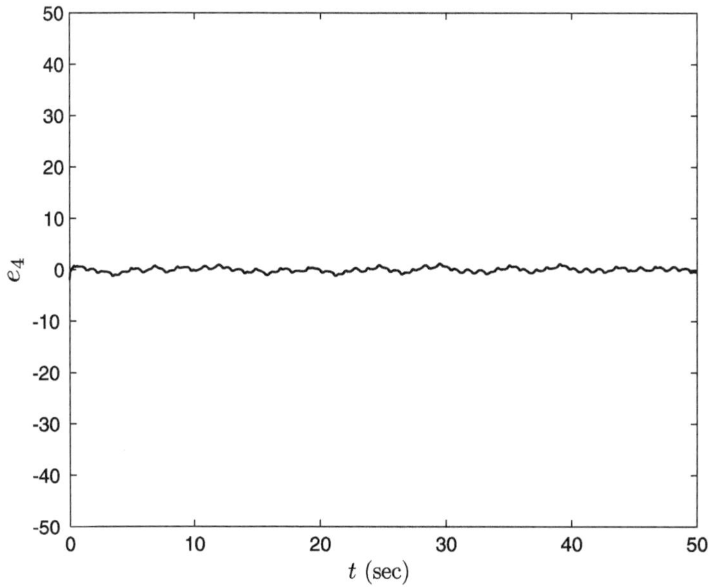

Fig. 8.11 Synchronization error $e_4(t)$ between two different hyperchaotic systems

8.4 Synchronizing Fuzzy Chaotic Systems with Time-Varying Delays

This section deals with the synchronization problem of chaotic systems with time-varying delays and unknown parameters. A fuzzy controller is developed and presented via T–S fuzzy models. By using an appropriate Lyapunov–Krasovskii functional, we derive synchronization conditions in the form of an LMI. The advantages of the approach are that the structurer of the fuzzy controller is simple and this method can be applied to most of the synchronization control of chaotic systems. A numerical example is given to demonstrate the validity of the proposed synchronization approach.

8.4.1 Basic Formulation

A T–S fuzzy system is described by a set of fuzzy IF–THEN rules where each rule locally represents a linear input–output realization of the system over a certain region of the state space. For the synchronization problem, the drive and response systems are designed such that the internal states of the drive and response systems are to be the same. Let the chaotic response system be described by a T–S fuzzy model with the following rule-base:

$$\begin{aligned} R^i:\ & \text{IF } z_1(t) \text{ is } M_{i1},\ldots, \text{ and } z_g(t) \text{ is } M_{ig}, \\ & \text{THEN } \dot{x}(t) = A_i x(t) + B_i x(t-\tau(t)) + \varepsilon(t) + u(t), \end{aligned}$$

$i = 1,2,\ldots,l$, where $x(t) \in \mathbb{R}^n$ is the state vector of the response system, $z(t) = (z_1(t), z_2(t), \ldots, z_g(t))^{\mathrm{T}}$ is a vector of the measurable premise variables, M_{ir} $(r = 1,2,\ldots,g)$ are fuzzy sets, l is the number of fuzzy rules, A_i and B_i are some constant matrices of compatible dimensions, $u(t) \in R^n$ is the control input vector, $\varepsilon(t)$ is the oscillated force or constant value in the chaotic dynamics, and $\tau(t)$ is a time-varying delay satisfying $0 < \tau(t) \le \tau_0$ and $0 \le \dot{\tau}(t) \le h < 1$.

Using the singleton fuzzifier, product fuzzy inference, and weighted average defuzzifier, the output of the above fuzzy response system is inferred as follows:

$$\dot{x}(t) = \sum_{i=1}^{l} v_i(z(t))[A_i x(t) + B_i x(t-\tau(t))] + \varepsilon(t) + u(t), \tag{8.21}$$

where

$$v_i(z(t)) = \frac{\omega_i(z(t))}{\sum_{i=1}^{l} \omega_i(z(t))},\ v_i(z(t)) \ge 0,\ \sum_{i=1}^{l} v_i(z(t)) = 1,$$

$$\omega_i(z(t)) = \prod_{j=1}^{g} M_{ij}(z(t)),\ \sum_{i=1}^{l} \omega_i(z(t)) > 0,\ \omega_i(z(t)) \ge 0,$$

and $M_{ij}(z(t))$ is the degree of membership of $z(t)$ in M_{ij}.

Similar to the fuzzy response system, the fuzzy drive system can be described by the following rules:

$$\begin{aligned} R^i: \ & \text{IF } \hat{z}_1(t) \text{ is } M_{i1}, \ldots, \text{ and } \hat{z}_g(t) \text{ is } M_{ig}, \\ & \text{THEN } \dot{y}(t) = \hat{A}_i y(t) + \hat{B}_i y(t-\tau(t)) + \varepsilon(t), \end{aligned}$$

$i = 1, \ldots, l$, where $y(t) \in \mathbb{R}^n$ is the state vector of the drive system, $\hat{z}(t) = [\hat{z}_1(t), \hat{z}_2(t), \ldots, \hat{z}_g(t)]^{\mathrm{T}}$ is a vector of the premise variables, and $\hat{A}_i$ and $\hat{B}_i$ are some unknown constant matrices of compatible dimensions. The overall fuzzy drive system can be inferred as

$$\dot{y}(t) = \sum_{i=1}^{l} \lambda_i(\hat{z}(t))[\hat{A}_i y(t) + \hat{B}_i y(t-\tau(t))] + \varepsilon(t). \tag{8.22}$$

Denote the synchronization error signal by $e(t) = x(t) - y(t)$. From (8.21) and (8.22), the error dynamics of $e(t)$ is arranged as

$$\begin{aligned} \dot{e}(t) =& \dot{x}(t) - \dot{y}(t) \\ =& \sum_{i=1}^{l} v_i(z(t)\,[A_i e(t) + B_i e(t-\tau(t))] \\ & + \left(\sum_{i=1}^{l} v_i(z(t)) - \sum_{i=1}^{l} \lambda_i(\hat{z}(t)) \right) [\hat{A}_i y(t) + \hat{B}_i y(t-\tau(t))] \\ & + \sum_{i=1}^{l} v_i(z(t))[(A_i - \hat{A}_i)y(t) + (B_i - \hat{B}_i)y(t-\tau(t))] + u(t) \\ =& \sum_{i=l}^{l} v_i(z(t))[A_i e(t) + B_i e(t-\tau(t))] \\ & + \sum_{i=1}^{l} v_i(z(t))[(A_i - \hat{A}_i)y(t) + (B_i - \hat{B}_i)y(t-\tau(t))] + \varpi(t) + u(t), \end{aligned} \tag{8.23}$$

where

$$\varpi(t) = \left(\sum_{i=1}^{l} v_i(z(t)) - \sum_{i=1}^{l} \lambda_i(\hat{z}(t)) \right) [\hat{A}_i y(t) + \hat{B}_i y(t-\tau(t))].$$

The purpose of this section is to design adaptive control law such that the drive and response systems are synchronized; that is,

$$\lim_{t \to \infty} e(t) = 0.$$

8.4.2 Main Result

In order to achieve the synchronization of the two systems, let us design the following fuzzy controller based on the PDC technique,

$$\begin{aligned} R^i:\ &\text{IF } z_1(t) \text{ is } M_{i1}, \ldots, \text{ and } z_g(t) \text{ is } M_{ig}, \\ &\text{THEN } u(t) = -K_i e(t) + (-A_i + C_i(t))y(t) + (-B_i + D_i(t))y(t-\tau(t)), \end{aligned}$$

$i = 1, 2, \ldots, l$, where K_i are the feedback gains to be designed later and $C_i(t)$ and $D_i(t)$ are adjustable control gains. The inferred control law is represented as

$$u(t) = \sum_{i=1}^{l} v_i(z(t))[-K_i e(t) + (-A_i + C_i(t))y(t) + (-B_i + D_i(t))y(t-\tau(t))]. \tag{8.24}$$

Substituting (8.24) into (8.23), we obtain the following error system:

$$\begin{aligned} \dot{e}(t) =& \dot{x}(t) - \dot{y}(t) \\ =& \sum_{i=1}^{l} v_i(z(t)) \left[(A_i - K_i)e(t) + B_i e(t-\tau(t))\right] \\ &+ \sum_{i=1}^{l} v_i(z(t)) \left[(C_i(t) - \hat{A}_i)y(t) + (D_i(t) - \hat{B}_i)y(t-\tau(t))\right] + \varpi(t). \end{aligned} \tag{8.25}$$

Theorem 8.2. For the error system (8.25), if there exist symmetric positive-definite matrices P and Q, a positive constant σ, feedback gain matrices K_i, and the parametric adaptive laws

$$\begin{cases} \dot{c}_{1jk}(t) = -v_k(z(t))\alpha_1 m_j y_k(t), \\ \quad \vdots \\ \dot{c}_{ljk}(t) = -v_k(z(t))\alpha_l m_j y_k(t), \end{cases} \tag{8.26}$$

$$\begin{cases} \dot{d}_{1jk}(t) = -v_k(z(t))\beta_1 m_j y_k(t-\tau(t)), \\ \quad \vdots \\ \dot{d}_{ljk}(t) = -v_k(z(t))\beta_l m_j y_k(t-\tau(t)), \end{cases} \tag{8.27}$$

where α_i and β_i are some given positive constants, m_j is the elements of $2e^{\mathrm{T}}P$, c_{ijk} and d_{ijk} are elements of C_i and D_i, $i = 1, 2, \ldots, l$, $j, k = 1, 2, \ldots, n$, such that

$$\begin{pmatrix} \Pi & PB_i & P & I \\ B_i^{\mathrm{T}}P & -(1-h)Q & 0 & 0 \\ P & 0 & -\sigma^2 I & 0 \\ I & 0 & 0 & -I \end{pmatrix} < 0, \tag{8.28}$$

where $\Pi = (A_i - K_i)^{\mathrm{T}}P + P(A_i - K_i) + Q$, then we obtain H_∞ performance given by

$$\int_0^{t_f} e^{\mathrm{T}}(t)e(t)\mathrm{d}t \le e^{\mathrm{T}}(0)e(0) + \sigma^2 \int_0^{t_f} \varpi^{\mathrm{T}}(t)\varpi(t)\mathrm{d}t.$$

Proof. Let us select the Lyapunov-Krasovskii functional as

$$V(t) = V_1(t) + V_2(t),$$

where

$$V_1(t) = e^{\mathrm{T}}(t)Pe(t) + \int_{t-\tau(t)}^{t} e^{\mathrm{T}}(s)Qe(s)\mathrm{d}s,$$

$$V_2(t) = \sum_{i=1}^{l}\sum_{j=1}^{n}\sum_{k=1}^{n}\left[\frac{1}{2\alpha_i}\left(c_{ijk}(t) - \hat{a}_{ijk}\right)^2 + \frac{1}{2\beta_i}(d_{ijk}(t) - \hat{b}_{ijk})^2\right],$$

where $\hat{a}_{ijk}$ and $\hat{b}_{ijk}$ are elements of unknown parameters matrices $\hat{A}_i$ and $\hat{B}_i$, respectively.

The derivative of $V(t)$ along the trajectory of (8.25) is

$$\begin{aligned}
\dot{V}_1(t) &= \dot{e}^{\mathrm{T}}(t)Pe(t) + e^{\mathrm{T}}(t)P\dot{e}(t) + e^{\mathrm{T}}(t)Qe(t) \\
&\quad - (1 - \dot{\tau}(t)e^{\mathrm{T}}(t-\tau(t))Qe(t-\tau(t))) \\
&\le \sum_{i=1}^{l} v_i(z(t))\left\{e^{\mathrm{T}}(t)[(A_i - K_i)^{\mathrm{T}}P + P(A_i - K_i) + Q]e(t)\right. \\
&\quad \left. + e^{\mathrm{T}}(t-\tau(t))B_i^{\mathrm{T}}Pe(t) + e^{\mathrm{T}}(t)PB_ie(t-\tau(t))\right\} \\
&\quad - (1-h)e^{\mathrm{T}}(t-\tau(t))Qe(t-\tau(t)) + \varpi^{\mathrm{T}}(t)Pe(t) + e^{\mathrm{T}}(t)P\varpi(t) \\
&\quad + \sum_{i=1}^{l} v_i(z(t))\left\{2e^{\mathrm{T}}(t)P(C_i(t) - \hat{A}_i)y(t)\right. \\
&\quad \left. + 2e^{\mathrm{T}}(t)P(D_i(t) - \hat{B}_i)y(t-\tau(t))\right\}
\end{aligned}$$

and

$$\begin{aligned}
\dot{V}_2(t) &= \sum_{i=1}^{l}\sum_{j=1}^{n}\sum_{k=1}^{n}\left[\frac{1}{\alpha_i}(c_{ijk}(t) - \hat{a}_{ijk})\dot{c}_{ijk}(t) + \frac{1}{\beta_i}(d_{ijk}(t) - \hat{b}_{ijk})\dot{d}_{ijk}(t)\right] \\
&= -\sum_{i=1}^{l}\sum_{j=1}^{n}\sum_{k=1}^{n} v_i(z(t))\Big[(c_{ijk}(t) - \hat{a}_{ijk})m_jy_k(t) \\
&\quad + (d_{ijk}(t) - \hat{b}_{ijk})m_jy_k(t-\tau(t))\Big] \\
&= -2e^{\mathrm{T}}(t)P\sum_{i=1}^{r} v_i(z(t))\left[(C_i(t) - \hat{A}_i)y(t) + (D_i(t) - \hat{B}_i)y(t-\tau(t))\right].
\end{aligned} \tag{8.29}$$

Using (8.26) and (8.29), we obtain

$$\begin{aligned}\dot{V}(t) &= \dot{V}_1(t) + \dot{V}_2(t) \\ &\leq \sum_{i=1}^{l} v_i(z(t)) \left\{ e^{\mathrm{T}}(t)[(A_i - K_i)^{\mathrm{T}}P + P(A_i - K_i) + Q]e(t) \right. \\ &\quad \left. + e^{\mathrm{T}}(t-\tau(t))B_i^{\mathrm{T}}Pe(t) + e^{\mathrm{T}}(t)PB_i e(t-\tau(t)) \right\} \\ &\quad - (1-h)e^{\mathrm{T}}(t-\tau(t))Qe(t-\tau(t)) + \varpi^{\mathrm{T}}(t)Pe(t) + e^{\mathrm{T}}(t)P\varpi(t). \end{aligned} \tag{8.30}$$

The H_∞ performance is satisfied as follows:

$$J = \int_0^{t_f} [e^{\mathrm{T}}(t)e(t) - \sigma^2 \varpi^{\mathrm{T}}(t)\varpi(t)]\mathrm{d}t, \tag{8.31}$$

where t_f denotes the terminal time of the control and σ is a disturbance attenuation value which denotes the effect of $\varpi(t)$ on $e(t)$. Using (8.30) and (8.31), we obtain

$$\begin{aligned} J &= \int_0^{t_f} \left[e^{\mathrm{T}}(t)e(t) - \sigma^2\varpi^{\mathrm{T}}(t)\varpi(t) + \dot{V}(t)\right]\mathrm{d}t - \int_0^{t_f} \dot{V}(t)\mathrm{d}t \\ &= \int_0^{t_f} \left[e^{\mathrm{T}}(t)e(t) - \sigma^2\varpi^{\mathrm{T}}(t)\varpi(t) + \dot{V}(t)\right]\mathrm{d}t - V(t_f) + V(0) \\ &\leq \int_0^{t_f} \sum_{i=1}^{l} v_i(z(t))\eta^{\mathrm{T}}(t)\Phi\eta(t)\mathrm{d}t + V(0), \end{aligned}$$

where $\eta(t) = (e(t)^{\mathrm{T}}, e(t-\tau(t))^{\mathrm{T}}, \varpi(t)^{\mathrm{T}})^{\mathrm{T}}$, $\Phi = \begin{pmatrix} \prod + I & PB_i & P \\ B_i^{\mathrm{T}}P & -(1-h)Q & 0 \\ P & 0 & -\sigma^2 I \end{pmatrix}$, and $\prod = (A_i - K_i)^{\mathrm{T}}P + P(A_i - K_i) + Q$.

If there exist symmetric positive-definite matrices $P > 0$ and $Q > 0$, a positive constant σ, and feedback gain matrices K_i such that

$$\begin{pmatrix} \prod + I & PB_i & P \\ B_i^{\mathrm{T}}P & -(1-h)Q & 0 \\ P & 0 & -\sigma^2 I \end{pmatrix} < 0, \tag{8.32}$$

we get

$$J = \int_0^{t_f} [e^{\mathrm{T}}(t)e(t) - \sigma^2\varpi^{\mathrm{T}}(t)\varpi(t)]\mathrm{d}t < V(0).$$

The matrix inequality (8.32) can be converted into (8.28) via the Schur complement. The proof is complete. □

8.4.3 Simulations

A numerical example is given to show the validity of the proposed synchronization method of T–S fuzzy chaotic systems with time-varying delays.

The fuzzy drive system [8] can be described as

$$\begin{aligned} R^i:\ &\text{IF } z_1(t) \text{ is } M_{i1},\ldots,\ z_g(t) \text{ is } M_{ig},\\ &\text{THEN } \dot{y}(t) = \hat{A}_i y(t) + \hat{B}_i y(t-\tau(t)) + \varepsilon(t), \end{aligned} \tag{8.33}$$

$i = 1,2$.

In this fuzzy model, the state vector $y = (y_1, y_2)^{\mathrm{T}}$ are taken as the premise variables. The membership functions of fuzzy sets are

R^1: IF $y_1(t)$ is M_{11}, THEN $\dot{y}(t) = \hat{A}_1 y(t) + \hat{B}_1 y(t-\tau(t)) + \varepsilon(t)$,

R^2: IF $y_1(t)$ is M_{21}, THEN $\dot{y}(t) = \hat{A}_2 y(t) + \hat{B}_2 y(t-\tau(t)) + \varepsilon(t)$,

where $y(t) = (y_1(t),\ y_2(t))^{\mathrm{T}}$. Let parameter matrices $\hat{A}_1$, $\hat{A}_2$, $\hat{B}_1$, $\hat{B}_2$, and $\varepsilon(t)$ be respectively selected as follows

$$\hat{A}_1 = \begin{pmatrix} 0 & 1 \\ -47.6 & -0.18 \end{pmatrix},\ \hat{A}_2 = \begin{pmatrix} 0 & 1 \\ 1.2 & -0.18 \end{pmatrix},\hat{B}_1 = \hat{B}_2 = \begin{pmatrix} 0 & 0 \\ 0.01 & 0.01 \end{pmatrix},$$

and

$$\varepsilon(t) = \begin{pmatrix} 0 \\ 25\cos(1.29t) \end{pmatrix}.$$

Let $\tau(t) = 0.125\sin(2t)$, the membership functions are selected as follows

$$M_{11}(y_1(t)) = \frac{y_1^2}{100},\ M_{21}(y_1(t)) = 1 - \frac{y_1^2}{100}.$$

Here, we suppose that the response chaotic system to be synchronized has the same structure as the drive system. The fuzzy response system can be described by the following rules:

R^1: IF $x_1(t)$ is F_{11}, THEN $\dot{x}(t) = A_1 x(t) + B_1 x(t-\tau(t)) + \varepsilon(t)$,

R^2: IF $x_1(t)$ is F_{21}, THEN $\dot{x}(t) = A_2 x(t) + B_2 x(t-\tau(t)) + \varepsilon(t)$,

where $x(t) = (x_1(t),\ x_2(t))^{\mathrm{T}}$,

$$A_1 = \begin{pmatrix} 0 & 1 \\ -39.6 & -0.1 \end{pmatrix},\ A_2 = \begin{pmatrix} 0 & 1 \\ 0.4 & -0.1 \end{pmatrix},\ B_1 = B_2 = \begin{pmatrix} 0 & 0 \\ 0.01 & 0.01 \end{pmatrix},$$

$\varepsilon(t)$ and $\tau(t)$ are the same definition as drive systems, and the membership functions are selected as follows:

$$M_{11}(x_1(t)) = \frac{x_1^2}{100},\ M_{21}(x_1(t)) = 1 - \frac{x_1^2}{100}.$$

We illustrate each of these synthesis procedures, in which the LMI Control Toolbox is used to compute the solutions of LMIs. From the assumption, we obtain the symmetric positive-definite matrices P and Q and the feedback gains K_1 and K_2 as follows:

$$P = \begin{pmatrix} 0.2299 & 0.0000 \\ 0.0000 & 0.2298 \end{pmatrix},\ Q = \begin{pmatrix} 1.5325 & 0.0000 \\ 0.0000 & 1.5325 \end{pmatrix},$$

$$K_1 = \begin{pmatrix} 8.0084 & -19.2968 \\ -19.2968 & 7.9097 \end{pmatrix}, \ K_2 = \begin{pmatrix} 8.0084 & 0.7000 \\ 0.7001 & 7.9097 \end{pmatrix}.$$

Given the initial state values of the drive system and the response system as $y(0) = (1, \ -1)^{\mathrm{T}}$ and $x(0) = (2, \ 3)^{\mathrm{T}}$ and $\sigma = 0.5$ as the prescribed value. The parametric adaptive laws are chosen as $\alpha_1 = 50$, $\alpha_2 = 10$, $\beta_1 = 2$, and $\beta_2 = 1$.

Let us assume that

$$C_1 = \begin{pmatrix} 0 & 0 \\ c_1^{21} & c_1^{22} \end{pmatrix}, \ C_2 = \begin{pmatrix} 0 & 0 \\ c_2^{21} & c_2^{22} \end{pmatrix},$$

where c_1^{21}, c_1^{22}, c_2^{21}, and c_2^{22} are uncertain parameters. The initial value of $C_1(t)$ and $C_2(t)$ are selected as follows

$$C_1(0) = \begin{pmatrix} 0 & 0 \\ c_1^{21}(0) & c_1^{22}(0) \end{pmatrix} = \begin{pmatrix} 0 & 0 \\ -0.4 & -2 \end{pmatrix},$$

$$C_2(0) = \begin{pmatrix} 0 & 0 \\ c_2^{21}(0) & c_2^{22}(0) \end{pmatrix} = \begin{pmatrix} 0 & 0 \\ -2 & -8 \end{pmatrix}$$

The synchronization errors and the estimated errors are illustrated as follows. The synchronization errors of each state variable are shown in Figs. 8.12 and 8.13. Changing parameters can be clearly seen in Fig. 8.14 by using the scheme proposed in Theorem 8.2. The adaptive synchronization is more rapidly achieved for uncertain drive system with the response system.

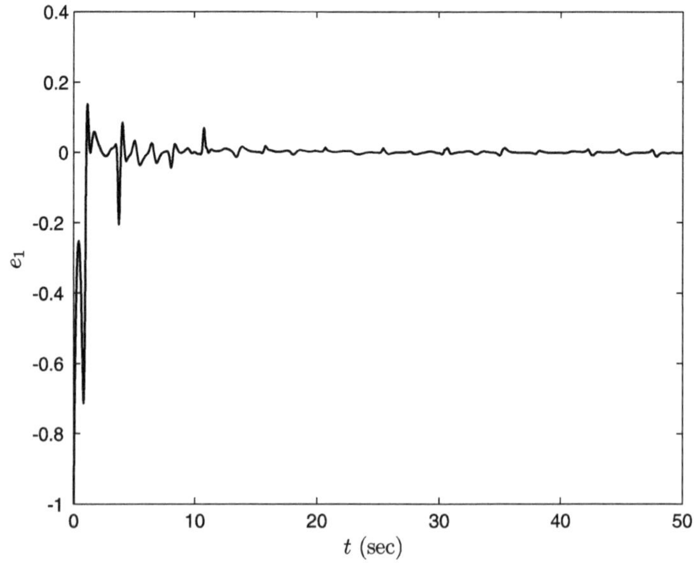

Fig. 8.12 The synchronization error e_1

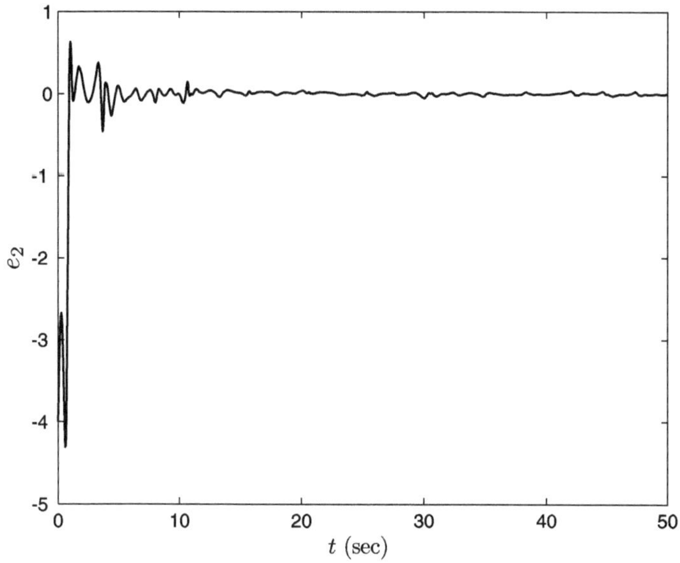

Fig. 8.13 The synchronization error e_2

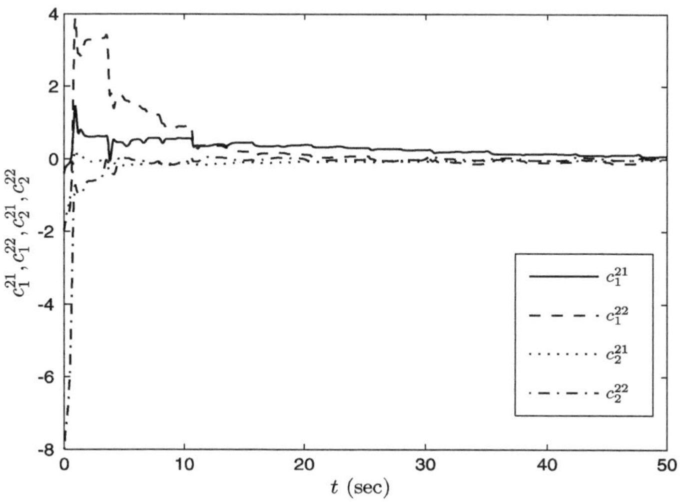

Fig. 8.14 Changing parameter curves of c_1^{21}, c_1^{22}, c_2^{21}, and c_2^{22}

8.5 Summary

In this chapter, first, we represented the typical chaotic and hyperchaotic systems via the T–S fuzzy models. Based on the fuzzy hyperchaotic models, simple fuzzy controllers were designed for synchronizing hyperchaotic systems. After that, a fuzzy controller was presented for the synchronization of chaotic systems based on the T–S fuzzy models with uncertain parameters and time-varying delays, which is based on the PDC technique, the T–S fuzzy models, and the adaptive control method. The theoretical analysis and simulation study show the validity of the proposed methods.

References

1. Chen G, Dong X (1998) *From Chaos to Order: Methodologies, Perspectives and Applications*. World Scientific, Singapore
2. Chen ZQ, Yang Y, Qi GY, Yuan ZZ (2007) A novel hyperchaos only with one equilibrium. Phys Lett A 360:696–701
3. El-Dessoky MM (2007) Synchronization and anti-synchronization of a hyperchaotic Chen system. Chaos Solitons Fractals 39:1790–1797
4. Elabbasy EM, Agiza HN, El-Dessoky MM (2006) Adaptive synchronization of a hyperchaotic system with uncertain parameter. Chaos Solitons Fractals 30:1133–1142
5. Gao TG, Chen ZQ, Yuan ZZ, Yu DC (2007) Adaptive synchronization of a new hyperchaotic system with uncertain parameters. Chaos Solitons Fractals 33:922–928
6. Jia Q (2007) Hyperchaos generated from the Lorenz chaotic system and its control. Phys Lett A 366:217–222
7. Jia Q (2007) Adaptive control and synchronization of a new hyperchaotic system with unknown parameters. Phys Lett A 362:424–429
8. Kim JH, Shin H, Kim E, Park M (2005) Synchronization of time-delayed TS fuzzy chaotic systems via scalar output variable. Int J Bifurc Chaos 15:2593–2601
9. Lian KY, Chiang TS, Chiu CS, Liu P (2001) Synthesis of fuzzy model-based designs to synchronization and secure communications for chaotic systems. IEEE Trans Syst Man Cybern B 31:66–83
10. Nikolov S, Clodong S (2006) Hyperchaos–chaos–hyperchaos transition in modified Rőssler systems. Chaos Solitons Fractals 28:252–263
11. Pecora LM, Carroll TL (1990) Synchronization in chaotic systems. Phys Rev Lett 64:821–824
12. Takagi T, Sugeno M (1985) Fuzzy identification of systems and its applications to modeling and control. IEEE Trans Syst Man Cybern 15:116–132
13. Tanaka K, Ikeda T, Wang HO (1998) A unified approach to controlling chaos via an LMI-based fuzzy control system design. IEEE Trans Circuits Syst I 45:1021–1040
14. Wang FQ, Liu CX (2006) Hyperchaos evolved from the Liu chaotic system. Chinese Phys 15:963–968
15. Wang GY, Zhang X, Zheng Y, Li YX (2006) A new modified hyperchaotic Lü system. Physica A 371:260–272
16. Wu XY, Guan ZH, Wu ZP (2008) Adaptive synchronization between two hyperchaotic systems. Nonlinear Anal 68:1346–1351
17. Yang DS, Zhang HG, Li AP, Meng ZY (2007) Generalized synchronization of two non–identical chaotic systems based on fuzzy model. Acta Phys Sinica 56:3121–3126
18. Yassen MT (2007) On hyperchaos synchronization of a hyperchaotic Lü system. Nonlinear Anal 68:3592–3600
19. Zhang HG, Cai LL, Bien Z (2000) A fuzzy basis function vector-based multivariable adaptive fuzzy controller for nonlinear systems. IEEE Trans Syst Man Cybern B 30:210–217

20. Zhang HB, Liao XF, Yu JB (2005) Fuzzy modeling and synchronization of hyperchaotic systems. Chaos Solitons Fractals 26:835–843
21. Zhang HG, Guan HX, Wang ZS (2007) Adaptive synchronization of neural networks with different attractors. Prog Nat Sci 17:687–695
22. Zhang HG, Huang W, Wang ZL, Chai TY (2006) Adaptive synchronization between different chaotic systems with unknown parameters. Phys Lett A 350:363–366
23. Zhang HG, Xie YH, Liu D (2006) Synchronization of a class of delayed chaotic neural networks with fully unknown parameters. Dyn Contin Discrete Impuls Syst B 13:297–308
24. Zhang HG, Yang DD, Chai TY (2007) Guaranteed cost networked control for T–S fuzzy systems with time delay. IEEE Trans Syst Man Cybern C 37:160–172

Chapter 9
Chaotification of Nonchaotic Systems

Abstract As a reverse process of suppressing or eliminating chaotic behaviors in order to reduce the complexity of an individual system or a coupled system, chaotification aims at creating or enhancing the system's complexity for some special applications. More precisely, chaotification is to generate some chaotic behaviors from a given system, which is nonchaotic or even is stable originally. In this chapter, we will investigate how to chaotify dynamical systems which are nonchaotic originally. In this chapter, first, we develop a simple nonlinear state feedback controller to chaotify the discrete-time fuzzy hyperbolic model (FHM) with uncertain parameters. Second, we use aan impulsive and nonlinear feedback control method to chaotify the continuous-time FHM. Finally, we chaotify two classes of continuous-time linear systems via a sampled data control approach.

9.1 Introduction

In this chapter, we will investigate how to chaotify dynamical systems which are nonchaotic originally. As a reverse process of suppressing or eliminating chaotic behaviors in order to reduce the complexity of an individual system or a coupled system, chaotification aims at creating or enhancing the system's complexity for some special applications. More precisely, chaotification is to generate some chaotic behaviors from a given system, which is nonchaotic or even is stable originally [22]. It is well known that chaotification of dynamical systems has many potential applications in electronic, mechanical, biological, and medical systems, etc. [9, 16, 20]. In recent years, many conventional control methods and special techniques were applied to the chaotification of discrete-time dynamical systems or continuous-time dynamical systems, such as linear feedback control [9, 12], time-delay feedback control [17, 19], nonlinear feedback control [11], adaptive inverse optimal control [25], impulsive control [21, 27], saturating control [13], switching control [1], and other control methods [22], etc. However, most of them are based on the assumption that the analytical representations of nonlinear dynamical systems to be chaotified are known exactly. For an unknown or uncertain nonlinear dynamical system, the above methods are ineffective. Recently, Zhang and Quan [24] and Margaliot et al. [14] successively proposed a novel fuzzy model named the hyperbolic tangent model (FHM). The FHM is a nonlinear model in nature which is very suitable for expressing nonlinear dynamical properties, and it is easy to design a stable and optimal controller based on the FHM than other models such as the fuzzy dynamic model, T–S model, etc. Moreover, the FHM has been shown to be a universal approximator, i.e., it can approximate an arbitrary nonlinear continuous function on a compact set with arbitrary accuracy [26]. This provides a theoretical foundation for using the FHM to represent complex nonlinear systems approximately. Therefore, the FHM can be regarded as an intermediate step for chaotifying an unknown nonlinear dynamical system. In this chapter, first, we develop a simple nonlinear state feedback controller to chaotify the discrete-time FHM with uncertain parameters. Second, we use an impulsive and nonlinear feedback control method to chaotify the continuous-time FHM. Finally, we chaotify two classes of continuous-time linear systems via a sampled data control approach.

9.2 Chaotification of Discrete-Time Fuzzy Hyperbolic Model with Uncertain Parameters

9.2.1 Basic Formulation

First, we define the fuzzy rule base for inferring the discrete-time FHM.

Definition 9.1 ([14, 23]). Given a plant with n state variables $x(k) = (x_1(k), \ldots, x_n(k))^{\mathrm{T}}$, we call the following fuzzy rule base a discrete-time hyperbolic-type fuzzy rule base if it satisfies the following conditions:

(i) Every fuzzy rule adopts the following form:

$$\begin{aligned} R^l: &\ \text{IF } x_1(k) \text{ is } F_{x_1} \text{ and } x_2(k) \text{ is } F_{x_2} \ \ldots \text{ and } x_n(k) \text{ is } F_{x_n}, \\ &\ \text{THEN } x_i(k+1) = \pm c_{x_1} \pm c_{x_2} \pm \cdots \pm c_{x_n}, \end{aligned}$$

$l = 1, 2, \ldots, 2^n$, where F_{x_i} $(i = 1, 2, \ldots, n)$ are fuzzy sets of $x_i(k)$ which include two linguistic values P (positive) and N (negative), c_{x_i} $(i = 1, 2, \ldots, n)$ is a positive constant corresponding to F_{x_i}, and '$\pm$' stands for either the plus or the minus sign. The actual signs in the THEN part are determined in the following manner: if the term characterizing F_{x_i} in the IF part is P, then c_{x_i} in the THEN part appears with a plus sign; otherwise c_{x_i} appears with a minus sign.

(ii) The constant term c_{x_i} in the THEN part corresponds to F_{x_i} in the IF part; that is, if there is an F_{x_i} term in the IF part, c_{x_i} must appear in the THEN part; otherwise c_{x_i} does not appear in the THEN part.

(iii) There are 2^n fuzzy rules in the rule base; that is, it includes all possible P and N combinations of state variables in the IF part and all sign combinations of constants in the THEN part. □

Theorem 9.1 ([14, 23]). Given a hyperbolic-type rule base, if we define the membership functions of P_{x_i} and N_{x_i} as

$$\mu_{P_{x_i}}(x_i(k)) = \exp\left[-\left(x_i(k) - s_{x_i}\right)^2/2\right],$$
$$\mu_{N_{x_i}}(x_i(k)) = \exp\left[-\left(x_i(k) + s_{x_i}\right)^2/2\right],$$

where $s_{x_i} > 0$, then we can derive the following model:

$$x(k+1) = A\tanh(Sx(k)),$$

where A is a constant matrix, $S = \operatorname{diag}\{s_{x_1}, \ldots, s_{x_n}\}$, and

$$\tanh(Sx(k)) := (\tanh(s_{x_1}x_1(k)), \ldots, \tanh(s_{x_n}x_n(k)))^{\mathrm{T}}.$$ □

This model is named the discrete-time FHM. In the sequel, s_{x_i} is denoted by s_i for simplicity.

Consider a general n-dimensional discrete-time autonomous system with the form

$$x(k+1) = g(x(k)), \tag{9.1}$$

where g is a C^1 nonlinear map. Denote $\bar{\mathbb{Z}} = \mathbb{N} \cup \{0\}$. Suppose that $t \in \bar{\mathbb{Z}}$ and denote g^t the t times compositions of g with itself. x^* is a p-periodic point of g ($p \in \mathbb{N}$) if $g^p(x^*) = x^*$ and $g^t(x^*) \neq x^*$ for $t < p$. When $p = 1$, x^* is a fixed point. Let $g'(x)$ and $\det(g'(x))$ be the Jacobian of g at the point x and its determinant, respectively. With this notation, we can give the definition of a snap-back repeller.

Definition 9.2 ([15]). Suppose that x^* is a fixed point of g with all eigenvalues of $g'(x^*)$ exceeding 1 in magnitude, and suppose that there exists a point $x^0 \neq x^*$ in a neighborhood[1] of x^* such that, for some positive integer m, $g^m(x^0) = x^*$ and $\det((g^m)'(x^0)) \neq 0$. Then, x^* is said to be a snap-back repeller of g. □

Lemma 9.1 ([15]). If system (9.1) has a snap-back repeller then the system is chaotic in the sense of Li–Yorke, namely,

(i) there exists a positive integer n such that, for every integer $p \geq n$, system (9.1) has p periodic points;
(ii) there exists a scrambled set (an uncountable invariant set S containing no periodic points) such that

a. $g(S) \subset S$,

---

[1] In [15], Marotto pointed out that identifying a repelling neighborhood of x^* is not in general a difficult task. In fact, since the local unstable manifold of x^* includes all points of $\mathbb{R}^n$ close to x^*, every closed ball $B(x^*, r)$ defined by the Euclidean norm must be so for sufficiently small $r > 0$.

b. for every $y \in S$ and any periodic point x of (9.1),

$$\lim_{k\to\infty} \sup \left\| g^k(x) - g^k(y) \right\| > 0,$$

c. for every $x, y \in S$ with $x \neq y$,

$$\lim_{k\to\infty} \sup \left\| g^k(x) - g^k(y) \right\| > 0;$$

(iii) there exists an uncountable subset $S_0 \subset S$ such that, for any $x, y \in S_0$,

$$\lim_{k\to\infty} \inf \left\| g^k(x) - g^k(y) \right\| = 0. \qquad \square$$

Lemma 9.2 (Gershgorin Disc Theorem [3]). Let A be a complex matrix whose ijth element is a_{ij}. For every $i = 1, 2, \ldots, n$, let $r_i := \sum_{j=1, j\neq i}^{n} \left|a_{ij}\right|$. Let D_i be a disc of radius r_i centered at a_{ii}. Then, all eigenvalues of A lie in the union of D_i for $i = 1, 2, \ldots, n$, i.e., $\lambda_i \in \bigcup_{j=1}^{n} D_j$ for $i = 1, 2, \ldots, n$. $\square$

9.2.2 Main Results

Now, consider the following uncertain discrete-time FHM:

$$x(k+1) = (A + \Delta A)\tanh(Sx(k)), \tag{9.2}$$

where ΔA represents the uncertain parameters. Assume that ΔA and S are bounded in the sense of infinite norm, i.e., there exist two positive constants γ and η such that $\|\Delta A\|_\infty \leq \gamma$ and $\|S\|_\infty \leq \eta$. Consequently, it can be inferred that $\|A + \Delta A\|_\infty \leq \rho$, where ρ is another positive constant.

Next, it will be proven that the controlled system is chaotic in the sense of Li–Yorke if we design the parameters of the following controller (9.3) appropriately. The relevant result is given as follows.

Theorem 9.2 ([28]). For the uncertain discrete-time FHM (9.2), given an arbitrarily small δ, there exists a sufficiently large constant β such that under the controller

$$\begin{aligned} u(k) &= \delta \sin\left(\frac{\beta\pi}{\delta} x(k)\right) \\ &= \left[\delta \sin\left(\frac{\beta\pi}{\delta} x_1(k)\right), \delta \sin\left(\frac{\beta\pi}{\delta} x_2(k)\right), \ldots, \delta \sin\left(\frac{\beta\pi}{\delta} x_n(k)\right)\right]^{\mathrm{T}}, \end{aligned} \tag{9.3}$$

the controlled uncertain discrete-time FHM

$$x(k+1) = (A + \Delta A)\tanh(Sx(k)) + u(k)$$

is chaotic in the sense of Li–Yorke.

Proof. Define

$$x(k+1) = (A+\Delta A)\tanh(Sx(k)) + \delta\sin\left(\frac{\beta\pi}{\delta}x(k)\right) := g(x(k)). \tag{9.4}$$

Obviously, $x^* = 0$ is a fixed point of g, which will be proven to be a snap-back repeller later. Define

$$\Phi(Sx) := \mathrm{diag}\left\{\frac{1}{\cosh^2(s_1x_1)},\ldots,\frac{1}{\cosh^2(s_nx_n)}\right\},$$
$$\cos\left(\frac{\beta\pi}{\delta}x\right) := \mathrm{diag}\left\{\cos\left(\frac{\beta\pi}{\delta}x_1\right),\ldots,\cos\left(\frac{\beta\pi}{\delta}x_n\right)\right\}.$$

Note that $\Phi(Sx^*) = \Phi(0) = I$ and $\cos\left(\frac{\beta\pi}{\delta}x^*\right) = \cos(0) = I$, where I denotes the identity matrix of appropriate dimensions. Differentiating (9.4) with respect to x at the fixed point $x^* = 0$ yields

$$g'(x^*) = (A+\Delta A)S + \pi\beta I.$$

According to Lemma 9.2, one can assert that all eigenvalues $\lambda_i(g'(x^*))$, $i = 1,2,\ldots,n$, lie in the union of all D_i of radius $\sum_{j=1,j\neq i}^{n} \left|(a_{ij}+\Delta a_{ij})s_j\right|$ centered at $(a_{ii}+\Delta a_{ii})s_i + \pi\beta$. Therefore, as long as one chooses an appropriate β, such that the following inequalities are met:

$$(a_{ii}+\Delta a_{ii})s_i + \pi\beta - \sum_{j=1,j\neq i}^{n} \left|(a_{ij}+\Delta a_{ij})s_j\right| > 1, \quad \text{for } i = 1,2,\ldots,n, \tag{9.5}$$

it is guaranteed that all $\lambda_i(g'(x^*))$ exceed 1 in magnitude for $i = 1,2,\ldots,n$. Indeed, according to the definition of infinite norm, and noting that S is a diagonal matrix, one has

$$\|(A+\Delta A)S\|_\infty = \max_i \left\{\sum_{j=1}^{n} \left|(a_{ij}+\Delta a_{ij})s_j\right|\right\}.$$

Therefore, the following inequalities are satisfied:

$$\sum_{\substack{j=1\\ j\neq i}}^{n} \left|(a_{ij}+\Delta a_{ij})s_j\right| \leq \|(A+\Delta A)S\|_\infty - |(a_{ii}+\Delta a_{ii})s_i|, \text{ for } i = 1,2,\ldots,n. \tag{9.6}$$

If $(a_{ii}+\Delta a_{ii})s_i \geq 0$, based on the inequalities (9.6), when

$$\beta > \frac{1}{\pi}\left(1 + \|(A+\Delta A)S\|_\infty\right),$$

one gets

$$(a_{ii}+\Delta a_{ii})s_i+\pi\beta-\sum_{j=1,j\neq i}^{n}\left|(a_{ij}+\Delta a_{ij})s_j\right|>1+2(a_{ii}+\Delta a_{ii})s_i,\ \text{for } i=1,2,\ldots,n.$$

Consequently, the inequalities (9.5) are satisfied.

Otherwise, for $(a_{ii}+\Delta a_{ii})s_i<0$, when $\beta>\frac{1}{\pi}(\,1+\|(A+\Delta A)S\,\|_{\infty})$, the inequalities (9.5) are satisfied directly. Then, according to the property of infinity norm, one has

$$\|(A+\Delta A)S\|_{\infty}\leq\|A+\Delta A\|_{\infty}\|S\|_{\infty}=\rho\eta.$$

Therefore, if $\beta>(1+\rho\eta)/\pi$, one can guarantee that all eigenvalues of $g'(x^*)$ exceed 1 in magnitude.

Next, it is shown that there exists a point $x^0\in B(x^*,r)$, with $x^0\neq x^*$, where $B(x^*,r):=\{x\colon \|x-x^*\|\leq r\}$ and r is a sufficiently small positive constant (that is, this r is small enough to guarantee that the closed ball $B(x^*,r)$ defined by Euclidean norm is a repelling neighborhood of x^*), such that $g(x^0)=0=x^*$ and $\det(g'(x^0))\neq 0$. In the sequel, for simplicity, $x<y$ denotes that every component of $x-y$ is negative and $x>y$ denotes that every component of $x-y$ is positive. First, we choose two points $x_1=\left(\frac{\delta}{2\beta},\ldots,\frac{\delta}{2\beta}\right)^{\mathrm{T}}$ and $x_2=\left(\frac{3\delta}{2\beta},\ldots,\frac{3\delta}{2\beta}\right)^{\mathrm{T}}$. Obviously, if $\beta>(3\sqrt{n}\delta)/(2r)$, then it can be guaranteed that $x_1\in B(x^*,r)$ and $x_2\in B(x^*,r)$ simultaneously.

Noting that $s_j>0$ (for $j=1,\ldots,n$), $x_1>0$, $x_2>0$, and $\tanh(x)<x$ for $x>0$, we can derive the following two inequalities:

$$\begin{aligned}g_i(x_1)&\geq-\sum_{j=1}^{n}\left(\left|a_{ij}+\Delta a_{ij}\right|\tanh\left(s_jx_{1j}\right)\right)+\delta\sin\left(\frac{\beta\pi}{\delta}x_{1i}\right)\\&\geq-\|A+\Delta A\|_{\infty}\|S\|_{\infty}\frac{\delta}{2\beta}+\delta\\&=-\frac{\rho\eta\delta}{2\beta}+\delta\end{aligned}\tag{9.7}$$

and

$$\begin{aligned}g_i(x_2)&\leq\sum_{j=1}^{n}\left(\left|a_{ij}+\Delta a_{ij}\right|\tanh(s_jx_{2j})\right)+\delta\sin\left(\frac{\beta\pi}{\delta}x_{2i}\right)\\&\leq\|A+\Delta A\|_{\infty}\|S\|_{\infty}\frac{3\delta}{2\beta}-\delta\\&=\frac{3\rho\eta\delta}{2\beta}-\delta,\end{aligned}\tag{9.8}$$

where $i=1,2,\ldots,n$. From (9.7) we can infer that if $\beta>(\rho\eta)/2$ then $g(x_1)>0$ can be guaranteed. From (9.8) we can infer that if $\beta>(3\rho\eta)/2$ then $g(x_2)<0$

can be guaranteed. Combining the above two results, it can be concluded that if $\beta > (3\rho\eta)/2$, then there exist two points $x_1 = \left(\frac{\delta}{2\beta}, \ldots, \frac{\delta}{2\beta}\right)^{\mathrm{T}}$ and $x_2 = \left(\frac{3\delta}{2\beta}, \ldots, \frac{3\delta}{2\beta}\right)^{\mathrm{T}}$, such that $g(x_1) > 0$ and $g(x_2) < 0$. Therefore, by the mean value theorem of calculus [10], there exists a point $x^0 \in B(x^*, r)$, $x_1 < x^0 < x_2$, such that $g(x^0) = 0 = x^*$. Then, as long as an appropriate β satisfying $\det(g'(x^0)) \neq 0$ is determined, we can conclude that x^* is a snap-back repeller of the map g defined in (9.4). In fact, considering that β can be chosen sufficiently large, we can always determine an appropriate constant β, such that $\det(g'(x^0)) = \det\left((A+\Delta A)S\Phi(Sx^0) + \pi\beta\cos\left(\frac{\beta\pi}{\delta}x^0\right)\right) \neq 0$. Then, it can be inferred that x^* is a snap-back repeller of the map g defined in (9.4).

To conclude, if an appropriate β is selected, such that $\beta > \bar{\beta}$ and $\det(g'(x^0)) \neq 0$ is satisfied, where $\bar{\beta} = \max\left\{\frac{1+\rho\eta}{\pi}, \frac{3\rho\eta}{2}, \frac{3\sqrt{n}\delta}{2r}\right\}$, then $x^* = 0$ is a snap-back repeller of the map g defined in (9.4). Then, according to Lemma 9.1, it can be concluded that the controlled system is chaotic in the sense of Li–Yorke. This completes the proof. □

Remark 9.1. On one hand, there exists the possibility that the selected β is not very appropriate, which makes the equality $\det(g'(x^0)) = 0$ met, though the probability of the event $\det(g'(x^0)) = 0$ is very small for a certain β. Then, the controlled system may be unable to generate chaos in this case. On the other hand, the exact value of r in $B(x^*, r)$ is not easy to determine. Based on the two facts above, a feasible method is presented here. Actually, we can regard β as an adjustable parameter, i.e., the value of β is increased gradually from the initial value 0. In this process, we can see that a series of period-doubling bifurcation behaviors occur continually. In the end, chaos will be generated in the controlled system. □

Remark 9.2. Assume that the original discrete-time system to be chaotified is fully unknown. However, if it can be approximated by the uncertain discrete-time FHM (9.2) with adequate accuracy, then we can believe that the controller for chaotification of the uncertain discrete-time FHM (9.2) can chaotify the original system at the same time. □

9.2.3 Simulations

To visualize the effectiveness of the theoretical analysis and controller design, an example is included for illustration. Consider an uncertain discrete-time FHM as follows:

$$x(k+1) = (A+\Delta A)\tanh(Sx(k)) + u(k),$$

where $x(k) = (x_1, x_2, x_3)^{\mathrm{T}}$, $A = \begin{pmatrix} 1 & 2 & 1.5 \\ 0 & 2 & 1 \\ 1.5 & 1 & 3 \end{pmatrix}$,

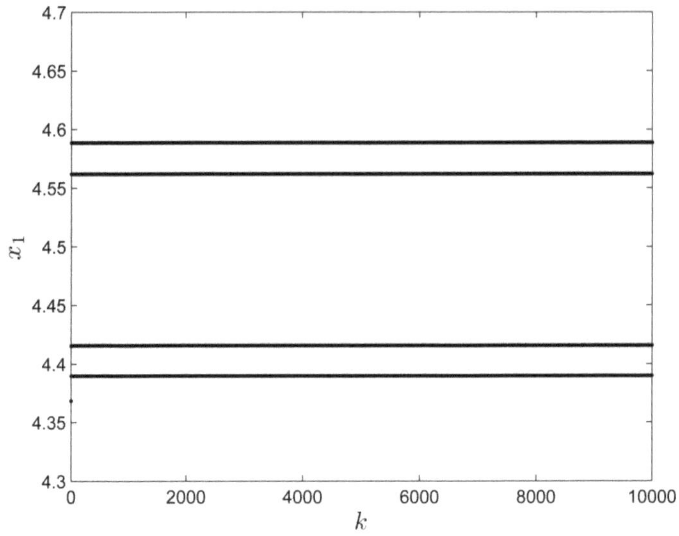

Fig. 9.1 The periodic orbits of x_1 with $\beta = 0$

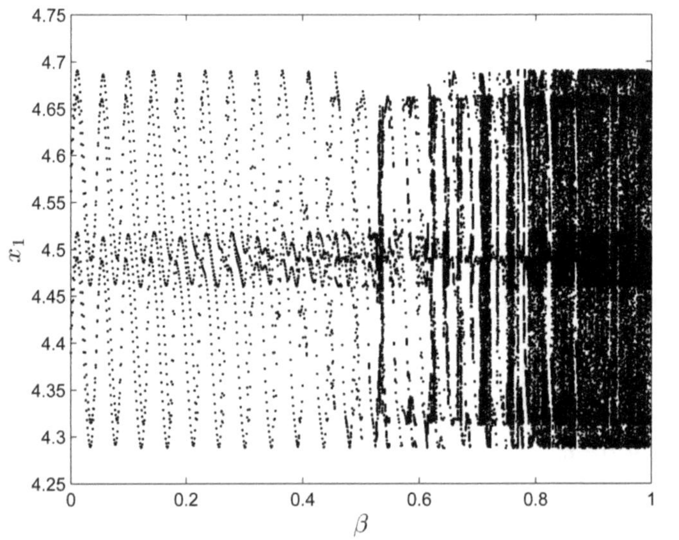

Fig. 9.2 The period-doubling bifurcation diagram of x_1

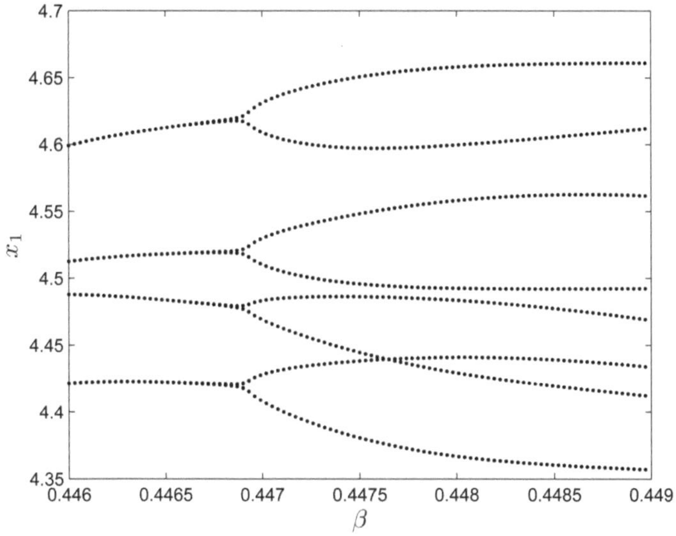

Fig. 9.3 The first period-doubling bifurcation diagram of x_1

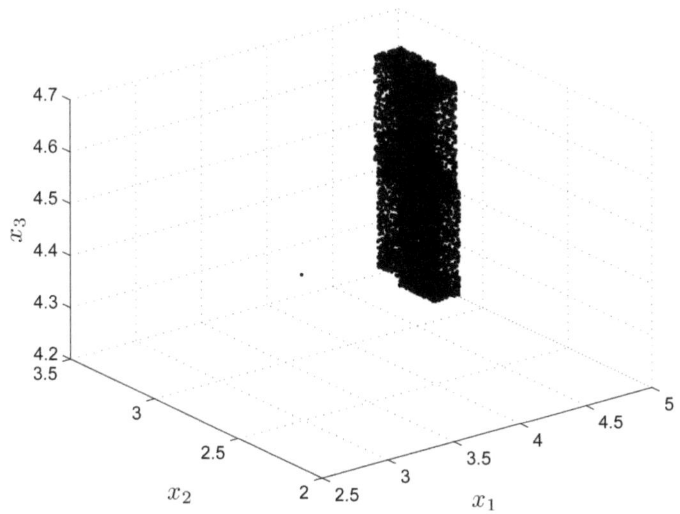

Fig. 9.4 The chaotic attractor of the controlled system with $\beta = 200$

$$\Delta A = \begin{pmatrix} 0.07\sin\left(\frac{k\pi}{2}\right) & 0.1\cos\left(\frac{k\pi}{2}\right) & 0 \\ 0.01\cos\left(\frac{k\pi}{2}\right) & 0 & 0.05\cos\left(\frac{k\pi}{2}\right) \\ 0.09\cos\left(\frac{k\pi}{2}\right) & 0.08\sin\left(\frac{k\pi}{2}\right) & 0 \end{pmatrix},$$

$S = \mathrm{diag}\{s_1, s_2, s_3\} = \mathrm{diag}\{1, 1, 1\}$, and

$$u(k) = \delta \sin\left(\frac{\beta\pi}{\delta} x(k)\right).$$

In the simulation, the magnitude of the control input is arbitrarily chosen as $\delta = 0.1$. The initial value of the controlled system is chosen as $x(0) = (0, 1, 2)^{\mathrm{T}}$. According to Theorem 9.2, β can be regarded as an adjustable control parameter. Fig. 9.1 presents the periodic orbits of an uncertain discrete-time FHM without control. When β takes values from 0 to 1 gradually, the period-doubling bifurcation[2] diagram of x_1 is shown in Fig. 9.2. For clarity, Fig. 9.3 describes the first period-doubling bifurcation of x_1. The bifurcation processes of x_2 and x_3 are similar to that of x_1, and are therefore omitted. The chaotic attractor of the controlled system is presented in Fig. 9.4. From these numerical results, it can be verified that the designed controller realizes the chaotification of the above uncertain discrete-time FHM.

9.3 Chaotification of Continuous-Time Fuzzy Hyperbolic Model Using Impulsive Control

9.3.1 Basic Formulation

First, we review some necessary preliminaries for Devaney's definition of chaos and the continuous-time FHM.

Definition 9.3 ([2]). A map $\phi : S \to S$, where S is a set, is chaotic if

(i) ϕ has sensitive dependence on initial conditions, i.e., for any $x \in S$ and any neighborhood $\tilde{N}$ of x in S, there exists a $\delta > 0$ such that $\|\phi^m(x) - \phi^m(y)\| > \delta$ for some $y \in \tilde{N}$ and $m > 0$, where ϕ^m is the mth-order iteration of ϕ, i.e., $\phi^m := \phi \circ \phi \circ \cdots \circ \phi$ (m times).
(ii) ϕ is topologically transitive, i.e., for any pair of subsets $U, V \subset S$, there exists an integer $m > 0$ such that $\phi^m(U) \cap V \neq \emptyset$.
(iii) The periodic points of ϕ are dense in S. □

---

[2] In mathematics, a period-doubling bifurcation in a dynamical system is a bifurcation in which the system switches to a new behavior with twice the period of the original system.

Remark 9.3. Definition 9.3 is for discrete-time systems. For continuous-time systems, we need to construct a Poincaré section so as to get a Poincaré map. Details will be provided later. □

Definition 9.4 ([24]). Given a plant with n input variables $x = (x_1(t), \ldots, x_n(t))^{\mathrm{T}}$ and an output variable $\dot{x}$, we call the fuzzy rule base a continuous-time hyperbolic-type fuzzy rule base if it satisfies the following conditions:

(i) Every fuzzy rule has the following form:

$$\begin{aligned} R^l: &\text{ IF } x_1 \text{ is } F_{x_1} \text{ and } x_2 \text{ is } F_{x_2} \ \ldots \text{ and } x_n \text{ is } F_{x_n}, \\ &\text{THEN } \dot{x}_i = \pm c_{x_1} \pm c_{x_2} \pm \cdots \pm c_{x_n}, \end{aligned}$$

$l = 1, 2, \ldots, 2^n$, where F_{x_i} $(i = 1, \ldots, n)$ is a fuzzy set of x_i, which includes two linguistic values P (positive) and N (negative), c_{x_i} $(i = 1, 2, \ldots, n)$ is a positive constant corresponding to F_{x_i}, and '$\pm$' stands for either the plus or the minus sign. The actual signs in the THEN part are determined in the following manner: if in the IF part the term characterizing F_{x_i} is P, then in the THEN part c_{x_i} appears with a plus sign; otherwise, c_{x_i} appears with a minus sign.

(ii) The constant term c_{x_i} in the THEN part corresponds to F_{x_i} in the IF part; that is, if there is an F_{x_i} term in the IF part, c_{x_i} must appear in the THEN part. Otherwise, c_{x_i} does not appear in the THEN part.

(iii) There are 2^n fuzzy rules in the rule base and these are all the possible P and N combinations of input variables in the IF part, and all the sign combinations of constants in the THEN part. □

The following theorem explains how a continuous-time FHM is constructed.

Theorem 9.3 ([24]). Given a hyperbolic-type fuzzy rule base, if we define the membership functions of P_{x_i} and N_{x_i} as

$$\mu_{P_{x_i}}(x_i) = \mathrm{e}^{-\frac{1}{2}(x_i - k_{x_i})^2}, \quad \mu_{N_{x_i}}(x_i) = \mathrm{e}^{-\frac{1}{2}(x_i + k_{x_i})^2},$$

where $k_{x_i} > 0$ $(i = 1, \ldots, n)$, then we can derive the following model:

$$\dot{x} = A \tanh(Kx), \tag{9.9}$$

where $K = \mathrm{diag}\{k_{x_1}, \ldots, k_{x_n}\}$, A is a constant matrix, and

$$\tanh(Kx) := (\tanh(k_1 x_1), \ldots, \tanh(k_n x_n))^{\mathrm{T}}.$$

This model is named continuous-time FHM. □

Definition 9.5 ([7]). A matrix A is said to be a Hurwitz diagonally stable matrix if there exists a diagonal matrix $Q = \mathrm{diag}\{q_i\} > 0$ $(i = 1, \ldots, n)$ such that

$$A^{\mathrm{T}} Q + QA < 0.$$ □

9.3.2 Main Results

Now, we consider the following control system:

$$\dot{x} = Af(x) + Bu, \tag{9.10}$$

where $x \in \mathbb{R}^n$, $f(x) = \tanh(Kx)$ is an n-dimensional vector function, $u = u(x(t)) \in \mathbb{R}^{m\times 1}$ is a vector function of state x, and $A \in \mathbb{R}^{n\times n}$ and $B \in \mathbb{R}^{n\times m}$ are both constant matrices.

Assumption 9.1. (A,B) is completely controllable. □

Under Assumption 9.1, we have the following lemma.

Lemma 9.3. For system (9.10), if the controller u is chosen as follows:

$$u(x(t)) = L\tanh(Kx(t)), \tag{9.11}$$

where $L \in \mathbb{R}^{m\times n}$ is a constant matrix such that $(A+BL)$ is a Hurwitz diagonally stable matrix, then the origin of the closed-loop system (9.10) is globally asymptotically stable.

Proof. Since $(A+BL)$ is a Hurwitz diagonally stable matrix, there exists a diagonal matrix $Q = \mathrm{diag}\{q_i\} > 0$ such that $(A+BL)^{\mathrm{T}}Q + Q(A+BL) < 0$. Choose the following Lyapunov function:

$$V(x) = 2\sum_{i=1}^{n}\int_0^{x_i} q_i f_i(\xi_i)\mathrm{d}\xi_i,$$

where $f_i(\cdot)$ is the ith element of $f(x)$; then $V(x)$ is positive definite and radially unbounded, namely, $V(x) > 0$ for all $x \neq 0$ and $V(x) \to +\infty$ as $\|x\| \to +\infty$. The time derivative of $V(x)$ along system (9.10) can be computed as follows:

$$\begin{aligned}\dot{V}(x) &= f^{\mathrm{T}}(x)Q\dot{x}(t) + \dot{x}^{\mathrm{T}}Qf(x)\\ &= f^{\mathrm{T}}(x)Q[Af(x)+Bu] + [Af(x)+Bu]^{\mathrm{T}}Qf(x).\end{aligned} \tag{9.12}$$

Substituting (9.11) into (9.12) and noting that $(A+BL)$ is a Hurwitz diagonally stable matrix, we get

$$\dot{V}(x) = f^{\mathrm{T}}(x)\left[(A+BL)^{\mathrm{T}}Q + Q(A+BL)\right]f(x) < 0, \quad \text{if } x \neq 0.$$

Moreover, the proof is independent of initial values, so the closed-loop system (9.10) is globally and asymptotically stable. □

Remark 9.4. For the purpose of deriving our main results, we only require the stability of closed-loop systems (9.10) and (9.11). It should be noted that there are other results for the stability of closed-loop systems (9.10) and (9.11). System (9.9) and the closed-loop systems (9.10) and (9.11) can be regarded as recurrent neural networks. They are also special cases of Lur'e systems [8]. □

Remark 9.5. Under the conditions of Lemma 9.3, solutions of the controlled system (9.10) are defined in a bounded region $D \subset \mathbb{R}^n$. □

Remark 9.6. If we let $\tilde{A} = A + BL$, system (9.10) with u given in (9.11) becomes

$$\dot{x} = \tilde{A}f(x). \tag{9.13}$$

Because of the form of $f(x)$, there exists a constant $\gamma > 0$ such that, for all different $x, y \in D$, $\|f(y) - f(x)\| \leq \gamma \|y - x\|$. That is to say, $f(x)$ is Lipschitzian on D. From the theory of ordinary differential equations, we know that system (9.10) has a unique continuous solution $\phi(t, t_0, x_0)$ through a given initial point, (t_0, x_0), which is also continuously dependent on x_0 [5]. □

Now, let us consider the following impulsive control system:

$$\begin{cases} \dot{x} = \tilde{A}f(x), & t \neq \tau_k, \\ \Delta x = I_k(x(t)), & t = \tau_k, \ k = 1, 2, 3, \ldots, \\ x(t_0^+) = x_0, & t_0 \geq 0, \end{cases} \tag{9.14}$$

where $\tilde{A}$ has been defined in (9.13), $I_k(x(t))$, $k = 1, 2, 3, \ldots$, are impulsive control signals defined in $D \subset \mathbb{R}^n$, $t_0 = \tau_0 < \tau_1 < \cdots < \tau_k$, $\tau_k - \tau_{k-1} = T$, $\lim_{k \to +\infty} \tau_k = +\infty$, $\Delta x = x(\tau_k^+) - x(\tau_k^-)$, and $x(\tau_k^+)$ and $x(\tau_k^-)$ are the right and left limits of $x(t)$ at $t = \tau_k$.

For system (9.14), we have the following lemma.

Lemma 9.4. If $I_k(x(t))$ is chosen as follows:

$$I_k(x(t)) = y_k - x(t^-), \quad t = \tau_k, \ k = 1, 2, 3, \ldots, \tag{9.15}$$

where $y_k = g(y_{k-1})$ and $g \colon Y \to Y$, $Y \subset D$, then, when $g(\cdot)$ is chaotic and satisfies Devaney's three criteria, the controlled system (9.14) is also chaotic and satisfies Devaney's criteria.

Proof. First, it should be emphasized that for system (9.14) to display chaotic dynamics, its phase space must be a finite region; that is to say, there exists an $M > 0$ such that the trajectory of system (9.14), $\phi(t, t_0, x_0)$, satisfies $\|\phi(t, t_0, x_0)\| \leq M$ for all t in the domain. In fact, for $t \in [\tau_0, \tau_1)$, system (9.14) is under the action of $\tilde{A}f(x)$. By Lemma 9.3, its trajectory, $\phi(t, t_0^+, x_0)$, is asymptotically stable. When $t = \tau_1$, $\tilde{A}f(x)$ turns off and $I_1(x(\tau_1))$ acts on system (9.14). The trajectory of system (9.14) jumps from $\phi(\tau_1^-, t_0^+, x_0)$ to x_0^1. We denote the trajectory at each instant $t = \tau_k$ as x_0^k, $k = 1, 2, 3, \ldots$, where $x_0^0 = x_0$. Since $I_1(x(\tau_1))$ is a finite impulse signal, x_0^1 is also finite. For $t \in (\tau_1, \tau_2)$, $I_1(x(\tau_1))$ turns off and $\tilde{A}f(x)$ turns on with the initial point (τ_1^+, x_0^1). The trajectory of system (9.14) in this time interval, denoted $\phi(t, \tau_1^+, x_0^1)$, is also asymptotically stable. The analysis in other time intervals, (τ_k, τ_{k+1}), $k = 2, 3, 4, \ldots$, is the same as that in (τ_1, τ_2), and situations at other time instants, τ_k, $k = 2, 3, 4, \ldots$, are similar to that at $t = \tau_1$. It is easy to see that although the trajectory may be discontinuous at $t = \tau_k$, $k = 1,2,3,\ldots$, the whole trajectory is indeed in a bounded region.

In the following we will prove that system (9.14) is chaotic and satisfies Devaney's three criteria.

To prove that a continuous-time system is chaotic, one method is to show that its Poincaré map is chaotic [4]. Suppose that $\phi(t,t_0^+,x_0)$ is a solution of system (9.13) with an initial value $x(t_0^+) = x_0$. According to Remark 9.6, the solution $\phi(t,t_0^+,x_0)$ is unique and, thus, $\phi^{-1}(t,t_0^+,x_0)$ exists. Define a set of Poincaré sections S_k as

$$S_k = \{(t,x)\colon x \in D, t = \tau_k\}, \;\; k = 1,2,3,\ldots.$$

The Poincaré map is defined as

$$P\colon V \to V, \;\; P = \psi \circ g \circ \psi^{-1},$$

where $V = \psi(Y)$ and $\psi(x) = \phi(\tau_k + T, \tau_k, x) = \phi(\tau_0 + T, \tau_0, x)$. We will first prove that P is extremely sensitive to initial values, i.e., there exists an $\varepsilon > 0$ and, for any two points $x_0, \bar{x_0} \in V$, there exists an $N > 0$, such that

$$\left\| P^N(x_0) - P^N(\bar{x_0}) \right\| > \varepsilon. \tag{9.16}$$

Since

$$P^N = \underbrace{(\psi \circ g \circ \psi^{-1}) \circ \cdots \circ (\psi \circ g \circ \psi^{-1})}_{N\text{th-order iteration of } P} = \psi \circ g^N \circ \psi^{-1}, \tag{9.17}$$

to prove that (9.16) holds is equivalent to prove that

$$\left\| \psi\left(g^N(y_0)\right) - \psi\left(g^N(\bar{y_0})\right) \right\| > \varepsilon \tag{9.18}$$

holds, where $y_0 = \psi^{-1}(x_0)$ and $\bar{y_0} = \psi^{-1}(\bar{x_0})$.

If (9.18) does not hold, i.e., for any ε_0 satisfying $\varepsilon \geq \varepsilon_0 > 0$ there exists an $n(\varepsilon_0) > 0$ such that

$$\|\psi(g^n(y_0)) - \psi(g^n(\bar{y_0}))\| \leq \varepsilon_0,$$

from the fact that $\psi(x)$ is continuous we know that, for the above ε_0, there exists a δ_0 such that

$$\|g^n(y_0) - g^n(\bar{y_0})\| < \delta_0 \;\text{ for all } n \geq n(\varepsilon_0).$$

But, this contradicts the fact that g is a chaotic map satisfying Devaney's definition. Therefore, P satisfies the first criterion of Devaney.

Next, we show that P is topologically transitive, that is, for any pair of subsets $E, F \subset V$, there exists an integer $N > 0$ such that

$$P^N(E) \cap F \neq \emptyset. \tag{9.19}$$

It is known that for any two subsets $\psi^{-1}(E)$ and $\psi^{-1}(F) \subset Y$, there exists an integer $N > 0$ such that

$$g^N\left(\psi^{-1}(E)\right) \cap \psi^{-1}(F) \neq \emptyset. \tag{9.20}$$

Acting on both sides of (9.20) by ψ, we get

$$\psi\left(g^N\left(\psi^{-1}(E)\right)\right)\cap F\neq\emptyset.$$

Noticing (9.17), we therefore proved (9.19).

Finally, we will prove that the periodic points of P are dense in V. Denote the set of periodic points of a map f as $\mathrm{Per}(f)$. Since $\overline{\mathrm{Per}(g)}=Y$ and ψ is one to one, we have $\overline{\mathrm{Per}(P)}=V$. This completes the proof. □

Summarizing the results above, we have the following theorem:

Theorem 9.4 ([27]). For the following system:

$$\begin{cases}\dot{x}=Af(x)+Bu, & t\neq\tau_k,\\ \Delta x=I_k(x(t)), & t=\tau_k,\\ x(t_0^+)=x_0, & t_0\geq 0,\quad k=1,2,3,\ldots,\end{cases}\tag{9.21}$$

where A and B satisfy Assumption 9.1, if u is chosen as in (9.11) and $I_k(x(t))$ is chosen as in (9.15), the system (9.21) will present chaotic dynamics and satisfy the three criteria of Devaney. □

Remark 9.7. The conclusion of Theorem 9.4 can also be kept if the impulsive control is chosen as $I_k(x(t))=h(y_k)-x(t^-)$, where $h\colon Y\to Y$ is a homeomorphism. A special case is to choose $h(y_k)=\Lambda y_k$, where $\Lambda=\mathrm{diag}\{\lambda_1,\ldots,\lambda_n\}$ with $\lambda_i>0$, $i=1,\ldots,n$, is a constant diagonal matrix. This means that we can chaotify the original system with arbitrarily small impulsive energy. □

9.3.3 Simulations

Suppose that we have the following fuzzy rule base:

R^1: IF x_1 is P_{x_1} and x_2 is P_{x_2}, THEN $\dot{x}_3=c_{x_1}+c_{x_2}$;
R^2: IF x_1 is N_{x_1} and x_2 is P_{x_2}, THEN $\dot{x}_3=-c_{x_1}+c_{x_2}$;
R^3: IF x_1 is P_{x_1} and x_2 is N_{x_2}, THEN $\dot{x}_3=c_{x_1}-c_{x_2}$;
R^4: IF x_1 is N_{x_1} and x_2 is N_{x_2}, THEN $\dot{x}_3=-c_{x_1}-c_{x_2}$;
R^5: IF x_1 is P_{x_1} and x_3 is P_{x_3}, THEN $\dot{x}_2=c_{x_1}+c_{x_3}$;
R^6: IF x_1 is N_{x_1} and x_3 is P_{x_3}, THEN $\dot{x}_2=-c_{x_1}+c_{x_3}$;
R^7: IF x_1 is P_{x_1} and x_3 is N_{x_3}, THEN $\dot{x}_2=c_{x_1}-c_{x_3}$;
R^8: IF x_1 is N_{x_1} and x_3 is N_{x_3}, THEN $\dot{x}_2=-c_{x_1}-c_{x_3}$;
R^9: IF x_2 is P_{x_2} and x_3 is P_{x_3}, THEN $\dot{x}_1=c_{x_2}+c_{x_3}$;
R^{10}: IF x_2 is N_{x_2} and x_3 is P_{x_3}, THEN $\dot{x}_1=-c_{x_2}+c_{x_3}$;
R^{11}: IF x_2 is P_{x_2} and x_3 is N_{x_3}, THEN $\dot{x}_1=c_{x_2}-c_{x_3}$;
R^{12}: IF x_2 is N_{x_2} and x_3 is N_{x_3}, THEN $\dot{x}_1=-c_{x_2}-c_{x_3}$.

Here, we choose membership functions P_{x_i} and N_{x_i} as follows:

$$\mu_{P_{x_i}}(x) = \mathrm{e}^{-\frac{1}{2}(x_i - k_i)^2}, \quad \mu_{N_{x_i}}(x) = \mathrm{e}^{-\frac{1}{2}(x_i + k_i)^2}.$$

Then, we have the following third-order model:

$$\dot{x} = Af(x) = A\tanh(Kx),$$

where $x = (x_1, x_2, x_3)^{\mathrm{T}}$, $A = \begin{pmatrix} 0 & c_{x_2} & c_{x_3} \\ c_{x_1} & 0 & c_{x_3} \\ c_{x_1} & c_{x_2} & 0 \end{pmatrix} = \begin{pmatrix} 0 & 1 & 3 \\ 2 & 0 & 3 \\ 2 & 1 & 0 \end{pmatrix}$, and $K = \mathrm{diag}\{k_1, k_2, k_3\} = \mathrm{diag}\{2, 3, 1\}$. According to Theorem 9.4, the controlled system is

$$\begin{cases} \dot{x} = (A + BL)f(x), & t \neq \tau_k, \\ \Delta x = I_k(x(t)), & t = \tau_k, \quad k = 1, 2, 3, \ldots, \\ x(t_0^+) = x_0. \end{cases} \tag{9.22}$$

We select $B = (1, 1, 1)^{\mathrm{T}}$ and $L = (-2, -1, -2)$. We calculate $Q = \mathrm{diag}\{q_1, q_2, q_3\} = \mathrm{diag}\{17.1186, 25.6779, 19.2584\}$. It is easy to verify that the matrix $A + BL$ is Hurwitz diagonally stable. The initial value is $x_0 = (5, 2, -1)^{\mathrm{T}}$ and the impulsive control is

$$I_k(x) = \Lambda g(y_{k-1}) - x(t^-),$$

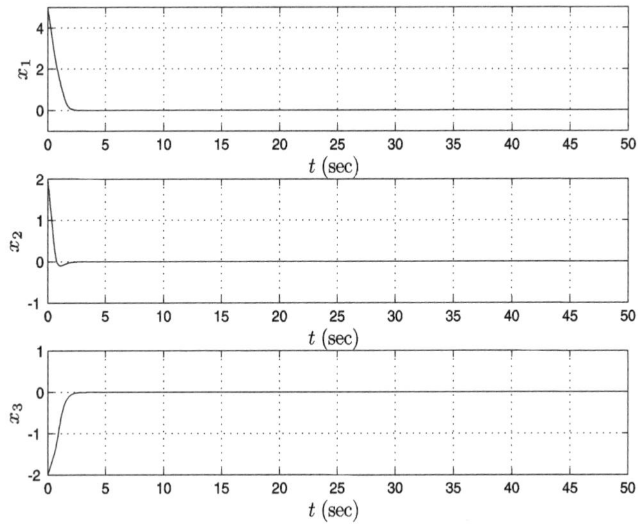

Fig. 9.5 State curves of system (9.23)

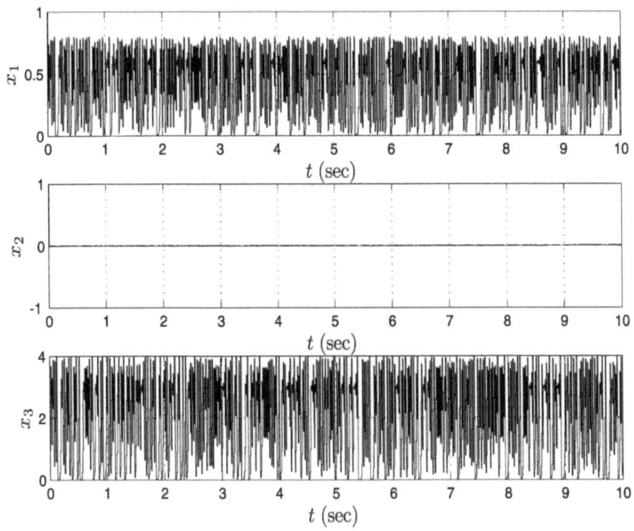

Fig. 9.6 State curves of system (9.22) when $\Lambda = \text{diag}\{0.8, 0, 4\}$

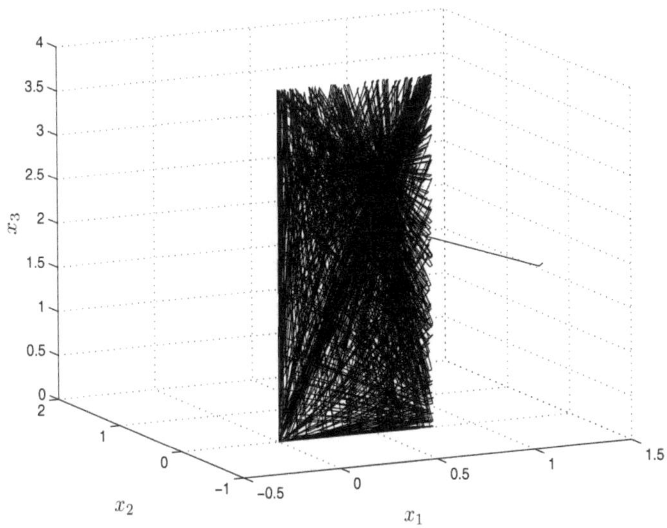

Fig. 9.7 Phase diagram of system (9.22) when $\Lambda = \text{diag}\{0.8, 0, 4\}$

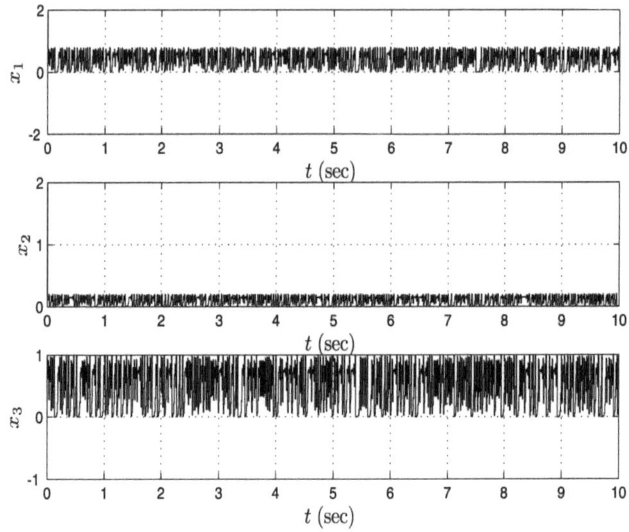

Fig. 9.8 State curves of system (9.22) when $\Lambda = \text{diag}\{0.8, 0.2, 1\}$

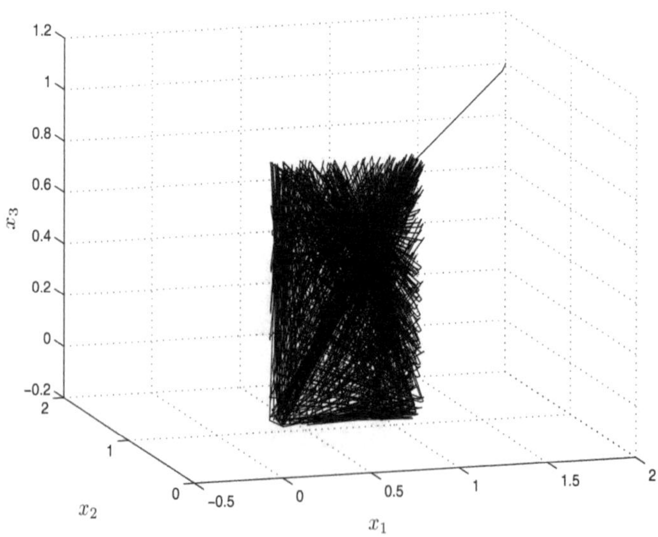

Fig. 9.9 Phase diagram of system (9.22) when $\Lambda = \text{diag}\{0.8, 0.2, 1\}$

where $\Lambda = \text{diag}\{\lambda_1, \lambda_2, \lambda_3\}$ and $g(y_{k-1}) = \left(g_1\left(y_{k-1}^1\right), g_2\left(y_{k-1}^2\right), g_3\left(y_{k-1}^3\right)\right)^{\mathrm{T}}$ with $g_i(\cdot)$ a logistic map:

$$y_k^i = g_i\left(y_{k-1}^i\right) = a y_{k-1}^i\left(1 - y_{k-1}^i\right), \quad i = 1,2,3.$$

When $a = 4.0$, the logistic map is chaotic.

Without impulsive control, the following system

$$\dot{x} = (A + BL) f(x) \tag{9.23}$$

is globally asymptotically stable, whose trajectories are shown in Fig. 9.5. With the impulsive control and $T = 0.01$, from Figs. 9.6–9.9, we see that the controlled system is in chaos for different Λ. These results verify the claims of Theorem 9.4 and Remark 9.7.

9.4 Chaotification of Linear Systems Using Sampled Data Control

9.4.1 Basic Formulation

Consider the following two classes of continuous-time linear systems. One class of systems is given as follows:

$$\dot{x} = Ax + Bu, \tag{9.24}$$

where $x \in \mathbb{R}^n$ is the state vector, $A \in \mathbb{R}^{n\times n}$ is a Hurwitz stable matrix whose eigenvalues are real and nonidentical to one another, i.e., $\lambda_i < 0$, $\lambda_i \neq \lambda_j$ $(i \neq j,$ $i,j = 1,2,\ldots,n)$, $B \in \mathbb{R}^{n\times m}$, $m \leq n$, $\text{rank}(B) = r$, $u \in \mathbb{R}^m$ satisfies

$$\|u\|_\infty \leq \varepsilon_0, \tag{9.25}$$

and ε_0 is an arbitrary positive constant. The aim is to design a controller satisfying (9.25) to chaotify system (9.24).

The other class of systems is described as follows:

$$\dot{x} = A'x + bu, \tag{9.26}$$

where $x \in \mathbb{R}^n$ is the state vector, and $A' \in \mathbb{R}^{n\times n}$ is a Hurwitz stable matrix whose eigenvalues are nonidentical to one another. Moreover, there is a pair of complex eigenvalues. i.e., $\lambda'_{1,2} = \alpha \pm \sqrt{-1}\beta$, $\lambda'_i \in \mathbb{R}$ $(i = 3,4,\ldots,n)$, $\alpha < 0$, $\lambda'_i < 0$, $\lambda'_i \neq \lambda'_j$ $(i \neq j,\ i,j = 3,4,\ldots,n)$, $b \in \mathbb{R}^{n\times 1}$, and u is an input variable satisfying

$$|u| \leq \varepsilon_0. \tag{9.27}$$

The aim is to design a controller satisfying (9.27) to chaotify the system (9.26).

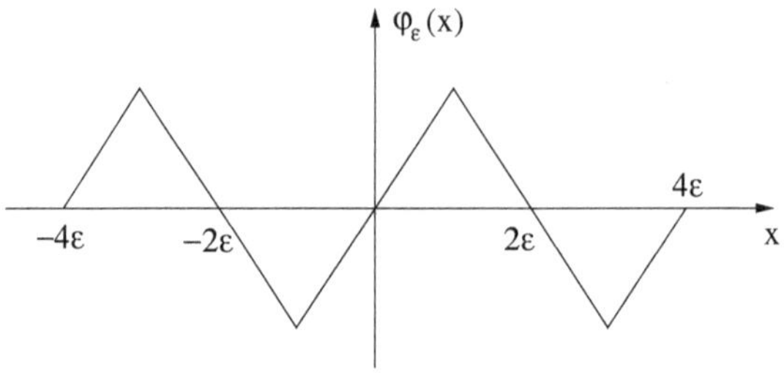

Fig. 9.10 The sawtooth function

To chaotify the above two classes of stable linear systems, a lemma is presented below.

Lemma 9.5 ([18]). Consider the following discrete-time linear systems:

$$x(k+1) = \Lambda x(k) + u(k), \tag{9.28}$$

where $x(k) = (x_1(k), x_2(k), \ldots, x_n(k))^{\mathrm{T}}$, and Λ is a Schur stable matrix, i.e., $\rho(\Lambda) < 1$, where $\rho(\cdot)$ denotes the spectral radius. If the controller $u(k)$ is chosen as

$$u(k) = \Phi_\varepsilon(\sigma x(k)) := (\varphi_\varepsilon(\sigma x_1(k)), \varphi_\varepsilon(\sigma x_2(k)), \ldots, \varphi_\varepsilon(\sigma x_n(k)))^{\mathrm{T}},$$

where $\sigma > \max\left\{3\|\Lambda\|_\infty, \sqrt{\|\Lambda\|_\infty + 1} + 1\right\}$, then the discrete-time linear system (9.28) is chaotic in the sense of Li–Yorke. Moreover, $\|u(k)\|_\infty \le \varepsilon$ $(k = 1, 2, \ldots)$, where ε is an arbitrarily small positive constant, and $\varphi_\varepsilon(x)$ is a sawtooth function shown in Fig. 9.10, which can be described as follows:

$$\varphi_\varepsilon(x) = (-1)^l (x - 2l\varepsilon), \;\; (2l-1)\varepsilon \le x \le (2l+1)\varepsilon, \;\; l \in \mathbb{Z}. \tag{9.29}$$

□

9.4.2 Main Results

First, we will develop a sampled data controller based on Lemma 9.5 to make the linear systems (9.24) chaotic.

Since the eigenvalues of A are nonidentical to one another, there exists a nonsingular matrix $Q \in \mathbb{R}^{n\times n}$ such that system (9.24) can be transformed into

$$\dot{\bar{x}} = \bar{A}\bar{x} + \bar{B}u, \tag{9.30}$$

where $\bar{x} = Q^{-1}x$, $\bar{A} = Q^{-1}AQ$ is a diagonal matrix whose diagonal elements are λ_i $(i = 1, 2, \ldots, n)$, $\bar{B} = Q^{-1}B := \left(\bar{b}_1^{\mathrm{T}}, \bar{b}_2^{\mathrm{T}}, \ldots, \bar{b}_n^{\mathrm{T}}\right)^{\mathrm{T}}$, and $\bar{b}_1, \bar{b}_2, \ldots,$ and $\bar{b}_n$ are row vectors of B.

Assuming that $\overline{b}_{i1}, \overline{b}_{i2}, \ldots, \overline{b}_{ir}$ are linearly independent columns of B, (9.30) can be rewritten as

$$\begin{cases} \dot{\overline{x}}_{i_1} = \lambda_{i_1}\overline{x}_{i_1} + \overline{b}_{i_1}u, \\ \dot{\overline{x}}_{i_2} = \lambda_{i_2}\overline{x}_{i_2} + \overline{b}_{i_2}u, \\ \quad\vdots \\ \dot{\overline{x}}_{i_r} = \lambda_{i_r}\overline{x}_{i_r} + \overline{b}_{i_r}u, \\ \quad\vdots \end{cases} \tag{9.31}$$

According to (9.31), introducing a new vector $\tilde{x}$, we can denote

$$\begin{cases} \tilde{x}_1 = \overline{x}_{i_1}, \\ \tilde{x}_2 = \overline{x}_{i_2}, \\ \quad\vdots \\ \tilde{x}_r = \overline{x}_{i_r}, \\ \quad\vdots \end{cases} \tag{9.32}$$

i.e., $\tilde{x} = R^{-1}\overline{x}$, where R^{-1} is a nonsingular transformation matrix corresponding to (9.32). Denoting $P = QR$, system (9.24) can be rewritten as

$$\dot{\tilde{x}} = \tilde{A}\tilde{x} + \tilde{B}u, \tag{9.33}$$

where $\tilde{x} = P^{-1}x$, $\tilde{A} = P^{-1}AP = \text{diag}\{\tilde{\lambda}_1, \tilde{\lambda}_2, \ldots, \tilde{\lambda}_n\}$, and $\tilde{B} = P^{-1}B$. Moreover, the first r row vectors of $\tilde{B}$ are linearly independent.

Since $\text{rank}(B) = \text{rank}(\tilde{B}) = r$, there exists a nonsingular matrix $C \in \mathbb{R}^{m\times m}$ satisfying

$$\tilde{B}C = \begin{pmatrix} I_{r\times r} & 0_{r\times(m-r)} \\ D_{(n-r)\times r} & 0_{(n-r)\times(m-r)} \end{pmatrix}, \tag{9.34}$$

where $I_{r\times r}$ is the identity matrix, $0_{r\times(m-r)}$ and $0_{(n-r)\times(m-r)}$ are zero matrices, and the matrix $D_{(n-r)\times r}$ can be represented as

$$D_{(n-r)\times r} = \begin{pmatrix} d_{11} & \ldots & d_{1r} \\ \vdots & \ddots & \vdots \\ d_{(n-r)1} & \ldots & d_{(n-r)r} \end{pmatrix}. \tag{9.35}$$

Choose

$$u = Cv,$$

where $v \in \mathbb{R}^{m\times 1}$ is a redefined control input. Therefore, the controlled system (9.33) is transformed into

$$\dot{\tilde{x}} = \tilde{A}\tilde{x} + \tilde{B}Cv. \tag{9.36}$$

Next, we will discretize system (9.36). Assume that the sampling interval $T > 0$ is sufficiently small. Cascading the sampler with a zeroth-order hold, system (9.36) is discretized to

$$\tilde{x}(k+1) = \tilde{A}(T)\tilde{x}(k) + \tilde{B}(T)v(k),$$

where $\tilde{A}(T) = \mathrm{e}^{\tilde{A}T} = \mathrm{diag}\left\{\mathrm{e}^{\tilde{\lambda}_1 T}, \mathrm{e}^{\tilde{\lambda}_2 T}, \ldots, \mathrm{e}^{\tilde{\lambda}_n T}\right\}$,

$$\begin{aligned}
\tilde{B}(T) &= \left(\int_0^{\mathrm{T}} \mathrm{e}^{\tilde{A}(t)}\mathrm{d}t\right)\tilde{B}C \\
&= \left(\int_0^{\mathrm{T}} \begin{pmatrix} \mathrm{e}^{\tilde{\lambda}_1 t} & \cdots & 0 \\ \vdots & \ddots & 0 \\ 0 & \cdots & \mathrm{e}^{\tilde{\lambda}_n t} \end{pmatrix} \mathrm{d}t\right)\tilde{B}C \\
&= \begin{pmatrix} \frac{1}{\tilde{\lambda}_1}(\mathrm{e}^{\tilde{\lambda}_1 T}-1) & \cdots & 0 \\ \vdots & \ddots & 0 \\ 0 & \cdots & \frac{1}{\tilde{\lambda}_n}(\mathrm{e}^{\tilde{\lambda}_n T}-1) \end{pmatrix} \begin{pmatrix} I_{r\times r} & 0_{r\times(m-r)} \\ D_{(n-r)\times r} & 0_{(n-r)\times(m-r)} \end{pmatrix} \\
&= \begin{pmatrix} \frac{1}{\tilde{\lambda}_1}(\mathrm{e}^{\tilde{\lambda}_1 T}-1) & \cdots & 0 & 0\cdots 0 \\ \vdots & \ddots & 0 & \vdots \ddots \vdots \\ 0 & \cdots & \frac{1}{\tilde{\lambda}_r}(\mathrm{e}^{\tilde{\lambda}_r T}-1) & 0\cdots 0 \\ \frac{d_{11}}{\tilde{\lambda}_{r+1}}(\mathrm{e}^{\tilde{\lambda}_{r+1} T}-1) & \cdots & \frac{d_{1r}}{\tilde{\lambda}_{r+1}}(\mathrm{e}^{\tilde{\lambda}_{r+1} T}-1) & 0\cdots 0 \\ \vdots & \ddots & \vdots & \vdots \ddots \vdots \\ \frac{d_{(n-r)1}}{\tilde{\lambda}_n}(\mathrm{e}^{\tilde{\lambda}_n T}-1) & \cdots & \frac{d_{(n-r)r}}{\tilde{\lambda}_n}(\mathrm{e}^{\tilde{\lambda}_n T}-1) & 0\cdots 0 \end{pmatrix},
\end{aligned}$$

$\tilde{x}(k) = \tilde{x}(kT)$, and $v(k) = v(kT)$ $(k = 0, 1, 2, \ldots)$, i.e., the discretized system is

$$\begin{cases}
\tilde{x}_1(k+1) = \mathrm{e}^{\tilde{\lambda}_1 T}\tilde{x}_1(k) + \frac{1}{\tilde{\lambda}_1}(\mathrm{e}^{\tilde{\lambda}_1 T}-1)v_1(k), \\
\qquad\vdots \\
\tilde{x}_r(k+1) = \mathrm{e}^{\tilde{\lambda}_r T}\tilde{x}_r(k) + \frac{1}{\tilde{\lambda}_r}(\mathrm{e}^{\tilde{\lambda}_r T}-1)v_r(k), \\
\tilde{x}_{r+1}(k+1) = \mathrm{e}^{\tilde{\lambda}_{r+1} T}\tilde{x}_{r+1}(k) + \frac{d_{11}}{\tilde{\lambda}_{r+1}}(\mathrm{e}^{\tilde{\lambda}_{r+1} T}-1)v_1(k) + \cdots \\
\qquad\qquad + \frac{d_{1r}}{\tilde{\lambda}_{r+1}}(\mathrm{e}^{\tilde{\lambda}_{r+1} T}-1)v_r(k), \\
\qquad\vdots \\
\tilde{x}_n(k+1) = \mathrm{e}^{\tilde{\lambda}_n T}\tilde{x}_n(k) + \frac{d_{(n-r)1}}{\tilde{\lambda}_n}(\mathrm{e}^{\tilde{\lambda}_n T}-1)v_1(k) + \cdots \\
\qquad\qquad + \frac{d_{(n-r)r}}{\tilde{\lambda}_n}(\mathrm{e}^{\tilde{\lambda}_n T}-1)v_r(k).
\end{cases} \tag{9.37}$$

Choose the control input as

$$v(k) = \begin{pmatrix} v_1(k) \\ \vdots \\ v_m(k) \end{pmatrix} = \begin{pmatrix} \frac{\tilde{\lambda}_1}{e^{\tilde{\lambda}_1 T}-1}\varphi_\varepsilon(\sigma\tilde{x}_1(k)) \\ \vdots \\ \frac{\tilde{\lambda}_r}{e^{\tilde{\lambda}_r T}-1}\varphi_\varepsilon(\sigma\tilde{x}_r(k)) \\ 0 \\ \vdots \\ 0 \end{pmatrix}, \tag{9.38}$$

where $\varphi_\varepsilon(x)$ is a sawtooth function. Then, the subsystem composed of the first r variables of (9.37) is given by

$$\begin{cases} \tilde{x}_1(k+1) = e^{\tilde{\lambda}_1 T}\tilde{x}_1(k) + \varphi_\varepsilon(\sigma\tilde{x}_1(k)), \\ \tilde{x}_2(k+1) = e^{\tilde{\lambda}_2 T}\tilde{x}_2(k) + \varphi_\varepsilon(\sigma\tilde{x}_2(k)), \\ \vdots \\ \tilde{x}_r(k+1) = e^{\tilde{\lambda}_r T}\tilde{x}_r(k) + \varphi_\varepsilon(\sigma\tilde{x}_r(k)), \end{cases}$$

which is equivalent to the following vector form:

$$\tilde{x}_R(k+1) = \tilde{A}_R(T)\tilde{x}_R(k) + \Phi_\varepsilon(\sigma\tilde{x}_R(k)), \tag{9.39}$$

where $\tilde{x}_R = \begin{pmatrix} \tilde{x}_1 \\ \vdots \\ \tilde{x}_r \end{pmatrix}$, $\tilde{A}_R(T) = \text{diag}\left\{e^{\tilde{\lambda}_1 T}, e^{\tilde{\lambda}_2 T}, \ldots, e^{\tilde{\lambda}_r T}\right\}$, and

$$\Phi_\varepsilon(\sigma\tilde{x}_R) = \begin{pmatrix} \varphi_\varepsilon(\sigma\tilde{x}_1) \\ \vdots \\ \varphi_\varepsilon(\sigma\tilde{x}_r) \end{pmatrix}.$$

It is obvious that $\|\Phi_\varepsilon(\sigma\tilde{x}_R)\|_\infty \leq \varepsilon$. To make sure that $\|u\|_\infty \leq \varepsilon_0$, denoting

$$\delta = \max\left\{\frac{\tilde{\lambda}_1}{e^{\tilde{\lambda}_1 T}-1}, \ldots, \frac{\tilde{\lambda}_r}{e^{\tilde{\lambda}_r T}-1}\right\}, \tag{9.40}$$

we can then choose

$$\varepsilon = \frac{\varepsilon_0}{\delta\|C\|_\infty}. \tag{9.41}$$

It can be seen that the last $n-r$ state variables $\tilde{x}_{r+1}, \tilde{x}_{r+2}, \ldots, \tilde{x}_n$ have no effect on the first r state variables. At the same time, it can be proven that $\tilde{x}_{r+1}, \tilde{x}_{r+2}, \ldots, \tilde{x}_n$ are bounded if the control input variables are bounded. In fact, the solution of the discrete-time system (9.37) can be given as

$$\tilde{x}(k) = \left(\tilde{A}(T)\right)^k \tilde{x}(0) + \sum_{i=1}^{k-1}\left(\tilde{A}(T)\right)^{k-1-i}\tilde{B}(T)v(i).$$

Denote $\left\|\tilde{A}(T)\right\|_\infty = \alpha$. Since $\tilde{\lambda}_i < 0$ $(i = 1,2,\ldots,n)$, it can be inferred that $\alpha < 1$. Furthermore, it can be seen that $\|v\|_\infty \leq \varepsilon\delta = \varepsilon_0/\|C\|_\infty$ from (9.38), (9.40), and (9.41). Therefore,

$$\begin{aligned}\lim_{k\to\infty} \|\tilde{x}(k)\|_\infty &\leq \lim_{k\to\infty} \alpha^k \|\tilde{x}(0)\|_\infty + \lim_{k\to\infty} \sum_{i=1}^{k-1} \alpha^{k-1-i} \left\|\tilde{B}(T)\right\|_\infty \|v(i)\|_\infty \\ &\leq \frac{\varepsilon_0 \left\|\tilde{B}(T)\right\|_\infty}{\|C\|_\infty} \lim_{k\to\infty} \sum_{i=1}^{k-1} \alpha^{k-1-i} \\ &= \frac{\varepsilon_0 \left\|\tilde{B}(T)\right\|_\infty}{\|C\|_\infty (1-\alpha)}. \end{aligned} \tag{9.42}$$

It is obvious that $\tilde{x}$ is chaotic if $\tilde{x}_R$ is chaotic. So, the next step is to chaotify system (9.39). Since $\tilde{\lambda}_i < 0$ $(i = 1,2,\ldots n)$, it is clear that $\mathrm{e}^{\tilde{\lambda}_i T} < 1$. Therefore, system (9.39) is stable without control (i.e., $v(k) \equiv 0$). According to Lemma 9.5, choosing

$$\sigma > \max\left\{3\left\|\tilde{A}_R(T)\right\|_\infty, \sqrt{\left\|\tilde{A}_R(T)\right\|_\infty + 1} + 1\right\}, \tag{9.43}$$

system (9.39) is chaotic in the sense of Li–Yorke, that is, system (9.37) is chaotic in the sense of Li–Yorke.

The fact that $\tilde{x}(k)$ in (9.37) is chaotic means that $\tilde{x}(t)$ in (9.33) and $x(t)$ in (9.24) are chaotic. Therefore, we can derive the following theorem.

Theorem 9.5 ([6]). For the n-dimensional continuous-time linear controlled system (9.24), if the nonlinear feedback controller is chosen as

$$u(t) = Cv(t) = C\begin{pmatrix} v_1(t) \\ \vdots \\ v_m(t) \end{pmatrix} = C\begin{pmatrix} \frac{\tilde{\lambda}_1}{\mathrm{e}^{\tilde{\lambda}_1 T}-1}\varphi_\varepsilon\left(\sigma(P^{-1}x(kT))_1\right) \\ \vdots \\ \frac{\tilde{\lambda}_r}{\mathrm{e}^{\tilde{\lambda}_r T}-1}\varphi_\varepsilon\left(\sigma(P^{-1}x(kT))_r\right) \\ 0 \\ \vdots \\ 0 \end{pmatrix}, \tag{9.44}$$

where $kT \leq t < (k+1)T$, $k = 0,1,2,\ldots$, then the controlled system (9.24) is chaotic in the sense of Li–Yorke, where P is the transformation matrix in (9.33), $(P^{-1}x(kT))_j$ $(j = 1,2,\ldots,r)$ is the jth component of the vector $P^{-1}x(kT)$, the matrix C is decided by (9.34), and ε and σ satisfy (9.41) and (9.43), respectively. □

Remark 9.8. In (9.35), denoting

$$D_i = \sum_{j=1}^{r} \left|d_{ij}\right|,$$

where $i = 1,2,\ldots,n-r$, we have

$$\left\|\tilde{B}(T)\right\|_{\infty} = \max\left\{\frac{1}{\tilde{\lambda}_1}(\mathrm{e}^{\tilde{\lambda}_1 T}-1),\ldots,\frac{1}{\tilde{\lambda}_r}(\mathrm{e}^{\tilde{\lambda}_r T}-1),\frac{D_1}{\tilde{\lambda}_{r+1}}(\mathrm{e}^{\tilde{\lambda}_{r+1} T}-1),\ldots,\right.$$
$$\left.\frac{D_{(n-r)}}{\tilde{\lambda}_n}(\mathrm{e}^{\tilde{\lambda}_n T}-1)\right\}.$$

From the above analysis, it can be seen that $\left\|\tilde{B}(T)\right\|_{\infty}$ will decrease with the decreasing of the sampling interval T. Therefore, it can be inferred from (9.42) that the chaotic region will shrink with the decreasing of the sampling interval T and the control upper bound ε_0. □

Remark 9.9. The control input in Theorem 9.5 satisfies $\|u\|_{\infty} \leq \varepsilon_0$, where ε_0 is an arbitrarily small positive constant. In fact, if the magnitude of the control input can be chosen arbitrarily large, then we can chaotify any linear system $\dot{x} = Ax + Bu$ by assigning appropriate poles and adopting the control input in the form of (9.44) as long as (A,B) is controllable completely. In this case, the controller consists of two parts: one is a linear state feedback controller and the other is a nonlinear state feedback controller. □

Next, we will investigate the problem of chaotification of the linear systems (9.26).

Since the eigenvalues of (9.26) are $\lambda'_{1,2} = \alpha \pm \mathrm{i}\beta$, $\lambda'_i \in \mathbb{R}$ $(i = 3,4,\ldots,n)$, there exists a nonsingular matrix P such that (9.26) can be transformed into the following Jordan standard form:

$$\dot{\tilde{x}} = \tilde{A}'\tilde{x} + \tilde{b}u, \tag{9.45}$$

where $\tilde{x} = P^{-1}x$, $\tilde{A}' = P^{-1}A'P = \begin{pmatrix} \tilde{A}'_1 & 0 \\ 0 & \tilde{A}'_2 \end{pmatrix}$ is a Jordan matrix, $\tilde{A}'_1 = \begin{pmatrix} \alpha & \beta \\ -\beta & \alpha \end{pmatrix}$, $\tilde{A}'_2 = \mathrm{diag}\left\{\lambda'_3,\ldots,\lambda'_n\right\}$, and $\tilde{b} = P^{-1}b := \left(\tilde{b}_1,\tilde{b}_2,\ldots,\tilde{b}_n\right)^{\mathrm{T}}$.

Discretizing (9.45) by a zeroth-order hold, we can derive the discretized linear systems as follows:

$$\tilde{x}(k+1) = \tilde{A}'(T)\tilde{x}(k) + \tilde{b}(T)u(k), \tag{9.46}$$

where T is the sampling interval,

$$\tilde{A}'(T) = \mathrm{e}^{\tilde{A}'T} = \begin{pmatrix} \tilde{A}'_1(T) & 0 \\ 0 & \tilde{A}'_2(T) \end{pmatrix},$$

$$\tilde{b}(T) = \left(\int_0^{\mathrm{T}} \mathrm{e}^{\tilde{A}'(t)}\mathrm{d}t\right)\tilde{b} = \frac{\mathrm{e}^{\alpha T}}{\alpha^2+\beta^2}\begin{pmatrix} \Lambda_1(T) & 0 \\ 0 & \Lambda_2(T) \end{pmatrix}\tilde{b},$$

$$\tilde{A}'_1(T) = \mathrm{e}^{\tilde{A}'_1 T} = \begin{pmatrix} \mathrm{e}^{\alpha T}\cos\beta T & -\mathrm{e}^{\alpha T}\sin\beta T \\ \mathrm{e}^{\alpha T}\sin\beta T & \mathrm{e}^{\alpha T}\cos\beta T \end{pmatrix},$$

$$\tilde{A}'_2(T) = \mathrm{diag}\left\{\mathrm{e}^{\lambda'_3 T},\ldots \mathrm{e}^{\lambda'_n T}\right\},$$

$$\Lambda_1(T) = \begin{pmatrix} \alpha\cos\beta T + \beta\sin\beta T - \alpha\mathrm{e}^{-\alpha T} & -(\alpha\sin\beta T - \beta\cos\beta T + \beta\mathrm{e}^{-\alpha T}) \\ \alpha\sin\beta T - \beta\cos\beta T + \beta\mathrm{e}^{-\alpha T} & \alpha\cos\beta T + \beta\sin\beta T - \alpha\mathrm{e}^{-\alpha T} \end{pmatrix},$$

$$\Lambda_2(T) = \text{diag}\{\sigma_3(T), \ldots, \sigma_n(T)\},$$

and

$$\sigma_i(T) = \frac{(\alpha^2+\beta^2)(\mathrm{e}^{\lambda_i' T}-1)}{\mathrm{e}^{\alpha T}\lambda_i'} \quad (i = 3, \ldots, n).$$

Choosing the sampling interval T as

$$T = T_0 = \frac{\pi}{\beta}, \tag{9.47}$$

we get $\tilde{A}_1'(T_0) = \text{diag}\{-\mathrm{e}^{\alpha T_0}, -\mathrm{e}^{\alpha T_0}\}$. Consequently,

$$\tilde{A}'(T_0) = \text{diag}\left\{-\mathrm{e}^{\alpha T_0}, -\mathrm{e}^{\alpha T_0}, \mathrm{e}^{\lambda_3' T_0}, \ldots, \mathrm{e}^{\lambda_n' T_0}\right\}.$$

Here,

$$\tilde{b}(T_0) = \frac{\mathrm{e}^{\alpha T_0}}{\alpha^2+\beta^2}\begin{pmatrix} -\alpha(1+\mathrm{e}^{-\alpha T_0}) & -\beta(1+\mathrm{e}^{-\alpha T_0}) & 0 \\ \beta(1+\mathrm{e}^{-\alpha T_0}) & -\alpha(1+\mathrm{e}^{-\alpha T_0}) & 0 \\ 0 & 0 & \Lambda_2(T_0) \end{pmatrix} \tilde{b} := \begin{pmatrix} \tilde{b}_1(T_0) \\ \vdots \\ \tilde{b}_n(T_0) \end{pmatrix},$$

where

$$\tilde{b}_1(T_0) = \frac{-\mathrm{e}^{\alpha T_0}}{\alpha^2+\beta^2}\left(\alpha(1+\mathrm{e}^{-\alpha T_0})\tilde{b}_1 + \beta(1+\mathrm{e}^{-\alpha T_0})\tilde{b}_2\right),$$

$$\tilde{b}_2(T_0) = \frac{-\mathrm{e}^{\alpha T_0}}{\alpha^2+\beta^2}\left(\alpha(1+\mathrm{e}^{-\alpha T_0})\tilde{b}_2 - \beta(1+\mathrm{e}^{-\alpha T_0})\tilde{b}_1\right),$$

$$\tilde{b}_i(T_0) = \frac{\tilde{b}_i}{\lambda_i'}\left(\mathrm{e}^{\lambda_i' T_0} - 1\right) \quad (i = 3, \ldots, n).$$

Therefore, (9.46) can be rewritten as follows:

$$\begin{cases} \tilde{x}_1(k+1) = -\mathrm{e}^{\alpha T_0}\tilde{x}_1(k) + \tilde{b}_1(T_0)u(k), \\ \tilde{x}_2(k+1) = -\mathrm{e}^{\alpha T_0}\tilde{x}_2(k) + \tilde{b}_2(T_0)u(k), \\ \tilde{x}_3(k+1) = \mathrm{e}^{\lambda_3' T_0}\tilde{x}_3(k) + \tilde{b}_3(T_0)u(k), \\ \quad\vdots \\ \tilde{x}_n(k+1) = \mathrm{e}^{\lambda_n' T_0}\tilde{x}_n(k) + \tilde{b}_n(T_0)u(k). \end{cases} \tag{9.48}$$

It is obvious that the state variables of (9.48), i.e., $\tilde{x}_1(k)$, $\tilde{x}_2(k)$, ..., $\tilde{x}_n(k)$, are independent. Since b is nonsingular, $\tilde{b}$ is nonsingular after a nonsingular transformation. So, all $\tilde{b}_i(T_0)$ are not zero. We consider the following one-dimensional systems:

$$\tilde{x}_1(k+1) = -\mathrm{e}^{\alpha T_0}\tilde{x}_1(k) + \tilde{b}_1(T_0)u(k). \tag{9.49}$$

According to Lemma 9.5, choosing the input $u(k)$ as

$$u(k) = \frac{\varphi_\varepsilon(\sigma\tilde{x}_1(k))}{\tilde{b}_1(T_0)}, \tag{9.50}$$

where

$$\varepsilon = \varepsilon_0 \left|\tilde{b}_1(T_0)\right| = \frac{2\varepsilon_0 e^{\alpha T_0}}{\alpha^2+\beta^2} \left|\alpha(1+e^{-\alpha T_0})\tilde{b}_1 + \beta(1+e^{-\alpha T_0})\tilde{b}_2\right|,$$
$$\sigma > \max\left\{3e^{\alpha T_0}, \sqrt{e^{\alpha T_0}+1}+1\right\},$$

the controlled system (9.49) is chaotic in the sense of Li–Yorke. Moreover, $|u(k)| \leq \varepsilon_0$.

With the control input $u(k)$ of (9.50), the solution of the closed-loop system (9.48) is given by

$$\tilde{x}(k) = \left(\tilde{A}'(T_0)\right)^k \tilde{x}(0) + \sum_{i=1}^{k-1} \left(\tilde{A}'(T_0)\right)^{k-1-i} \tilde{b}(T_0)u(k).$$

Denote $\left\|\tilde{A}'(T_0)\right\|_\infty = \rho$. Since $\alpha < 0$ and $\lambda_i' < 0$ $(i = 3,4,\ldots,n)$, it is obvious that $\rho < 1$. Therefore,

$$\begin{aligned}\lim_{k\to\infty} \|\tilde{x}(k)\|_\infty &\leq \lim_{k\to\infty} \rho^k \|\tilde{x}(0)\|_\infty + \lim_{k\to\infty} \sum_{i=1}^{k-1} \rho^{(k-1-i)} \left\|\tilde{b}(T_0)\right\|_\infty |u(k)| \\ &\leq \frac{\varepsilon_0 \left\|\tilde{b}(T_0)\right\|_\infty}{1-\rho},\end{aligned}$$

i.e., the state variables of system (9.48) are bounded. For the n-dimensional system (9.48), the fact that $\tilde{x}_1(k)$ is chaotic means that $\tilde{x}_2(k)$, ..., $\tilde{x}_n(k)$ are chaotic.

The fact that the discretized system (9.48) is chaotic means that the continuous-time systems (9.45) and (9.26) are chaotic. Therefore, when $\tilde{b}_1(T_0) \neq 0$, we can adopt the following control input:

$$u(t) = \frac{\varphi_\varepsilon(\sigma\tilde{x}_1(k))}{\tilde{b}_1(T_0)} = \frac{\varphi_\varepsilon(\sigma(P^{-1}x(kT_0))_1)}{\tilde{b}_1(T_0)},$$

where $kT_0 \leq t < (k+1)T_0$, $k = 0,1,2,\ldots$, to chaotify the system (9.26).

For $\tilde{x}_2,\ldots,\tilde{x}_n$, we can derive similar conclusions. Therefore, we have the following theorem.

Theorem 9.6 ([6]). For the n-dimensional continuous-time linear systems (9.26), when $\tilde{b}_i(T_0) \neq 0$ $(i = 1,2,\ldots,n)$, we can adopt the following control input to chaotify the continuous-time linear systems (9.26)

$$u(t) = \frac{\varphi_\varepsilon(\sigma(P^{-1}x(kT_0))_i)}{\tilde{b}_i(T_0)}, \tag{9.51}$$

where $kT_0 \le t < (k+1)T_0$, $k = 0,1,2,\ldots$, $|u(t)| \le \varepsilon_0$, P is the nonsingular matrix which transforms (9.26) into (9.45), $(P^{-1}x(kT))_i$ $(i = 1,2,\ldots,n)$ denotes the ith component of the vector $P^{-1}x(kT)$, T_0 is determined by (9.47), $\varepsilon = \varepsilon_0 \left|\tilde{b}_i(T_0)\right|$,

$$\sigma > \max\left\{ 3\mathrm{e}^{\alpha T_0}, \sqrt{\mathrm{e}^{\alpha T_0}+1}+1, 3\mathrm{e}^{\lambda_i' T_0}, \sqrt{\mathrm{e}^{\lambda_i' T_0}+1}+1 \right\} \quad (i = 3,4,\ldots,n),$$

$\tilde{b}_i(T_0)$ is given by

$$\begin{cases} \tilde{b}_1(T_0) = \dfrac{-\mathrm{e}^{\alpha T_0}}{\alpha^2+\beta^2}\left(\alpha(1+\mathrm{e}^{-\alpha T_0})\tilde{b}_1 + \beta(1+\mathrm{e}^{-\alpha T_0})\tilde{b}_2\right), \\ \tilde{b}_2(T_0) = \dfrac{-\mathrm{e}^{\alpha T_0}}{\alpha^2+\beta^2}\left(\alpha(1+\mathrm{e}^{-\alpha T_0})\tilde{b}_2 - \beta(1+\mathrm{e}^{-\alpha T_0})\tilde{b}_1\right), \\ \tilde{b}_i(T_0) = \dfrac{\tilde{b}_i}{\lambda_i'}\left(\mathrm{e}^{\lambda_i' T_0}-1\right) \quad (i = 3,\ldots,n), \end{cases} \tag{9.52}$$

and $\left(\tilde{b}_1,\tilde{b}_2,\ldots,\tilde{b}_n\right)^{\mathrm{T}} = P^{-1}b$. □

Remark 9.10. The control input $u(k)$ in Theorem 9.6 satisfies $|u| \le \varepsilon_0$, where ε_0 is an arbitrarily small positive constant. If, without this restraint, i.e., the magnitude of the control input $u(k)$ can be given arbitrarily large, then we can chaotify any linear systems $\dot{x} = Ax+bu$ as long as (A,b) is completely controllable. Here, we can assign the poles of the closed-loop system appropriately by state feedback control first, and then add a nonlinear feedback control (9.51) to the system; then, we can chaotify any linear systems $\dot{x} = Ax+bu$. In this case, the controller also consists of two parts: one is a linear state feedback controller and the other is a nonlinear state feedback controller. □

9.4.3 Simulations

In this subsection, two illustrative examples are presented to demonstrate the effectiveness of the proposed chaotification approach. First, let us consider the following three-dimensional continuous-time linear system:

$$\begin{pmatrix} \dot{x}_1 \\ \dot{x}_2 \\ \dot{x}_3 \end{pmatrix} = \begin{pmatrix} -29/18 & 17/18 & 1/6 \\ 7/18 & -19/18 & 1/6 \\ -5/18 & 11/18 & -5/6 \end{pmatrix} \begin{pmatrix} x_1 \\ x_2 \\ x_3 \end{pmatrix} + \begin{pmatrix} 1 & 0 \\ 1 & 1 \\ 0 & 1 \end{pmatrix} \begin{pmatrix} u_1 \\ u_2 \end{pmatrix}. \tag{9.53}$$

By calculation, the eigenvalues of the system matrix A is given by $\lambda_1 = -1/2$, $\lambda_2 = -1$, and $\lambda_3 = -2$. Choosing the nonsingular matrix P as $P = \begin{pmatrix} 1 & 1 & 2 \\ 1 & 1 & -1 \\ 1 & -2 & 1 \end{pmatrix}$, the transformation is given by $\tilde{x} = P^{-1}x$. System (9.53) is transformed into the following diagonal form:

$$\begin{pmatrix} \dot{\tilde{x}}_1 \\ \dot{\tilde{x}}_2 \\ \dot{\tilde{x}}_3 \end{pmatrix} = \begin{pmatrix} -1/2 & 0 & 0 \\ 0 & -1 & 0 \\ 0 & 0 & -2 \end{pmatrix} \begin{pmatrix} \tilde{x}_1 \\ \tilde{x}_2 \\ \tilde{x}_3 \end{pmatrix} + \begin{pmatrix} 2/3 & 8/9 \\ 1/3 & -2/9 \\ 0 & -1/3 \end{pmatrix} \begin{pmatrix} u_1 \\ u_2 \end{pmatrix}.$$

It can be verified that the first two row vectors of $\tilde{b}$ are linearly independent. So, we get $C^{-1} = \begin{pmatrix} 2/3 & 8/9 \\ 1/3 & -2/9 \end{pmatrix}$, i.e., $C = \begin{pmatrix} 1/2 & 2 \\ 3/4 & -3/2 \end{pmatrix}$. It can be calculated easily that $\|C\|_\infty = 2.5$. According to (9.43), we can choose $\sigma = 3$. Assume that $\varepsilon_0 = 0.2$ and the initial values are chosen as $(0.01, 0.01, 0)^{\mathrm{T}}$ and the sampling intervals are chosen as $T = 5$, $T = 1$, and $T = 0.2$, respectively. According to (9.40), δ is presented as $\delta = 1.0068$, $\delta = 1.5820$, and $\delta = 5.5167$, respectively. According to (9.41), ε is presented as $\varepsilon = 0.0795$, $\varepsilon = 0.0506$, and $\varepsilon = 0.0145$. The chaotic response curves are shown in Figs. 9.11–9.13 corresponding to various sampling intervals, respectively. From the simulation results we can see that the scales of the chaotic region become smaller and smaller with the decrease of sampling intervals.

Second, let us consider the following three-dimensional continuous-time linear system:

$$\begin{pmatrix} \dot{x}_1 \\ \dot{x}_2 \\ \dot{x}_3 \end{pmatrix} = \begin{pmatrix} -10/9 & -1/9 & -25/18 \\ 5/18 & -13/18 & -5/18 \\ 2/3 & 2/3 & -1/6 \end{pmatrix} \begin{pmatrix} x_1 \\ x_2 \\ x_3 \end{pmatrix} + \begin{pmatrix} 1 \\ 1 \\ 1 \end{pmatrix} u. \tag{9.54}$$

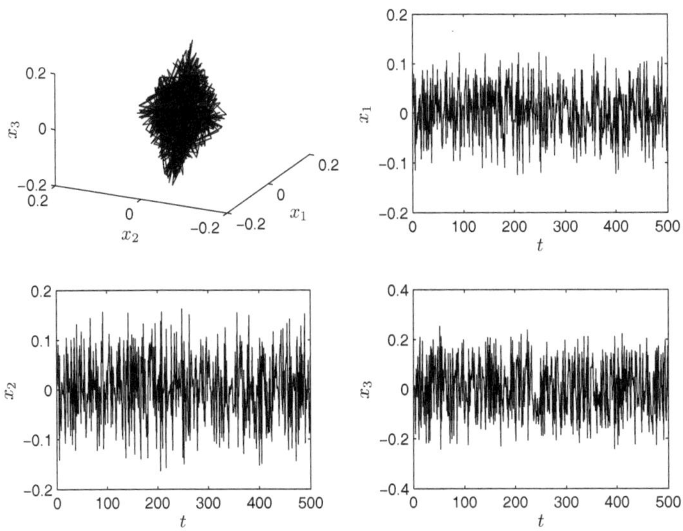

Fig. 9.11 Chaotic orbit in phase space and state variables of system (9.53) with $T = 5$

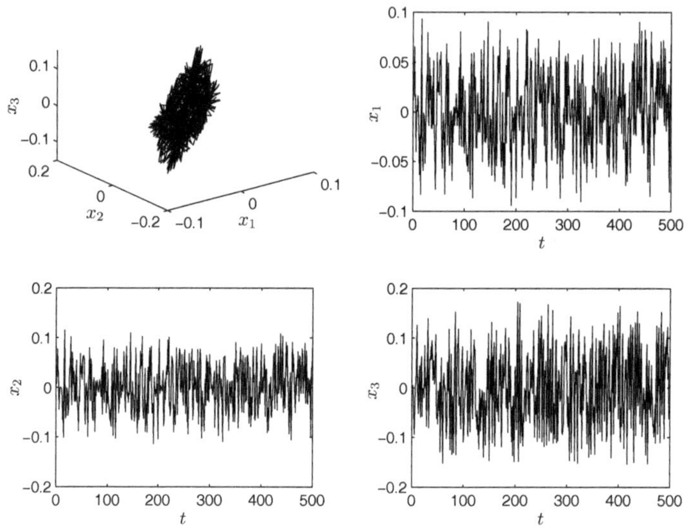

Fig. 9.12 Chaotic orbit in phase space and state variables of system (9.53) with $T = 1$

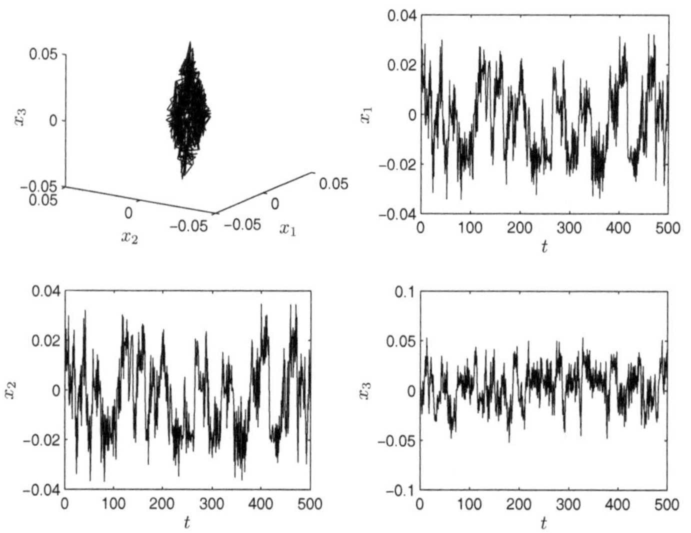

Fig. 9.13 Chaotic orbit in phase space and state variables of system (9.53) with $T = 0.2$

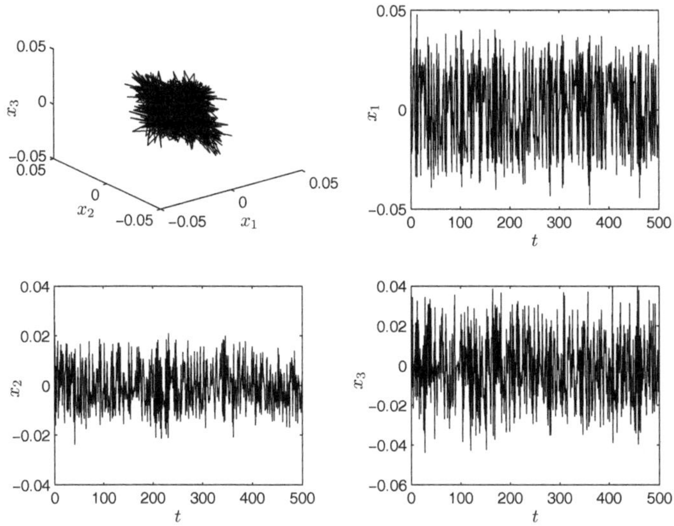

Fig. 9.14 Chaotic attractor and state variables of the controlled system (9.54)

The eigenvalues of the state matrix A' are given by $\lambda'_{1,2} = -1/2 \pm \mathrm{i}$, $\lambda'_3 = -1$. Choosing a nonsingular matrix $P = \begin{pmatrix} 1/9 & 5/9 & 1/3 \\ 2/9 & 1/9 & -1/3 \\ 1/3 & -1/3 & 0 \end{pmatrix}$, the linear system (9.54) is transformed into

$$\begin{pmatrix} \dot{\tilde{x}}_1 \\ \dot{\tilde{x}}_2 \\ \dot{\tilde{x}}_3 \end{pmatrix} = \begin{pmatrix} -1/2 & 1 & 0 \\ -1 & -1/2 & 0 \\ 0 & 0 & -1 \end{pmatrix} \begin{pmatrix} \tilde{x}_1 \\ \tilde{x}_2 \\ \tilde{x}_3 \end{pmatrix} + \begin{pmatrix} 4 \\ 1 \\ 0 \end{pmatrix} u.$$

It is easy to see that $\tilde{b}_1 = 4$, $\tilde{b}_2 = 1$, and $\tilde{b}_3 = 0$. According to (9.47), we get $T_0 = \pi$. According to (9.52), we can derive $\tilde{b}_1(T_0) = \frac{2}{5}(\mathrm{e}^{-\pi/2} + 1) \approx -0.4832$. Assuming that $\varepsilon_0 = 0.2$, according to Theorem 9.6, we get $\varepsilon \approx 0.0996$. Choosing $\sigma = 3$ and the initial states $x_0 = (0.01, 0.01, 0.02)^{\mathrm{T}}$, the linear system (9.54) is chaotified and is shown by Fig. 9.14.

9.5 Summary

In this chapter, we have investigated the problem of chaotification of both linear systems and nonchaotic nonlinear systems. In Sect. 9.2, we developed a simple sinusoidal nonlinear feedback controller to chaotify the discrete-time FHM with uncertain parameters. In Sect. 9.3, a composite controller based on nonlinear state feedback control and the impulsive control method was designed to chaotify the continuous-time FHM. In Sect. 9.4, two classes of continuous-time linear systems were chaotified based on sampled data control. It is verified that all the chaos generated by the above methods satisfies either Li–Yorke's definition or Devaney's definition through mathematical analysis. We believe that these chaotification methods can be used not only for FHM and the above two classes of linear systems, but also for a wide class of nonchaotic systems.

References

1. Chen QF, Zhong QH, Hong YG, Chen GR (2007) Generation and control of spherical circular attractors using switching schemes. Int J Bifurc Chaos 17:243–253
2. Devaney RL (1989) *An Introduction to Chaotic Dynamical Systems*, 2nd edn. Addison-Wesley, New York
3. Golub GH, van Loan CF (1983) *Matrix Computations*. Johns Hopkins University Press, Baltimore
4. Guckenheimer J, Holmes P (1983) *Nonlinear Oscillations, Dynamical Systems, and Bifurcations of Vector Fields*. Springer, New York
5. Hirsch MW, Smale S (1974) *Differential Equations, Dynamical Systems, and Linear Algebra*. Academic Press, New York
6. Huang W (2005) A study of synchronization of chaotic systems and chaotification of nonchaotic systems. Ph.D. dissertation, Northeastern University, Shenyang
7. Kaszkurewicz E, Bhaya A (2000) *Matrix Diagonal Stability in Systems and Computation*. Birkhäuser, Boston
8. Khalil HK (2002) *Nonlinear Systems*, 3rd edn. Prentice Hall, New Jersey
9. Lai DJ, Chen GR (2005) Chaotification of discrete-time dynamical systems: an extension of the Chen–Lai algorithm. Int J Bifurc Chaos 15:109–117
10. Lang S (1997) *Undergraduate Analysis*, 2nd edn. Springer, New York
11. Li Z, Park JB, Chen GR, Joo YH, Choi YH (2002) Generating chaos via feedback control from a stable TS fuzzy system through a sinusoidal nonlinearity. Int J Bifurc Chaos 12:2283–2291
12. Lu JG (2005) Generating chaos via decentralized linear state feedback and a class of nonlinear functions. Chaos Solitons Fractals 25:403–413
13. Lu JG (2006) Chaotic behavior in sampled-data control systems with saturating control. Chaos Solitons Fractals 30:147–155
14. Margaliot M, Langholz G (2003) A new approach to fuzzy modeling and control of discrete-time systems. IEEE Trans Fuzzy Syst 11:486–494
15. Marotto FR (2005) On redefining a snap-back repeller. Chaos Solitons Fractals 25:25–28
16. Schiff SJ, Jerger KD, Duong H, Chang T, Spano ML, Ditto WL (1994) Controlling chaos in the brain. Nature 370:615–620
17. Wang XF, Chen GR, Yu XH (2000) Anticontrol of chaos in continuous-time systems via time-delay feedback. Chaos 10:1–9
18. Wang XF, Chen GR (2000) Chaotifying a stable LTI system by tiny feedback control. IEEE Trans Circuits Syst I 47:410–415
19. Wang XF, Chen GR, Man KF (2001) Making a continuous-time minimum-phase system chaotic by using time-delay feedback. IEEE Trans Circuits Syst I 48:641–645
20. Yang W, Ding M, Mandell AJ, Ott E (1995) Preserving chaos: control strategies to preserve complex dynamics with potential relevance to biological disorders. Phys Rev E 51:102–110

21. Yang L, Liu Z, Chen GR (2002) Chaotifying a continuous-time system via impulsive input. Int J Bifurc Chaos 12:1121–1128
22. Yang RT, Hong YG, Qin HS, Chen GR (2005) Anticontrol of chaos for dynamic systems in p-normal form: a homogeneity-based approach. Chaos Solitons Fractals 25:687–697
23. Zhang HG, Quan YB (2000) Modeling and control based on fuzzy hyperbolic model. Acta Autom Sinica 26:729–735
24. Zhang HG, Quan YB (2001) Modeling, identification and control of a class of non-linear system. IEEE Trans Fuzzy Syst 9:349–354
25. Zhang HG, Wang ZL, Liu D (2004) Chaotifying fuzzy hyperbolic model using adaptive inverse optimal control approach. Int J Bifurc Chaos 14:3505–3517
26. Zhang HG, Wang ZL, Li M, Quan YB, Zhang MJ (2004) Generalized fuzzy hyperbolic model: a universal approximator. Acta Autom Sinica 30:416–422
27. Zhang HG, Wang ZL, Liu D (2005) Chaotifying fuzzy hyperbolic model using impulsive and nonlinear feedback control approaches. Int J Bifurc Chaos 15:2603–2610
28. Zhao Y, Zhang HG, Zheng CD (2008) Anticontrol of chaos for discrete-time fuzzy hyperbolic model with uncertain parameters. Chinese Phys B 17:529–535

Index

MIX
Papier aus verantwortungsvollen Quellen
Paper from responsible sources
FSC® C105338

If you have any concerns about our products,
you can contact us on
ProductSafety@springernature.com

In case Publisher is established outside the EU,
the EU authorized representative is:
Springer Nature Customer Service Center GmbH
Europaplatz 3, 69115 Heidelberg, Germany

Printed by Libri Plureos GmbH
in Hamburg, Germany